DATE DUE

2			
MR 10 '05			
JY 19 '05			
1 1 08			

DEMCO 38-296

Mass Extinctions and Their Aftermath

Mass Extinctions and Their Aftermath

A. Hallam
University of Birmingham

and

P.B. Wignall
University of Leeds

OXFORD • NEW YORK • TOKYO

OXFORD UNIVERSITY PRESS

1997

)xford OX2 6DP

Athens Auckland Bangkok Bogota Bombay
Buenos Aires Calcutta Cape Town Dar es Salaam
Delhi Florence Hong Kong Istanbul Karachi
Kuala Lumpur Madras Madrid Melbourne
Mexico City Nairobi Paris Singapore
Taipei Tokyo Toronto Warsaw

and associated companies in
Berlin Ibadan

Oxford is a trade mark of Oxford University Press

Published in the United States
by Oxford University Press, Inc., New York

© A. Hallam and P. B. Wignall, 1997

A catalogue record for this book is available from the British Library

Library of Congress Cataloging in Publication Data
Hallam, A. (Anthony), 1933–
Mass extinctions and their aftermath / A. Hallam and P. B. Wignall.
Includes bibliographical references and index.
1. Extinction (Biology)   2. Paleoecology.   3. Catastrophes
(Geology) I. Wignall, P. B.   II. Title
QE721.2.E97H35 1997   576.8'4–dc21   97–22349

ISBN 0 19 854917 2 (Hbk)
      0 19 854916 4 (Pbk)

Typeset by Hewer Text Composition Services, Edinburgh
Printed in Great Britain by Bookcraft (Bath) Ltd, Midsomer Norton, Avon

Preface

Since the spectacular discovery of an iridium anomaly at the Cretaceous–Tertiary boundary in Italy, giving rise to the interpretation by Luis Alvarez and his colleagues of a giant meteorite impact leading, among other things, to the demise of the dinosaurs, an immense interest has developed in the subject of mass extinctions. This has resulted in the publication of a huge number of scientific papers since 1980, the date of the seminal Alvarez paper, and also many books. Some of these books have been aimed at a wider audience than the scientific community with specialised knowledge and are therefore lacking in technical detail, others have been volumes with many authors of widely varying expertise, usually based on a scientific symposium. Yet others have focused on only one mass extinction event, usually that at the Cretaceous–Tertiary boundary. What has been lacking is a comprehensive review of Phanerozoic mass extinctions as a whole, with sufficient technical detail to satisfy the geological and palaeontological community, and abundant references to the primary literature. In particular it is of prime importance to put the all-too-celebrated end-Cretaceous event into perspective. This is what we have attempted to do in this book. Where illustrations have been copied from previously published work, permission has been sought from the authors and publishers concerned.

No such wide-ranging effort, which involves having some understanding not just of palaeontology but also a number of other sub-disciplines of the Earth sciences, could have been achieved without considerable help from a number of colleagues, who have kindly offered helpful comments and criticisms on various chapters. These are, in alphabetical order, Mike Benton, Martin Brasier, Pat Brenchley, Andy Gale, Thor Hansen, Don Prothero and Richard Twitchett. To these people we offer our heartfelt thanks. In addition, we should like to acknowledge with gratitude the help provided by June Andrews for her secretarial services, Jackie Stokes for drafting diagrams and Gail Radcliffe for supplying photographs.

Birmingham   A.H.
Leeds   P.B.W.
October 1996

Contents

1 *The study of mass extinctions*

Introduction

Having been largely ignored throughout most of the history of palaeontological and geological research, the subject of mass extinctions has emerged within the last couple of decades as one of the most lively and contentious issues in the whole of science. Not only the Earth sciences community but also physical and biological scientists have involved themselves in the debate, and the media have exhibited intense interest.

Since the term '*mass extinction*' means different things to different people it is desirable at the start to attempt some sort of definition. That of Sepkoski (1986) is particularly useful. He defines mass extinction as any substantial increase in the amount of extinction (that is, lineage termination) suffered by more than one geographically widespread higher taxon during a relatively short interval of geological time, resulting in at least temporary decline in their standing diversity. This is a general definition purposefully designed to be somewhat vague. An equally vague but more concise one offered here is that a mass extinction is an extinction of a significant proportion of the world's biota in a geologically insignificant period of time. The vagueness about extinctions can be dealt with fairly satisfactorily in particular cases by giving percentages of taxa, but the vagueness about time is more difficult to deal with. A significant question about mass extinctions is how catastrophic they were, so we also require a definition of catastrophe in this context. According to Knoll (1984), it is a biospheric perturbation that appears instantaneous when viewed at the level of resolution provided by the geological record.

It is important to appreciate the limits of time resolution that can be established from the stratigraphic record. In the view of Dingus (1984), it is unlikely that one can distinguish, for the most discussed extinction event across the Cretaceous–Tertiary boundary, episodes of extinction lasting 100 years or less from episodes lasting as long as 100 000 years. In favourable circumstances this upper age limit may be reducible to a few thousand years. Sampling problems also make the determination of the magnitude of a catastrophe difficult, as will be discussed later. One should not quibble too much, however, about whether an event was catastrophic or not if it can be reduced to a few thousand years. Working with a time frame of many millions of years, virtually all geologists would concur that it was catastrophic in this context, if the effects were sufficiently considerable.

Raup (1992) has given us some idea of the vast geographic area that must be affected to produce a crisis worthy of the designation 'mass extinction'. He maintains that more than half the Earth's surface needs to be environmentally affected to produce mass extinctions of the magnitude found in the fossil record. His work is supported by Flessa and Jablonski's (1995) analysis of Recent bivalve genera. Their simulation results suggest that substantial levels of generic extinction can only be produced by geographically extensive environmental changes. For example, even the elimination of the entire Indo-Pacific fauna at all latitudes would result in the extinction of only slightly more than half of the global fauna.

Historical background

The earlier history of mass extinctions study has been reviewed by Hallam (1989b), and a more condensed version is presented here.

The great French anatomist and palaeontologist Georges Cuvier was the first person to recognise that certain organisms, such as the mammoth and giant sloth, had become extinct. He went on to argue for episodes of mass extinction and in effect became the father of catastrophist thought in the early nineteenth century. Under the influence of Charles Lyell, however, English geologists tended to reject Cuvier's catastrophism so that Charles Darwin's mid-century view, expressed in *The origin of species*, that dramatic faunal turnovers in the stratigraphic record reflected major stratigraphic hiatuses, provoked no dissent. 'The old notion of all the inhabitants of the earth having been swept away by catastrophes at successive periods is generally given up.' Thus the marked

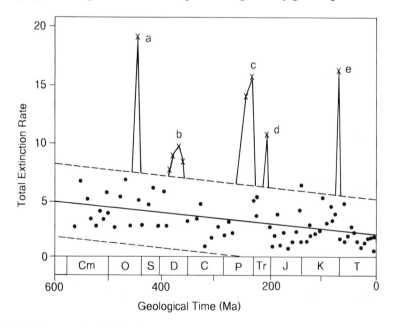

Fig. 1.1 Extinction rates (families/m.y.) of marine animals during the Phanerozoic showing the 'big five' mass extinctions as clear peaks with crosses standing above the enveloped background extinction level: (a) Late Ordovician; (b) Late Devonian; (c) Late Permian; (d) Late Triassic; (e) Late Cretaceous. Cm represents Cambrian; O, Ordovician; S, Silurian; D, Devonian; C, Carboniferous; P, Permian; Tr, Triassic; J, Jurassic; K, Cretaceous; T, Tertiary. Adapted from Raup and Sepkoski (1982).

differences between Palaeozoic, Mesozoic and Cenozoic faunas, recognised by John Phillips as early as 1840, did not elicit much interest in terms of extinction. The subject was largely ignored until early in the twentieth century, when Chamberlin (1909) proposed that major faunal changes through time, which provided the basis for stratigraphic correlation, were under the ultimate control of epeirogenic movements of the continents and ocean basins. Chamberlin's was an isolated article that did not stimulate a major research programme, and the study of extinction events remained dormant for many years.

The modern study of mass extinctions began with a series of papers in the 1950s and 1960s principally focused on the end-Permian event (Schindewolf 1954; Beurlen 1962; Newell 1962). This culminated with Newell's (1967) overview of Phanerozoic extinctions pertinently entitled 'Revolutions in the history of life'. Besides the major extinction events at the end of the Palaeozoic and Mesozoic eras, Newell recognised four other events in the Phanerozoic record of marine families which were also dramatically sudden in the context of the much longer time intervals preceding and following them; these six events took place at the end of the Cambrian, Ordovician, Permian, Triassic, Cretaceous, and near the end of the Devonian. Raup and Sepkoski's (1982) much later statistical analysis of extinction rate, expressed as families going extinct per stratigraphic stage, confirmed the last five as major episodes of mass extinction (Fig. 1.1) and these have generally been accepted since as the 'big five'; they also show up as abrupt decreases in diversity (Fig. 1.2).

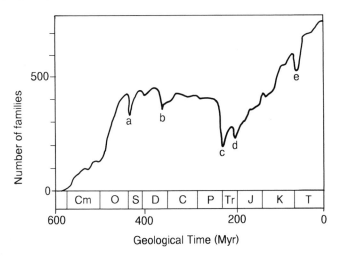

Fig. 1.2 Marine family diversity during the Phanerozoic showing the long-term increase punctuated by diversity crashes caused by the 'big five' mass extinction events. Diversity decreases correspond to the following percentage losses of families: (a) 12%; (b) 14%; (c) 52%; (d) 12%; (e) 11%. Abbreviations as in Fig. 1.1. After Raup and Sepkoski (1982).

Research in recent years, involving many additions, corrections, and reinterpretations of both taxonomy and stratigraphy, has only served to sharpen these five major events (Sepkoski 1993; Benton and Storrs 1994; Jablonski 1994; Benton 1995). Sepkoski's family – and genus – level compendia have allowed the species-level extinction intensities of the five major mass extinctions to be calculated (Table 1.1). On the basis of his compendium of marine genera, Sepkoski (1986) recognised no fewer than 27 Phaner-ozoic extinction events, which can be called major and minor mass extinctions (Fig. 1.3).

Table 1.1 Extinction intensities at the five major mass extinctions in the fossil record: species-level estimates based on a rarefaction technique

Mass extinction	Families		Genera	
	Observed Extinction (%)	Calculated species loss (%)	Observed Extinction (%)	Calculated species loss (%)
End-Ordovician	26	84	60	85
Late Devonian	22	79	57	83
End-Permian	51	95	82	95
End-Triassic	22	79	53	80
End-Cretaceous	16	70	47	76

Source: simplified from Jablonski (1994).

The fossil record of marine invertebrates generally has the advantage of abundant specimens, good stratigraphic control, closely spaced samples, uniform preservation quality and broad geographic distribution. The quality, however, varies substantially from group to group and for some, notably the echinoderms, the fossil record is rather poor (Benton and Simms 1995). The record of vertebrate tetrapods known from continental environments is generally incomplete. Thus most dinosaur genera are known only from a single stratigraphic stage, which would suggest, on a literal reading of the record, that dinosaurs suffered total generic extinction 24 or 25 times during their history (Padian and Clemens 1985). However, at the family level there is only one end-Cretaceous event, because dinosaur families generally had stratigraphic ranges exceeding a stage.

To counter the problem of assessing the relative incompleteness of the tetrapod record, Benton (1989) has proposed a simple completeness metric (SCM) which compares the number of families known to be present with the numbers that ought to be present, that is, which span several stages but are not known from every stage. His SCM percentage figures for different vertebrate groups are as follows: birds, 57; bats, 76; frogs and salamanders 42; lizards and snakes, 49; mammal-like reptiles, 94; placentals, 87. He recognised four mass extinctions and two minor Cenozoic extinctions from the terrestrial tetrapod record, with the relative magnitude being given in terms of percentages of families that disappeared: Early Permian (Sakmarian–Artinskian), 58; Late Permian–Early Triassic (Tatarian–Scythian), 49; Late Triassic (Carnian–Rhae-tian), 22; Late Cretaceous (Maastrichtian), 14; early Oligocene (Rupelian), 8; late Miocene (Tortonian–Messinian), 2. Interestingly, neither the major Early Permian event nor the minor Cenozoic events correspond to marine extinctions.

Periodicity of mass extinctions

The possibility that regularly spaced extinction events have punctuated the Earth's history was raised by Georges Cuvier in the early days of palaeontology, but the real debate on periodicity did not begin until a century and a half later when Fischer and

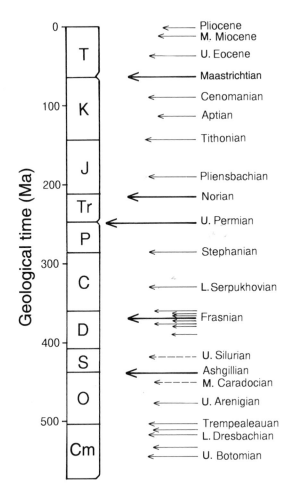

Fig. 1.3 Summary of the temporal distribution of extinction events through the Phanerozoic marine record. The lengths of the arrows, which indicate the events, show very approximately their relative magnitudes. After Sepkoski (1986).

Arthur (1977) proposed that a 32 m.y. cyclicity occurred in post-Palaeozoic extinctions. Based on a small data set, by the later standards of Jack Sepkoski, they claimed to see periodic decreases in the diversity of ammonoid genera and globigerinid species associated with a changeover in the dominant top-predators in the oceans. Despite the similarity to the later claims of Raup and Sepkoski, discussed below, Fischer and Arthur's contribution cannot be considered to mark the start of the current periodicity debate. Many of their extinction events have not been subsequently confirmed (for example, the Bathonian and Valanginian events) while their two Tertiary extinctions, in the Oligocene and Pliocene, do not even appear to be shown in their own data (Fischer and Arthur 1977: Fig. 1). They do, however, deserve credit for being the first to recognise the end-Cenomanian mass extinction. Several of the so-called 32 m.y. cycles are also a product of the curious time-scale used. For example, Fischer and Arthur's 222 Ma date

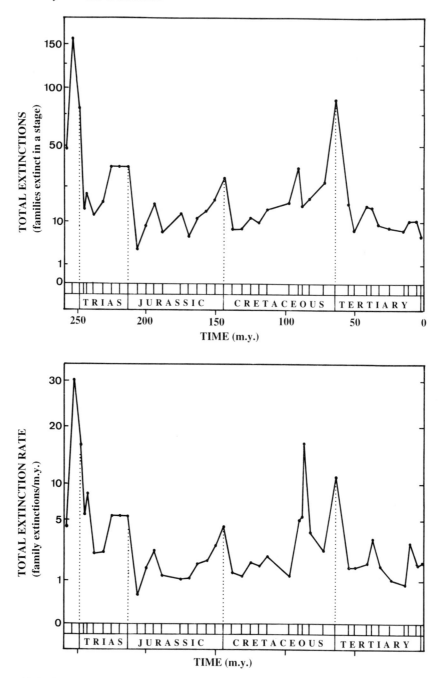

Fig. 1.4 Extinction intensity over the past 250 m.y. for marine animal families calculated using the four extinction metrics discussed in the text. A 26 m.y. periodicity is apparent, although the mid-Jurassic, Aptian and Tertiary events are subdued or missing. A square-root scale has been used as the ordinate in order to reduce the variation at higher extinction intensities. This ensures that the huge end-Permian extinction can be depicted on the same plot as the lesser post-Palaeozoic extinctions. After Sepkoski and Raup (1986).

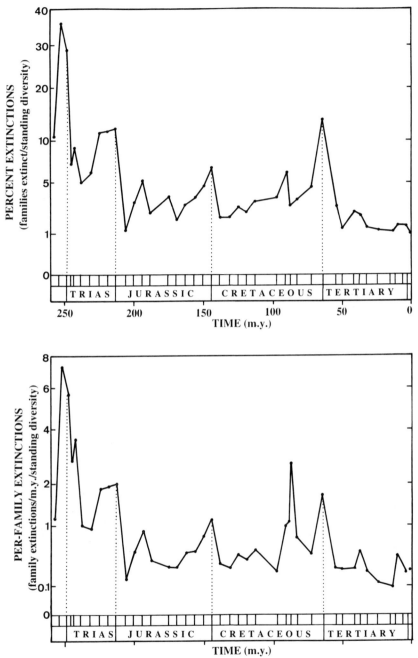

for the Permo-Triassic boundary is too young by nearly 25 m.y. compared to the then accepted age (it is 30 m.y. too young compared to the currently accepted age), but this exceptionally young age is required if the end-Permian and end-Triassic extinctions are to be separated by 32 m.y.

After the false start of the Fischer–Arthur hypothesis, the periodicity debate got under way following Sepkoski's impressive literature compilation of the known ranges for all marine families (Sepkoski 1982a). Ironically, while Sepkoski's initial assessment of his data revealed 15 mass extinctions in the Phanerozoic, including the long-recognised big five, he concluded that: 'Waiting times between mass extinctions were extremely variable and appear not to conform to . . . a cyclic incidence' (Sepkoski 1982b: p. 283). The periodicity debate truly entered the arena in a series of papers published by Sepkoski and his Chicago colleague Dave Raup between 1984 and 1986 (Raup and Sepkoski 1984, 1986; Sepkoski and Raup 1986) in which they found that in the past 250 m.y. mass extinctions have exhibited a regular 26 m.y. periodicity. The extinctions included three of the big five, the end-Permian, end-Triassic and end-Cretaceous, as well as several more minor and new extinctions (Fig. 1.4).

This startling conclusion, potentially the most significant discovery in the Earth sciences since the plate-tectonic revolution, was based on Sepkoski's data set of 3500 marine family ranges from the mid-Permian to the Recent of which 970 are extinct – a figure which includes nearly 250 extinct cephalopod families. In their original paper, Raup and Sepkoski (1984) considered only extinct families in order to eliminate the distorting effect of the 'pull of the Recent'. This effect, documented by Raup (1978), is caused by the extending of 'last' occurrences for extant taxa. All range data from the fossil record are defined by first and last occurrences but, for extant taxa, the 'last' occurrence is the present day, which may be considerably later than their last fossil occurrence, thus 'pulling' the range to the Recent. An extreme example is seen for the class Monoplacophora; their fossil range is Cambrian to Devonian, but their living representatives extend this range forward by 400 m.y.

Raup and Sepkoski also excluded single-datum occurrences from their analyses: a considerable proportion of these were soft-bodied taxa encountered in Lagerstätten. The exceptional preservation and rare occurrence of such deposits ensure that the true range of Lagerstätten taxa is unlikely to be even approximated by the fossil record, therefore their exclusion from the analyses is clearly justified. Raup and Sepkoski were forced to manipulate their data set further in order to incorporate some of the more imprecise ranges. For the majority of families their ranges could be resolved to stage level but, for many, only series-level ranges were known. The series-level extinctions were therefore apportioned evenly among the relevant stage-level extinctions.

Having modified their data set somewhat, Raup and Sepkoski devised four extinction metrics:

Total extinction = number of families going extinct in a stage.

This is clearly the simplest measure of extinction but it does not account for the number of families at risk of extinction. They therefore also calculated:

$$\text{Percent extinction} = \frac{\text{total extinction}}{\text{number of families extant during the stage}}$$

However, the variability of stage durations adds a considerable source of error to this metric; an exceptionally long stage could generate a mass extinction simply by accumulating background extinctions. Therefore a third metric normalises for time:

$$\text{Total extinction rate} = \frac{\text{total extinction}}{\text{stage duration}}$$

Unfortunately, this metric probably contains an even greater source of error as stage durations are poorly known, probably highly variable (cf. Hoffman 1985) and have to be averaged between a few reliable radiometric dates. Thus, a stage, which is of considerably longer duration than other stages within its general interval, may generate a pseudo-mass extinction. This effect can potentially be tested because origination rates should also be correspondingly high.

The most ideal extinction metric calculates a probability of extinction:

$$\text{Per-family extinction rate} = \frac{\text{total extinction rate}}{\text{standing familial diversity}}$$

Once again, the uncertainty of stage durations is incorporated in this measure.

Clearly all extinction metrics have their problems, and it has been general practice to compare extinction patterns using all four measures. It is testimony to the robust nature of the signal from the fossil record that the 26 m.y. periodicity is shown by all four metrics (Fig. 1.4).

The intense critical analysis that followed the initial flurry of the Raup and Sepkoski papers over a period of about five years falls into three distinct categories: whether the data set and the manipulations of it were valid; whether the apparent periodicity of the curve was a statistical reality; and whether the minor mass extinctions, that are fundamental to the definition of periodicity, are in fact true mass extinctions, as defined in the opening section of this chapter.

In an early exegesis, Hoffman (1985) strongly criticised Raup and Sepkoski's (1984) removal of extant families from their analysis as it greatly exaggerated the younger extinction events, particularly in the Tertiary where there are so few extinct families that the chance clustering of a few extinct families can produce an 'event'. Thus, Hoffman (1985) observed that the so-called mid-Miocene mass extinction only corresponds to the loss of five families; background extinction rates in the Mesozoic are commonly substantially higher than this. However, in their later analyses Raup and Sepkoski included extant families and periodicity was still found (Raup and Sepkoski 1986; Sepkoski and Raup 1986).

Hoffman's criticism also highlighted the problem of Raup and Sepkoski's *de facto* definition of mass extinctions as peaks on their extinction curves. They are thus local maxima 'within their local neighborhood' (Sepkoski and Raup 1986: p. 10); a definition that ignores the possibility that two or more mass extinctions may be closely spaced in time but have a different cause.

In a direct attack on the validity of Sepkoski's taxonomic data base, Patterson and Smith (1987) claimed that a large proportion of the families used in the analyses were not true monophyletic clades but, rather, consisted of polyphyletic or paraphyletic taxa. This raised the possibility that many 'pseudo-extinctions' – simple name-changing due to the vagaries of 'systematists' judgement' (Patterson and Smith 1987: p. 250) – were being included. Patterson and Smith went on to demonstrate, for their own specialist groups, fish and echinoderms, that no mass extinctions could be seen in a cladistically truthful data base of monophyletic taxa. In his immediate retort, Sepkoski (1987) noted that families are only a proxy for groups of species, however defined, and that a family extinction must record extinction at a lower taxonomic level. The classic example considers the two orders of dinosaurs and the birds which together form a monophyletic clade. Clearly, it would be cladistic facetiousness to claim that dinosaurs did not go extinct because birds are still extant. Sepkoski's (1987) further point, that the vagaries of family designations is unlikely to generate a 26 m.y. periodicity, is also justified.

Patterson and Smith's (1987) critique could also be further criticised for their choice of taxa. Fish have proved extraordinarily immune to all mass extinction events, with the exception of the Late Devonian crises. Echinoderms, on the other hand, have suffered mass extinctions, notably in the Late Permian, but, due to the rarity of complete, articulated specimens, their fossil record is as poor as that for terrestrial vertebrates (Benton and Simms 1995). Indeed, Boucot (1990) considered that all echinoderm range data should be eliminated from any mass extinction study due to the overriding control of 'taphonomic noise' on their distribution. It is therefore not so surprising that Patterson and Smith found neither mass extinctions nor periodicity in their fish and echinoderm data.

The periodicity claim has also been subjected to a variety of statistical analyses with somewhat contradictory results. For example, Hoffman and Ghiold (1985) have argued that a pseudo-periodicity could be produced in a random time series which would be expected to produce a peak, on average, at every fourth step. As average post-Palaeozoic stage durations are somewhere between 5.5. and 6.4 m.y. the source of the periodicity becomes apparent. Following this line of argument, the more variable and longer average duration of the Palaeozoic stages may account for the often overlooked absence of periodicity in this era. However, Harper (1987) showed that random models should have peaks at every third step and Sepkoski (1986) noted that the actual extinction peaks occur after every fifth interval. Overall, diverse statistical approaches seem to confirm the existence of periodicity (Sepkoski 1989).

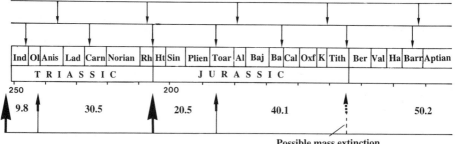

Possible mass extinction

Fig. 1.5 Occurrence of post-Palaeozoic mass extinctions plotted against the Gradstein *et al.* (1994) Mesozoic time-scale and, for the Cenozoic, the Harland *et al.* (1989) time-scale. The best fit occurs with a periodicity of 20.5 m.y. centred on the end-Triassic and Early Toarcian extinction events. This produces a 50% success rate; only 6 of a predicted 12 mass extinctions occur. A fit to a 28.5 m.y. periodicity centred on the end-Cretaceous and end-Cenomanian events is also shown for comparison: this most closely approximates to Raup and Sepkoski's original proposed 26 m.y. periodicity although only four of a predicted nine mass extinctions occur near the correct time.

An alternative, evolutionary argument, propounded by Stanley (1990) and Lutz (1987), suggested that the periodicity was a consequence of an embedded recovery interval. Thus, faunas in the post-extinction interval were said to be initially dominated by a few, hardy, extinction-resistant taxa, with mass extinction-prone communities only evolving a considerable time later. There is a strong element of the much older concept of 'perched' faunas in this argument (Johnson 1974), but it receives little support from the fossil record. With the exception of the huge end-Permian extinction (Fig. 1.2), pre-extinction diversity levels are generally attained within one or two stages after the event (Sepkoski 1989).

The main argument against mass extinction periodicity is the number of events that are 'missing' from the curves. Thus, the mid-Miocene event is far too insignificant to be considered a mass extinction. The Late Eocene event is also a very minor one and it is considerably different from many others in being a rather protracted, gradual affair (cf. Chapter 10). The Aptian mass extinction is a negligible event, while the end-Jurassic is a clear peak, but it is probably the product of an over-reliance on the European data base; thus, it records changes in only one region (Hallam 1986; Chapter 6). A mid-Jurassic extinction is predicted around the Bathonian–Callovian boundary but it is not seen. Therefore, out of a total of ten mass extinctions predicted by the periodicity hypothesis, only five or six occurred with certainty.

Hallam (1984b) has argued that the periodicity may be an artefact of the time-scale employed. This was rejected on the grounds that the random error encountered in all time scales is unlikely to produce a periodic signal (Raup and Sepkoski 1986; Sepkoski 1989). This notion can be tested using the latest and most rigorously assessed Mesozoic time-scale proposed by Gradstein *et al.* (1994), supplemented by the Harland *et al.* (1989) time-scale for the Cenozoic. Our only quibble is with Gradstein *et al.*'s age for the Permo-Triassic boundary. This is based on the 251.2 Ma age that Claoué-Long *et al.* (1991) obtained for the boundary clay in the Meishan section in southern China. Gradstein *et al.* correctly highlighted the fact that the boundary clay occurs high in the latest Permian conodont zone rather than precisely at the boundary; they therefore

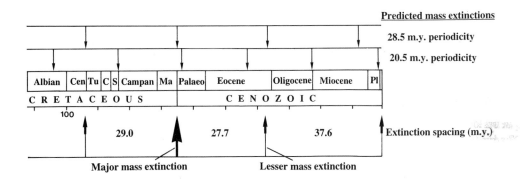

assigned a 248.2 Ma age to the boundary. However, the boundary clay is only 14 cm beneath the conodont-defined boundary; even allowing for slow sedimentation rates, this amount of strata is unlikely to have accumulated in 3 m.y. We therefore suggest that 251 Ma is probably a better estimate for the boundary.

Using this latest time-scale it can be seen that the periodicity is even less apparent (Fig. 1.5). The average mass extinction spacing is 28.5 m.y. but, using this value as the period, only four out of a predicted nine mass extinctions occur at the right time, even allowing for a 5 m.y. error. The end-Triassic and early Toarcian extinctions have a spacing of 20.5 m.y. and, using this period centred around these two mass extinctions, generates the best hit-to-miss ratio (Fig. 1.5): 6 out of 12! This suggests that both the time-scale and the choice of period length do indeed affect the appearance of periodicity. Like Fischer and Arthur (1977) before them, Raup and Sepkoski's most significant error was their choice of a young age for the Permo-Triassic boundary. A consequence of the more realistic older age is that a further mass extinction is predicted to occur somewhere in the Anisian–Ladinian interval of the Middle Triassic (Fig. 1.5). Like many other predicted mass extinctions, this event is missing.

As well as consistently and vigorously defending the periodicity claim, Sepkoski has continuously updated his marine family compendium (Sepkoski 1992a) and produced a compilation for marine genera (Sepkoski 1986). Calculation of the extinction metrics for genera reveals the previously detected mass extinction events with even sharper clarity (Sepkoski 1989), as predicted by Sepkoski (1987). This led him to conclude that: 'These [mass extinctions] are strong signals that show through the noise of imperfect data' (Sepkoski 1993: p. 45). A small Aptian extinction 'appears' in this generic data but the Bathonian–Callovian event obstinately refuses to materialise.

A new, independent compilation of all fossil families has recently appeared (Benton 1993). This is the product of 90 specialists and contains range data for 7186 families including, importantly, terrestrial organisms, from which Benton (1995) has calculated the various extinction metrics. An important finding is that the end-Jurassic 'extinction' is much more important for terrestrial than marine organisms – a fact attributed to the Lagerstätten effect of the Solnhofen Limestone which immediately precedes this extinction (Benton 1995: p. 55). Curiously, Benton (1995) found Campanian and Albian extinction peaks but neither the Aptian nor the end-Cenomanian peaks. The Campanian peak was attributed to a back-smearing (see the next section for a discussion of this effect caused by random range truncations) from the end-Cretaceous mass extinction. The Albian event may be a consequence of the good preservation of Albian plant and insect beds in North America relative to the Cenomanian. 'In addition, the imprecise dating of extinctions of many insect families as simply 'Early Cretaceous' leads to the artifact of build-up of . . . extinctions' (Benton 1995: p. 55).

A minor, end-Early Triassic peak was also found; Raup and Sepkoski had tentatively indentified this event in their data but had excluded it from consideration as it was primarily composed of cephalopod extinctions. Benton's (1995) analysis revealed that it is also a modest extinction event for tetrapods, particularly amphibians. This peak, if a genuine one (cf. p. 116), is far too soon after the end-Permian extinction to be included among the 26 m.y. peaks. Benton (1995: p. 57), in a cogent conclusion, with which we concur, noted that: 'The present data do not lend strong support to this idea [of periodic mass extinctions], because only six, or perhaps seven, of the events are evident.'

Mass extinctions: terminology and analysis

As with any newly evolving field, mass extinction studies have rapidly generated a new list of terms. This argot is designed as a shorthand for recurrent phenomena associated with biotic crises, but it has also served to focus attention on some of the more salient and exigent issues of debate.

The typical sequence of events in a mass extinction is initiated by an extinction phase, during which diversity falls rapidly, followed by a survival or lag phase of minimal diversity and then a recovery phase of rapid diversity increase (Fig. 1.6). In practice, the distinction between the extinction phase and the survival phase is rather an arbitrary one as extinctions can continue into the survival phase. For example, the majority of articulate brachiopods became extinct in the Late Permian but several genera persisted into the earliest Triassic only to become extinct in this survival phase. Such taxa, which outlive the majority of their clade, are known as **holdover taxa**: their fate is commonly an evanescent one (Fig. 1.7). **Progenitor taxa**, on the other hand, appear during the extinction or survival phase and rapidly radiate during the recovery phase. The bivalve *Claraia* and the plant *Isoetes* are two examples of progenitor taxa from the earliest Triassic (cf. Chapter 5) while *Mytiloides* is a further bivalve example from the Cenomanian–Turonian event. The ecological preferences of progenitor and holdover taxa can provide important clues as to the nature of environmental conditions during the extinction and survival phases.

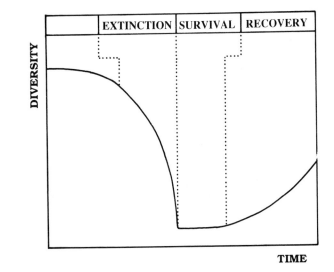

Fig. 1.6 Phases of a mass extinction crisis. After Kauffman (1986), based on information in Donovan (1989).

The survival phase is also characterised by **disaster taxa**, usually long-ranged species of opportunists, whose presence in swarm abundances is a sure sign of elevated environmental stresses. Disaster taxa, and opportunists generally, are typically small, morphologically simple forms: the tube-like morphology of the small foraminifer *Earlandia* is a typical example found in huge numbers immediately following both the end-Devonian and end-Permian extinctions.

The recovery interval is marked by the radiation of both progenitors and the taxa that survived the mass extinction (Fig. 1.7). It is also marked by the reappearance of taxa that had disappeared from the fossil record during the extinction phase several million years earlier. This significant phenomenon of many mass extinctions was documented by Batten (1973) and Waterhouse and Bonham-Carter (1976) and was subsequently named the Lazarus effect by Jablonski (1986a) after the biblical character who was brought back from the dead by Jesus. The absence of **Lazarus taxa** from the fossil record, known as their **outage**, is one of the more intriguing aspects of the survival phase. The importance of Lazarus taxa depends, in part, on their taxonomic ranking, although this has rarely been explored. Kauffman and Erwin (1995) explicitly stated that Lazarus taxa were species, but somewhat earlier Harries and Kauffman (1990) had rather confusingly written of new species arising from Lazarus lineages. In fact all documented Lazarus taxa are either genera or families; no Lazarus species are known to us. This is important because, if Lazarus species did appear after a lengthy absence, it would imply that species-level evolution had somehow ceased during their outage.

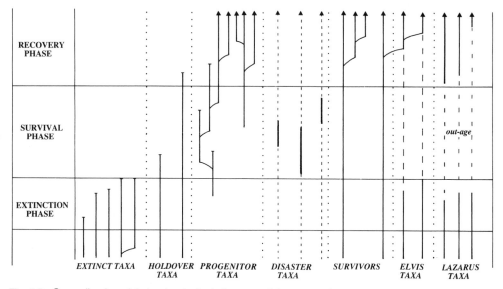

Fig. 1.7 Generalised model showing typical phases and the range of responses of species during and after a mass extinction. Based, in part, on Kauffman and Erwin (1995).

In his original definition, Jablonski (1986a) considered that Lazarus taxa were important because they could be used to monitor the quality of the fossil record by assessing the relative proportion of Lazarus taxa present at any one time – Paul's (1982) and Benton's (1989) simple completeness metric is a comparable measure. In contrast, Kauffman and Erwin (1995) considered that the abundance of Lazarus taxa was a measure of the importance of **refugia** at times of crisis. In their definition, refugia are sanctuaries, such as oceanic islands and 'various deep-sea habitats', where small populations are able to 'emigrate and hide' from the vicissitudes of their normal habitats. Only after the crisis has passed do the Lazarus taxa return to their old habitats. Refugia are implied to be small havens or sites not normally sampled in the geological record. However, quite apart from the problem of how such emigrants could displace resident taxa which were presumably well adapted to their habitat, one is forced to ask why the Lazarus taxa had not already colonised such habitats at times of less stress and larger population sizes. In fact, in the only detailed treatise on the subject, Vermeij (1986) concluded that the recognition of traditional refugia is far from straightforward. For example, the terrestrial fauna of modern, oceanic islands shows some evidence of belonging to relict, primitive lineages, but only at the order level. As Darwin noted long ago, islands and small refuges are sites of rampant speciation and are therefore unlikely to preserve ancient genera or species. Even more importantly, Vermeij (1986: p. 239) observed that 'no molluscan examples of relict marine species . . . confined to oceanic islands are known at present; in fact . . . quite the reverse seems to be the case: relict taxa appear to have continental [shelf] distributions'.

A simpler explanation for Lazarus taxa might lie in the reduction of population sizes at times of extinction. While many species' populations may decline below the minimum viable population sizes and thus go extinct, a certain proportion of species will presumably become very rare but just manage to avoid extinction. They will thus become correspondingly rare fossils that are unlikely to be collected. In this alternative, the proportion of Lazarus taxa becomes just as important a measure of the extinction crisis as the various extinction metrics. That this is the case is indicated by the fact that the most significant Lazarus effects occur after two of the biggest extinctions, the end-Ordovician and end-Permian.

Erwin and Droser (1993) tackled the problem of Lazarus taxa from a different angle. Using the mid-Triassic reef faunas as their example, they suggested that, although the various sponge and algal taxa resembled Late Permian examples, this was a result of convergent evolution. The younger taxa were therefore exhibiting homoplasy and were termed **Elvis taxa** 'in recognition of the many Elvis impersonators who have appeared since the death of the King.' (Erwin and Droser 1993: p. 623). Clearly Elvis taxa tell a different story than Lazarus taxa and it is imperative that their relative proportions are distinguished.

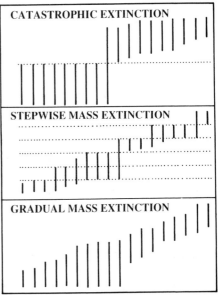

Fig. 1.8 Three end-members for the extinction phase. Vertical lines represent stratigraphic ranges. After Kauffman (1986).

The precise pattern of extinction during the geologically brief crisis has proved to be a further subject of debate, one which has seen the introduction of statistical analysis to range charts for the first time. Kauffman (1986) identified three distinct extinction patterns: abrupt, stepped and gradual (Fig. 1.8). Clearly there can be overlap between these types, but attempts at categorising extinctions are an integral part of several kill mechanisms. Thus, stepped extinction is predicted for the multiple cometary impact kill mechanism (Hut *et al.* 1987), while an abrupt extinction is held as characteristic (but by no means diagnostic) of a single, large bolide impact. Unfortunately, extinction patterns are as a much controlled by the sampling methodology as by the nature of the event itself – hence the need for statistical rigour.

Range charts record the series of point occurrences of fossil species against a stratigraphic column. Such charts give a feel for the relative frequency of occurrence of species; more common species will have shorter gaps between their occurrences than rarer ones. An important point to be remembered when interpreting extinctions in range charts is that the highest occurrence of a fossil species is unlikely to represent the very last individual of that species (Signor and Lipps 1982; Raup and Sepkoski 1982). Thus, all species ranges are artificially truncated to some extent. This will tend to make mass extinctions appear more gradual and slightly earlier in the stratigraphic record: an effect known both as the **Signor–Lipps effect** and **backsmearing**. The effect is readily reproduced by applying a series of random range truncations to an ideal abrupt extinction event (Fig. 1.9; Springer 1990). As Signor and Lipps (1982: p. 291) cogently noted, this effect means that 'gradual extinction patterns prior to mass extinction do not necessarily eliminate catastrophic extinction hypotheses'.

The Signor–Lipps effect is least important for the commonest species, which tend to have short gap lengths (Koch and Morgan 1988). Thus, the extinction level of several common species provides the best approximation of the level of mass extinction

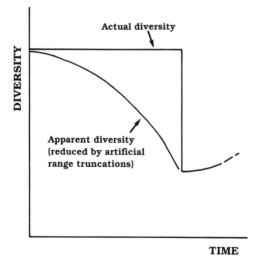

Fig. 1.9   The Signor–Lipps effect: the effect of random range truncations on an abrupt mass extinction. Artificial range truncations (likely to be the norm in the fossil record) produce an apparently gradual extinction. After Signor and Lipps (1982) and Springer (1990).

horizons. The difference in gap length between common and rare species is the fundamental attribute used in several attempts to assign error bars to a species' last (and first) occurrence. Strauss and Sadler (1989) were among the first to determine confidence intervals; however, their calculations require that fossilisation is random, sedimentation rate and facies are constant, and collecting intensity is uniformly distributed throughout the study section. None of these criteria is ever likely to be met. Therefore, in an attempt to provide a more utilitarian method, Marshall (1994) produced a statistical technique which made no assumptions about gap-length distribution beyond that it was continuous. Constant sedimentation rate is no longer a prerequisite in this method, but the requirement that each species has at least seven occurrences (six gap lengths) excludes the rarer species from analysis. In an alternative approach, Raup (1989) applied bootstrapping simulations to gap lengths in order to assign confidence intervals, thereby allowing error bars to be calculated for the position of the end-Cretaceous extinction.

Essentially, confidence intervals provide little more than a 'feel' for the veracity of extinction patterns as the overly restrictive assumptions for their application are never met. In an attempt to marry statistical rigour with the real world of the geological record, Holland (1995) considered the range of fossils, assumed to have a depth-related Gaussian distribution, within the framework of the Exxon sequence stratigraphic model. This showed that a gradual mass extinction can be made to appear abrupt by rapid water-depth changes. One of the most significant of such changes in the sequence stratigraphic model occurs at the maximum flooding surface; the rapid decline of sedimentation rates beneath this surface further concentrates last occurrences, thereby exacerbating the abruptness. An important conclusion of Holland's work is that, for those mass extinctions associated with rapid transgression (and there are a number), the fossil record must be searched in the expanded, nearshore sections because the condensed, offshore sections may give an abrupt pattern of last occurrences that is merely the result of low sedimentation rates. A similar but lesser effect may also be produced by sequence boundaries, although the decline in sedimentation rates beneath such erosive surfaces may not be as significant.

The main contribution of the statistical assessments has been to demonstrate that, where the extinction phase is of short duration (i.e. less than 1 m.y.), it is not possible to distinguish between the three alternatives of Figure 1.8. Holland's (1995) work has also emphasised that fossil occurrences are not data points in a void; the control of shifting facies belts and sedimentation rate fluctuations must also be considered.

Mass extinctions: different in kind?

Originations and extinctions have occurred throughout the fossil record, with the continual increase of global diversity indicating that the former have, on average been more common than the latter (Benton 1995). The only major reversals to this trend have occurred during mass extinctions, with the huge end-Permian mass extinction in particular resetting the diversity 'clock' back to Early Ordovician levels (Fig. 1.2; Sepkoski 1984). The current interest in mass extinctions has engendered a debate as to whether these events are different from background extinctions or whether they are simply the product of normal extinction processes with the volume turned up. This debate has in turn spilled over into a longer-running one on the ultimate cause of Phanerozoic diversity increase (see, for example, Van Valen 1973, 1984; Signor 1982; Sepkoski 1984; Raup and Boyajian 1988; Gilinsky and Good 1991; Courtillot and Gaudemer 1996).

Statistical analysis initially revealed that the big five mass extinctions were quantitatively too large to be considered the tail end of a continuum of background extinction intensities (Raup and Sepkoski 1982; Fig. 1.1). This conclusion has failed the test of further analyses (McKinney 1987; Hoffman 1989b,c; Hubbard and Gilinsky 1992) although, as Raup (1986: p. 1529) noted, 'this does not deny the possibility that [background and mass] extinctions are qualitatively different from each other'.

Dave Jablonski, a Chicago colleague of both Raup and Sepkoski, has undertaken the most persistent attempt to distinguish qualitatively background and mass extinctions, primarily using a bivalve and gastropod data set from the Late Cretaceous–Palaeocene of North America (Jablonski 1986a,b, 1989, 1991). He has noted that, during background times, extinction rates are lower for those genera with planktotrophic larvae, broad geographic range of constituent species and high species richness. During the end-Cretaceous mass extinction none of these survival traits appears to have operated and only those genera (but not species) with broad geographic ranges preferentially survived. This observation is presumably related to the global nature of the end-Cretaceous crisis, although it is surprising that broadly distributed species did not fare better. However, in a trenchant analysis of these patterns, Hoffman (1989a) noted that Jablonski's data really only pertained to one province, making conclusions concerning geographic distribution somewhat suspect. Looking at similar data from western Europe, Hoffman could find no distinction between background and mass extinctions. Neither was Erwin (1989) able to detect any differences in gastropod extinction patterns before and during the end-Permian crisis. Jablonski's observations may not, therefore, constitute general 'rules' for mass extinctions.

Tropical taxa are commonly considered to be preferentially affected during mass extinctions (see, for example, Waterhouse and Bonham-Carter 1976; Copper 1977; Sheehan 1979; S.M. Stanley 1988; Banerjee and Boyajian 1996) – an occurrence that might relate to the greater endemicity of tropical taxa rather than selection for cold-tolerant taxa *per se* (Jablonski 1986b). In contrast, Raup and Boyajian's (1988) demonstration that reef taxa (the quintessential tropical taxa) extinction rates are not proportionally elevated relative to level-bottom taxa during mass extinction intervals argues against latitudinal bias. For several mass extinctions the qualitative pattern is probably sufficient to infer a tropical bias to extinction (for example, the end-Ordovician extinction) but it has rarely been quantitatively demonstrated. A problem with interpreting latitudinal selectivity is the much higher standing diversity of the tropics relative to higher latitudes. With more taxa to kill, simple metrics such as total extinction will inevitably be higher in the tropics. A metric normalising for regional diversity must be applied to test truly for latitudinal bias.

Mass extinctions have not yet been shown to be either qualitatively or quantitatively different from background extinction patterns and, in his characteristically polemical attack on Jablonski, Hoffman (1989a: p. 257) argued that 'mass extinctions could not rightly be called mass extinctions at all for they . . . represent artificial collections of causally-independent phenomena instead of causally coherent events'. That some mass extinctions may have multiple causes is of course true, but Hoffman missed the point; mass extinctions are unique in recording global crises that, by virtue of their magnitude, are capable of removing dominant taxa and destroying whole communities. They thus 'provide opportunities for diversification of taxa that had been minor constituents of the pre-extinction biota' (Jablonski 1986b: p. 313). All extinctions affect the course of history but mass extinctions cause irrevocable, fundamental and unpredictable changes in the large-scale evolutionary and faunal dominance trends of background times.

The biological significance of mass extinctions

Darwin was firmly of the opinion that biotic interactions, such as competition for food and space – the 'struggle for existence' – were of considerably greater importance in promoting evolution and extinction than changes in the physical environment. This is clearly brought out by this quotation from *The origin of species*:

Species are produced and exterminated by slowly acting causes . . . and the most important of all causes of organic change is one which is almost independent of altered . . . physical conditions, namely the mutual relation of organism to organism – the improvement of one organism entailing the improvement or extermination of others.

This Darwinian view has been accepted uncritically by generations of evolutionary biologists, but it is not in accord with the Phanerozoic record of extinctions. Simultaneous extinctions of groups of different biology and habitat, on a global or regional scale, clearly imply adverse changes in the physical environment, on a scale that could not have been adapted for by classic Darwinian microevolution during more normal times. In other words, to use the laconic language of Raup (1991), their extinctions could have been a matter of bad luck rather than bad genes. This theme has also been taken up by the decidedly unlaconic Stephen Jay Gould (1989), who has emphasised the dependence of evolution on a succession of historical contingencies and opportunistic responses. Mass extinctions have not operated randomly. Features evolved adaptively during normal times have proved insufficient to prevent extinction at times of catastrophic change.

What Darwin had in mind when he discussed the 'struggle for existence' was the eventual success of adaptively superior organisms, a concept that has received a modern formulation by Van Valen (1973) with his well-known Red Queen hypothesis. This type of competition implies that later arrivals on the evolutionary scene can displace the inhabitants of given ecological niches and is hence appropriately termed 'displacive competition' (Hallam 1990c). On the other hand, success may favour the incumbents. The earlier species occupant of a niche would stay there until some physical disturbance caused its elimination. Only after the niche had been so vacated could another species, which could be a direct evolutionary descendant, come to reoccupy it. This alternative phenomenon can be termed 'pre-emptive competition' (Hallam 1990c) and implies that the prime motor of evolutionary change is, contrary to Darwin's belief, the physical not the biotic environment. Indeed, evolution could conceivably grind to a halt in the absence of abiotic change, a thought that has led Stenseth and Maynard Smith (1984) to formulate the so-called stationary model of evolution, as opposed to the Red Queen model characterised by continuous biotic interactions.

It has become increasingly apparent that in general the fossil record of both vertebrates and invertebrates supports the pre-emptive model (Benton 1987; Hallam 1990c). This is most evident from the record of mass extinctions and subsequent radiations, for example, the end-Mesozoic extinction of dinosaurs and early Cenozoic radiation of the mammals, but it is clear also from a host of lesser extinction and radiation events. Rosenzweig and McCord (1991) believe that the most significant phenomenon for long-term evolutionary progress may be the process of what they call 'incumbent replacement'. New clade species acquire a key adaptation that gives them a higher competitive speciation than the old clade sources of replacement of extinct species. Incumbent replacement proceeds at a rate limited by the extinction rate and often seems to be linked to mass extinction events.

Knoll (1984) and Traverse (1988) have maintained that the turnover of plant taxa with time exhibits a fundamentally different pattern from that of animals, with plants being less vulnerable to mass extinction episodes and with displacive competition playing a major role. However, Dimichele and Aronson (1992) dispute this for the Carboniferous–Permian vegetational transition, which they consider to be replacive rather than displacive. Taxa that originated in peripheral, drier habitats in the Late Palaeozoic tend to be subgroups of seed plants. The life histories of seed plants suggest a priori a greater resistance to extinction than most groups of 'lower' vascular plants, such as ferns, lycopsids, and sphenopsids. This prediction is confirmed by the nearly continuous expansion of seed-plant diversity since the Palaeozoic, at the expense of lower vascular plants (Niklas *et al.* 1985).

Palaeoenvironmental analysis

In order to determine as far as possible the relationship between mass extinctions and environmental change the maximum effort should be made to extract relevant facies and geochemical information from the stratigraphic record. Little need be written here about the use of sedimentology and palaeoecology in facies analysis, which is thoroughly dealt with in a number of textbooks, beyond recommending facies analysis within a framework of sequence stratigraphy (Walker and James 1992). It is worth summarising, however, the use of certain isotopes and trace elements in geochemical analysis, because this has provided some key evidence bearing on environmental change at and across mass extinction horizons.

Carbon isotopes

The isotopic ratio of ^{13}C to ^{12}C has proved probably the most informative, considering the Phanerozoic as a whole. Marine biomass is depleted by about $25 \pm 5‰$ in ^{13}C compared with marine bicarbonate, the principal reservoir of inorganic carbon in the environment. The average $\delta^{13}C$ value for oceanic bicarbonate ranges from about $-1‰$ to $0‰$, while that for oceanic carbonate ranges from -0.5 to $1‰$. Therefore precipitated carbonate monitors dissolved bicarbonate in seawater quite closely (Holser *et al.* 1988). As the isotopic ratio in organic matter (reduced carbon reservoir) is considerably lower than in the carbon dioxide-carbonate (oxidised) reservoir, sequestration of organic matter is one of the prime controls on the ^{13}C to ^{12}C ratio in seawater. Other things being equal, the greater the rate of organic carbon burial, the higher the ratio of ^{13}C to ^{12}C in a well-mixed world ocean will be. Thus an oceanic anoxic event is likely to be recorded as a positive excursion of δ^{13}C carbonate (Arthur *et al.* 1987).

In the present ocean system the calcareous skeletons of plankton are characterised by positive δ^{13}C values because the dissolved bicarbonate in surface waters is relatively depleted in ^{12}C due to photosynthesis. On the other hand, bottom waters are relatively enriched in ^{12}C due to respiration of isotopically light organic carbon, and benthic foraminifera are characterised by this. For this reason a positive surface- to bottom-water gradient develops. If for some reason there is a drastic decline in organic productivity in surface waters, the gradient disappears, and δ^{13}C carbonate will record a strong negative shift. Such a phenomenon at the Cretaceous–Tertiary boundary led Hsü and McKenzie (1985) to propose the term Strangelove Ocean, after the mad scientist character memorably played by Peter Sellers in the eponymous film of Stanley Kubrick. The kind of catastrophe explored with excessive enthusiasm by Dr Strangelove was a nuclear holocaust, but the mass destruction of life could have another cause, such as extraterrestrial impact.

Negative shifts of δ^{13}C in both carbonate and organic matter can be more extended through the stratigraphic succession and need not necessarily signify a catastrophic fall in productivity. Thus there could be a net relative flux of carbon from the total organic reservoir to the total carbonate reservoir, either by decrease of burial and storage of new organic carbon (relative to carbonate) or by increase in oxidation of older organic carbon, for instance through erosion of coal beds during times of regression (Magaritz *et al.* 1992).

Sulphur isotopes

As with carbon isotopes, the principal fractionation is biologically mediated and is faithfully recorded in sediments. The decisive control in the ocean is sulphate reduction by bacteria. The δ^{34} of seawater sulphate is about 20%, while the δ^{34}‰S of bacteriogenic sulphides ranges between -40–0‰. Long-term carbon and sulphur isotope curves through the Phanerozoic have been plotted (Holser *et al.* 1988) but the latter depends on the isotopic composition of sulphates, which are only sporadically distributed through the stratigraphic record. Since short-term changes are of interest in the study of mass extinctions, it is very desirable to study the isotopic composition of the ubiquitous iron sulphide mineral pyrite, but unfortunately so far very little such research has been done.

Sulphides in sediments and anoxic waters are commonly depleted in ^{34}S by 45–70‰ relative to seawater sulphate, far beyond the apparent capabilities of sulphate-reducing bacteria. This can be explained by cycles of fractionation accompanying bacterial disproportionation of sulphur, followed by sulphide oxidation. Repeated cycles of oxidation to sulphur and subsequent disproportionation should produce sulphides much more depleted in ^{34}S than those produced in the initial reduction of sulphate to sulphide (Canfield and Thamdrup 1994). This has an obvious bearing on the use of sulphur isotope ratios in establishing anoxic conditions.

Oxygen isotopes

The use of oxygen isotope ratios, expressed as δ^{18}O values, is familiar in palaeotemperature studies, most notably in the Cenozoic, with negative shifts indicating higher temperatures, but has figured little in mass extinction studies because of the serious problem of post-depositional alteration in diagenesis or contact with meteoric ground-waters. Consequently, oxygen isotope ratios are recorded through the Phanerozoic much less faithfully than carbon or sulphur isotope ratios. Even for the younger Cenozoic, in comparatively unaltered material from deep-sea cores, it is difficult to disentangle a purely temperature effect from a salinity effect, because the melting of isotopically light polar ice, which is frozen freshwater, effects an isotopic change in seawater, whose salinity is slightly lowered. Fortunately for the study of climatic cycles and isotope stratigraphy, relatively warm intervals correlate with ice melting, and so the isotope shifts are in the same direction.

Strontium isotopes

In the modern ocean the ratio of ^{87}Sr to ^{86}Sr is predominantly the result of mixing of two major strontium fluxes, continental weathering (average river values 0.712) and inter-action of seawater with basalt (0.703) at mid-oceanic ridges. Strontium has a long residence time in seawater (about 4 m.y.) compared with present ocean mixing times, and consequently ^{87}Sr values are homogenous in the modern ocean (Veizer and Compston 1974; Burke *et al.* 1982).

The isotopic values of calcareous fossils and marine carbonates correspond to average coeval seawater values. They have been shown to vary considerably through time and are proving increasingly valuable for stratigraphic correlation, at least in the younger part of the column (Veizer 1989). The ^{87}Sr is radiogenic strontium generated by decay with time of ^{87}Rb and derives from weathering of continental rocks. The marked rise in the ratio within the last 100 m.y. results from an overall fall of sea level through this time, with consequential increase of continental area and hence runoff. Indeed, runoff seems to be the key factor in controlling the ratio (Tardy *et al.* 1989).

Iridium

Ever since the celebrated discovery of a pronounced iridium anomaly at the Cretaceous–Tertiary boundary at Gubbio, Italy by Alvarez *et al.* (1980) there has been an intense interest in this trace element as an indicator of bolide (asteroid or comet) impact. Because platinum-group elements (PGEs) such as iridium are greatly enriched in meteorites compared with the Earth's crustal rocks, an extraterrestrial source has been widely invoked for abnormal concentrations of iridium at a number of extinction horizons (in practice, iridium is the element most frequently studied because of its amenability to neutron activation analysis).

A number of cautionary remarks are required, however. To make a persuasive case for bolide impact, iridium values of at least several parts per billion need to be established, and appropriate chondritic ratios of such platinum-group elements as iridium, asmium, platinum, and gold should be determined if possible (Orth *et al.* 1990). It has become evident from study of modern ocean sediments that many of the small PGE 'spikes' found in the stratigraphic record can be accounted for by purely terrestrial processes involving element redistribution in diagenesis, at and across redox boundaries (Colodner *et al.* 1992). Even if the PGE concentrations are extraterrestrial in origin, they could be derived from the more or less continuous rain of micrometeorite dust, at condensed horizons such as those associated with hardgrounds and the basal parts of transgressive systems tracts, to use the jargon of sequence stratigraphy (see, for example, Wallace *et al.* 1991). Iridium concentrations in excess of 1 ppb are known from such horizons, which are not usually associated with mass extinctions (Hallam 1992).

2 Extinctions in the early history of the Metazoa

Although this book is, for obvious reasons, concerned essentially with the fate of Phanerozoic organisms, claims have been made for mass extinction of Vendian faunas immediately prior to the explosive Cambrian radiation, and these must be addressed. It will therefore be necessary initially to say something about correlation of strata across the Proterozoic–Cambrian boundary, a subject of considerable contention for many years. Important episodes of mass extinction also took place towards the end of the Early Cambrian, and a succession of extinctions among the dominant trilobite fauna in the Late Cambrian will also be considered.

Correlation of the Vendian and Lower Cambrian

After intensive research by many workers it appears that no single stratigraphic section provides a complete view of the succession across the Proterozoic–Cambrian boundary and through the Lower Cambrian (Brasier 1989). A consensus has emerged that the best sections globally are in Siberia, Newfoundland, and China. The Siberian Platform is important in providing a link between archaeocyaths and the so-called small shelly fossils (SSFs), and between these and trilobites. The archaeocyaths have the best record for the Lower Cambrian and provide the basis for the Russian stage names Tommotian, Atdabanian, Botomian, and Toyonian (Table 2.1). A section in south-east Newfoundland links SSFs to both trilobites and trace fossils and is demonstrably above the Vendian. Sections in southern China provide a good reference stratigraphy for the Palaeotethys margin of Gondwana.

Table 2.1 Subdivisions and correlation of Vendian and Lower Cambrian

System	Series (Newfoundland)	Stage (Russia)	Stage (China)
Cambrian (Lower)	Branchian	Toyonian	Maozhuangian + Longwangmiaoan
		Botomian	Canglangpuian
		Atdabanian	Qionzhusian
	Placentian	Tommotian	
		Nemakit-Daldynian (or Manykaian)	Meishucunian
Vendian		Kotlinian	Dengyingxian
		Redkinian	

Source: based on Landing (1994) and Brasier (1995).

Problems in correlation involve the absence of trilobites in SSF assemblages, diachronous first appearances and condensation, reworking, and hiatuses. After much discussion it has been resolved by the International Stratigraphic Commission that the boundary be taken at the base of the *Phycodes pedum* zone at Fortune Head, Newfoundland (Landing 1994). The proposal to use trace fossils to define the base of a system developed from the recognition that they do not have the environmentally restricted distributions characteristic of post-Cambrian time, and include ichnogenera and species with a limited stratigraphic range but broad geographic distribution. There is evidently an upward replacement world-wide of the low-diversity, simple, subhorizontal sediment-feeder traces of the uppermost Proterozoic by complex feeding, escape and dwelling burrows. This means that trace fossils have the same significance for zonation and correlation of the boundary interval as body fossils have in younger intervals. Trace fossils are present to the near exclusion of shelly fossils in many siliciclastic-dominated sequences and the ichnofossil zonation in the Proterozoic–Cambrian boundary interval is evidently practical for interregional correlation. On the other hand, boundary-interval SSFs are markedly provincial, and geographically widespread taxa often have long stratigraphic ranges. Furthermore, diverse Lower Cambrian skeletal fossil assemblages are restricted to very shallow marine facies.

In Newfoundland it now appears that diversity increase of trace fossils took place before the appearance of diverse skeletonised metazoans, with the important implication that substantial stratigraphic gaps exist below the basal Cambrian stages in both Siberia and China (Landing 1994: Fig. 1). This introduces an important caveat to the use of carbon isotopes in correlation. For the purpose of such isotopic study carbonate-rich strata are required, and the ones best constrained biostratigraphically are in Siberia. Magaritz *et al.* (1986) and Magaritz (1989) recorded a marked negative shift of $\delta^{13}C$ at the regional Proterozoic–Cambrian boundary immediately following an equally pronounced positive excursion. With the revised placing of the boundary, however, this marked change took place within the basal Cambrian, at the boundary of the Nemakit-Daldynian and Tommotian stages (Brasier 1995; Table 2.1). The comprehensive review of carbon isotope correlation by Brasier *et al.* (1990) indicates a far from straightforward picture. Landing (1994) notes that the dramatic negative excusion in $\delta^{13}C$ values in the Siberian lowermost Tommotian and the positive excursion (Dahai maximum) of Meishucunian B in southern China coincide with abrupt faunal replacements at apparent unconformities, suggesting that the abruptness of the changes might reflect missing stratigraphic section. Evidently carbon isotope stratigraphy cannot be treated as a panacea for correlation at this time interval.

It is convenient to summarise here the principal results of isotope studies involving strontium and sulphur as well as carbon, for Vendian and Cambrian time (Fig. 2.1). The strontium isotope results are based on marine carbonates and the sulphur isotope results on marine evaporites and are important only for establishing broad trends through time. Only the carbon isotope results allow the possibility of monitoring temporal changes through brief stratigraphic intervals (see, for example, Brasier *et al.* 1994). Figure 2.1 shows a Middle Vendian peak in $\delta^{13}C$ and two others in the Early and Late Cambrian, the former preceded by a pronounced trough. In more detail the Lower Cambrian contains multiple peaks and troughs. The $^{87}Sr/^{86}Sr$ ratio rises sharply in the Middle Vendian and less sharply in the Late Vendian–Nemakit-Daldynian interval. There is a pronounced drop in ratio at the base of the Tommotian, followed by a rise towards the Middle Cambrian (Derry *et al.* 1994). The δ^{34}S curve shows a rise through the Vendian to a peak in the Early Cambrian, followed by a slight fall and rise through the rest of the period. The possible environmental significance of these changes will be considered in due course.

Vendian extinctions

Extinctions have been claimed for three distinct groups of organisms: acritarchs, trace fossils, and the celebrated Ediacara fossils.

Most acritarchs probably represent the cyst stage in the life cycle of planktonic algae. Proterozoic acritarch biotas attained a maximum of diversity and morphological complexity during the Vendian, with the youngest assemblages containing distinctively Proterozoic, large, process-bearing cysts occurring in rocks stratigraphically above the Varangian glacials but below Ediacaran organisms. The latest Proterozoic biotas are dominated by simple, leiosphaerid acritarchs and small, spiny microhystrids (Knoll

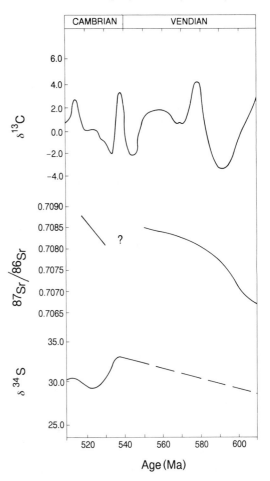

Fig. 2.1 Carbon, sulphur, and strontium isotopic changes in the Vendian and Lower Cambrian. Based on Donnelly *et al.* (1990), Kaufman *et al.* (1993) and Derry *et al.* (1994).

1989). Large acanthomorph acritarchs, that had survived the Varangian glaciation, evidently became extinct during the Vendian because they are entirely absent from well-sampled Lower Cambrian rocks (Mendelson and Schopf 1992). Originally Vidal and Knoll (1982) inferred a short-term, sharp fall in diversity in the latest Proterozoic, but Schopf (1992) questions this, arguing on the basis of a large subsequent increase in data that the Late Proterozoic diversity decline was gradual, after reaching a maximum about 950–900 Ma. This view is clearly at variance with that of Knoll (1989), who maintained that the diversity peaked much later, in the Vendian, but even in this case there seems to be no clear evidence of a mass extinction episode close to the era boundary. At any rate, the main diversity decline seems to have preceded the appearance of the Ediacara fauna (Knoll and Walter 1992).

According to Crimes (1989), a number of trace fossils are restricted to the Vendian, namely the ichnogenera *Bilinichnus*, *Harlaniella*, *Nenoxites*, *Vimenites*, *Intrites*, *Vendichnus*, and *Palaeopascichnus*, but an equal number survived into the Phanerozoic, so that any claim for mass extinction would be unjustified. It could, of course, be argued that ichnotaxa represent behavioural groups, of which very few have become extinct since the beginning of the Phanerozoic.

The most plausible case for mass extinction at the end of the Proterozoic concerns the Ediacara fauna, elements of which have been discovered in almost every continent (Brasier 1989). This 'fauna' is unquestionably enigmatic and has been subjected to widely different interpretations, with doubt even being thrown on whether it is a fauna at all. The conventional interpretation has been that the principal components are impressions of cnidarian polyps and medusoids, annelids, and subordinate anthropods, but with distinctive body plans that disappeared from the record before the start of the Cambrian (Glaessner 1984; Runnegar and Fedonkin 1992). However, Seilacher (1984, 1989, 1992) has challenged this interpretation, arguing that few if any of the Ediacaran fossils can be referred to living phyla. He proposed instead a new animal 'kingdom', the Vendobionta, in which foliate shapes, large sizes, and the necessary compartmentalisation were achieved by quilting of the skin rather than by multicellularity. Thus the genus *Spriggina*, which has been interpreted as having many of the characters that might be expected in a primitive trilobite, has been reoriented by Seilacher (1989) so that the purported head shield becomes a holdfast, and 'appendages' a foliate body. Such reorientation is mandatory if *Spriggina* is to be accommodated within the extinct Vendobionta. This characteristically original and ingenious interpretation, which of course implies an end-Proterozoic extinction, has, however, received a sceptical response from other experts in the field (Runnegar and Fedonkin 1992; Conway Morris 1993). It does seem, however, that the peculiar 'quilted' morphology of many 'petal organisms' and the unusual concentric symmetries of many so-called medusoids are unique to the Vendian, and such apparently soft-bodied organisms of large size and surface area are virtually unknown from younger strata.

Interpretation of the Ediacaran fossils has been thrown into further confusion by Retallack's (1994a) equally original interpretation of them as lichens, on grounds of morphology, growth, internal structures, habitats, and preservation. In particular, they are believed to have possessed exceptional resistance to compaction, comparable to logs but not jellyfish and other Metazoa. Furthermore, the microscopic structure of permineralised specimens is held to provide compelling evidence of lichen affinities.

Retallack makes the cogent point that many of the Ediacaran fossils, with a high surface-to-volume ratio, are unusually large for early metazoans, because certain examples of these are only a few millimetres in diameter, whereas 'vendobionts' can reach 1 m.

With three radically different interpretations to choose from, it is difficult for an uncommitted observer to come to definitive conclusions, except perhaps to accept that a group of distinctive organisms disappeared rather dramatically at or near to the end of the Vendian, in a manner that is at least consistent with mass extinction. The disappearance cannot be dismissed as a function of taphonomy because well-preserved soft-bodied organisms are known from the Chenjiang (Lower Cambrian, China) and Burgess Shale (Middle Cambrian, Canada) faunas (Brasier 1989). These rich faunas are almost totally different from those in the Vendian, though Conway Morris (1993) claims that a few Burgess Shale fossils do indeed resemble Ediacaran taxa.

Provisionally accepting, therefore, a major extinction event, the question arises as to whether it was relatively gradual or sudden, and what kind of stratigraphic gap exists between the last Vendian and the first undoubted Cambrian fossils. According to Brasier (1989), the evidence from Avalonia (including Newfoundland), Baltica and north-western Canada clearly points to the disappearance of the Ediacaran fossils prior to the appearance of the *Phycodes pedum* assemblages that heralds the diversification of more penetrative burrows and arthropod limbs. The interregnum is marked by a presumed medusoid assemblage of low diversity, by shallow horizontal traces of *Hartaniella* and *Palaeopaschichus*, and by levels of abundant organic remains of *Vendotaenia* and *Tyrasotaenia*. In most sections, unfortunately, there is a clear stratigraphic hiatus associated with unconformity at the base of the Cambrian. Nevertheless, Brasier considers that the most complete successions of Canada and Siberia suggest that the so-called petalonanean fauna disappeared some time before the first appearance of skeletal fossils and diverse trace fossils of Nemakit-Daldynian type. New biostratigraphic and radiometric data reported from Namibia, however, suggest that the Ediacara fauna survived at least until close to the base of the Nemakit-Daldynian (Grotzinger *et al.*, 1995). These authors argue that there was a greater overlap of Ediacaran and Cambrian organisms than previously recognised. Thus it is claimed that skeletonised organisms first appeared below the Precambrian–Cambrian boundary and that Ediacaran organisms could have existed concurrently with other major invertebrate groups, including macrophagous predators. This latter claim, at least, is likely to prove controversial.

One of the best sections for studying an apparently more or less continuous fossiliferous succession from the Vendian to the Cambrian is in the Wernecke Mountains of the Yukon, Canada (Narbonne and Hoffman 1987). Ediacaran macrofossils occur in the upper part of the Windermere Supergroup. This is disconformably overlain by the Vampire Formation, which contains in its basal beds small shelly fossils including *Anabarites* and *Protohertzina*, together with the trace fossils *Cruziana* and *Rusophycus*, which can be clearly assigned to the Lower Cambrian. The strata with Ediacaran fossils comprise alternating beds of carbonate and fine siliciclastic units, mainly siltstone and shale, with thin sandstone storm beds. Soft-bodied macrofossils are found only in these siliciclastic units, which are thought to signify a shallow sublittoral environment. A total of 14 species of metazoan, one species of metaphyte and five trace fossil taxa are recorded, with 11 of the 14 metazoan species being referred to nine 'medusoid' genera; among the ichnofossils only *Planolites* is common. The highest Ediacaran soft-bodied faunas occur in siltstone unit 1, which is overlain by carbonates of the Risky Formation, with only the trace fossil *Palaeophycus* recorded. Taking into account Narbonne and Hoffman's detailed stratigraphic logging, it could be misleading to infer, as Conway Morris (1993) does, that the Ediacaran fauna disappeared below the Proterozoic–Cambrian boundary, because the fossil disappearance evidently coincides with the loss of siliciclastic facies, implying a regional environmental control.

The Early Cambrian radiation

In a book devoted to mass extinctions and their aftermath it would be a gross distortion to refer to the celebrated Cambrian radiation as a mere recovery after the preceding Vendian extinctions. Instead it must rank as probably the most spectacular event in the whole fossil record. As such it has been treated fully elsewhere (see, for example, Brasier 1979; McMenamin and McMenamin 1990; Bengtson 1994) and much discussed in terms of the exploitation of evolutionary opportunities consequent upon environmental changes. Accordingly it need only be considered briefly here. Before this is done it should be noted that there is still some doubt as to whether the Ediacaran and Cambrian radiations are one and the same, or two separate radiations.

The pattern of metazoan generic diversity in the Upper Vendian and Cambrian (Fig. 2.2) illustrates clearly how dramatically rapid the Early Cambrian radiation event was, spanning a mere 15 m.y. or so in the Tommotian and Atdabanian; later Phanerozoic radiations were much slower (Sepkoski 1992b). The archaeocyath peak in Fig. 2.2 reflects a massive radiation of this reef-building group in the late Early Cambrian,

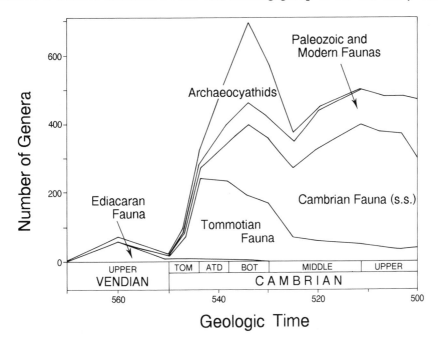

Fig. 2.2 Diversity of metazoan genera in the upper Vendian and Cambrian. After Sepkoski (1992b).

representing about 240 genera in the Lower Botomian, approximately half the diversity of all other metazoans at this time. Some 120 genera of mostly soft-bodied organisms are recorded from the Burgess Shale alone, compared with 70 Ediacaran genera world-wide, a fact which puts into perspective the relative diversity of Vendian and Cambrian faunas.

The radiation involved both skeletal and trace fossils, so cannot be attributed solely to a biomineralisation phenomenon. All well-mineralised phyla living today are present in the Cambrian record, which suggests that most or all of living 'soft-bodied' phyla were also in existence then. In the Burgess Shale only about 20% of metazoan species had normally preservable hard parts. Mineralised shells appeared independently at the Vendian–Cambrian boundary in at least four groups: arthropods, molluscs, hyoliths, and brachiopods. The oldest sponge spicules, of hexactinellid type, also occur in the oldest Cambrian. Calcareous, phosphatic, and siliceous skeletons all appeared at about the same time in an astonishing diversity of shapes: spicules, shells, sclerites, teeth and spine-like processes, cones, and cap- and plate-shaped structures (Bengtson 1992). It is believed by Knoll *et al.* (1993) that the widespread and dramatic appearance of calcified microbes near the Proterozoic–Cambrian boundary reflects the emergence of skeletons as the principal agents of carbonate deposition. Supersaturation levels of seawater are inferred to have decreased sharply, curtailing the distribution and abundance of non-skeletal precipitates. In consequence, cyanobacterial sheaths were more commonly deposited under conditions where they served as sites for $CaCO_3$ nucleation induced by bacterial decay. There was also a major radiation of acritarchs at the start of the Cambrian (Knoll and Walter 1992).

The large expansion in diversity of trace fossils is recorded by Crimes (1989). No fewer than 21 ichnogenera appear in the Tommotian and Atdabanian, including such familiar taxa as *Cruziana*, *Rusophycus*, *Palaeodictyon*, *Rhizocoralliuon*, *Diplocraterion*, *Chondrites*, *Nereites*, and *Asteriacites*. They testify to a mixture of shallow and deep burrowing as well as crawling traces, implying a significant increase in bioturbation compared with the Vendian.

According to Landing (1994), the earliest Cambrian radiation was a two-stage process featuring replacement of uppermost Proterozoic trace producers by Phanerozoic-aspect assemblages and a subsequent appearance of diverse skeletonised metazoans. Among the latter, Brasier (1990) recognises the following sequence.

1. Phosphatic and calcareous tubes and small phosphatic jaw-like protoconodonts, in the lower parts of the Nemakit-Daldynian stage.

2. In the upper Nemakit-Daldynian or lower part of an extended Tommotian, tubes and protoconodonts are joined by cap-shelled and coiled molluscs, hyolith tubes and phosphatic sclerites.

3. Trilobites appear in the Atdabanian/Qiongzhusian, implying delayed calcification. There was a general switch to calcitic skeletons from Tommotian times onwards (Brasier 1995).

Brasier believes that the strong association between skeletal preservation and phosphogenic events may have been largely controlled by taphonomic factors, in that the first widespread appearance of small shell spaces and abundant faecal/organic substrates could have provided suitable traps for the accumulation of grainstone phosphorites for the first time in the stratigraphic record.

Late Early Cambrian extinctions

A perusal of Fig. 2.2 indicates a marked fall of diversity to a low in the early Middle Cambrian. This largely reflects a precipitous decline in the number of archaeocyath genera, with the group almost going extinct, though it survived at a low diversity level until the Late Cambrian. The extinctions also affected other groups, including trilobites (especially redlichiids), hyoliths, anabaritids, and tommotiids (Debrenne 1991; Sepkoski

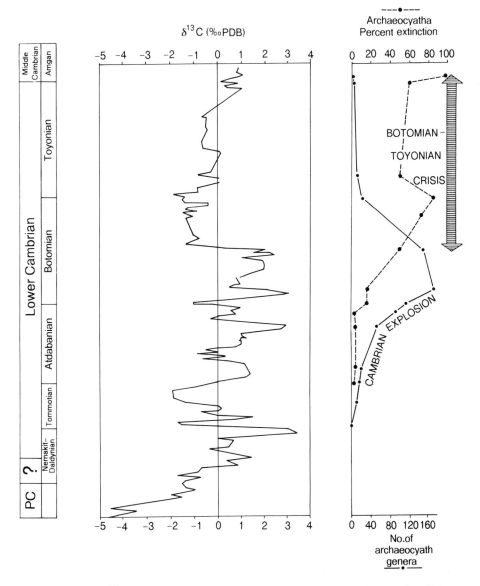

Fig. 2.3 Composite δ^{13}C profile for Siberian Platform and exponential diversification of reef biota followed by mass extinction during Botomian–Toyonian crisis. Simplified from Brasier *et al.* (1994).

1992b). Thus many of the most distinctive Early Cambrian faunas had disappeared by the end of the subperiod.

Further details of the archaeocyath extinctions, based on the well-studied Siberian sections, are given by Brasier *et al.* (1994). As indicated in Fig. 2.3, there was a sharp rise in diversity until the Early Botomian, followed by a fall to a very low level in the Toyonian, with the percentage of genera going extinct reaching maxima at about the Botomian–Toyonian and Lower–Middle Cambrian boundaries. Figure 2.3 also shows that this so-called Botomian–Toyonian crisis correlates with a time of largely negative δ^{13}C.

Zhuravlev (1996) and Zhuravlev and Wood (1996) consider that the extinctions can be resolved globally into two events (cf. Brasier 1995). The earlier and more severe, mid-Botomian event, named the Sinsk Event, is associated with the rapid decline of tropical shallow-water benthos, but coeval temperate and deep-water faunas remained unchanged. This, and the younger, Early Toyonian, Hawke Bay Event, eliminated or drastically reduced much of the so-called Cambrian biota, including reef-associated forms such as archaeocyaths, radiocyaths, cribicyaths, and coralomorphs. Subsequent reef-building in the Middle Cambrian was dominated by calcified cyanobacterial communities, especially *Epiphyton* and *Renalcis*, with no metazoan component. *Renalcis* itself disappeared at the beginning of the Late Cambrian, leaving only thrombolitic-stromatolitic communites in shallow tropical environments. Metazoans did not resume a role in reef-building until the latest Cambrian–Early Ordovician.

Possible causes of the Vendian–Early Cambrian biotic events

Although the evidence is not as decisive as for events within the Phanerozoic, it seems reasonable to accept that there was a mass extinction event affecting the Ediacaran organisms at the end of the Proterozoic (Brasier 1989). While it has been tentatively suggested that they may have gone extinct because they lacked defences against newly emergent Cambrian predators (McMenamin and McMenamin 1990), this is unlikely, because the general view has been that the organisms in question had apparently disappeared by about the end of the Redkinian stage of the Vendian, well before the appearance of diverse skeletal and trace fossils (Brasier 1989), although this has been challenged recently, as already noted, by Grotzinger *et al.* (1995). Most plausibly though, some event or events in the physical environment should be sought.

Sea-level changes and associated anoxia bound up with a so-called regression–transgression couplet, leading to drastic reductions in shallow marine habitat area (cf. Hallam 1989a), appear to be the likeliest candidates (Brasier 1989). There are clear indications of regression probably linked to major plate-tectonic changes that affected parts of the Laurentian and Baltic cratons (with rifting) and Avalonia and Armorica (with accretion; Brasier 1982). In many parts of the world, including Siberia and south China (Landing 1994), there is an unconformity between the Vendian and Lower Cambrian, which is consistent with a global sea-level fall. For Siberia, however, Brasier (1995) claims an episode of widespread emergence followed rapidly by a massive transgression at the Nemakit-Daldynian–Tommotian boundary, just within the Lower Cambrian, rather than at the Vendian–Cambrian boundary. Black shales, associated with a subsequent sea-level rise, were not widespread until latest Redkinian–Kotlinian to

Nemakit-Daldynian times; the associated heavy $\delta^{34}S$ values (Fig. 2.1) suggest extensive bacterial sulphate reduction on the sea floor (Brasier 1995). The global maximum of $\delta^{13}C$ could be due to a phase of increased carbon burial during an oceanic anoxic event. However, in Brasier (1995: Fig. 4) the isotope maximum is seen to coincide with a time of low sea level. The overall picture is consistent with that of Cook and Shergold (1984), who argued that a major episode of global phosphogenesis in the latest Proterozoic–Early Cambrian should have been preceded by a widespread anoxic phase in the oceans.

Only one attempt has been made to associate the end-Proterozoic extinction event with bolide impact. Hsü *et al.* (1985) reported an iridium anomaly associated with a light $\delta^{13}C$ spike above a metalliferous horizon within a black shale succession in China. The carbon isotope results were held to signify a possible 'Strangelove Ocean' effect caused by massive reduction in organic productivity following impact, comparable to what has been claimed for the Cretaceous–Tertiary boundary. Unfortunately for this interpretation, the horizon in question is not at the Vendian–Cambrian boundary but well within the Lower Cambrian, and there are no associated extinctions (Brasier 1989). It is likely, therefore, that the iridium anomaly is associated with a redox boundary effect (cf. Colodner *et al.* 1992) and has nothing to do with impact.

The Early Cambrian radiation was of such a scale and character as to dwarf the importance of any preceding extinctions. This unique event in Earth history has been extensively reviewed by a multitude of authors, most recently in Bengtson (1994). It was clearly due to a combination of interacting physical and biotic factors. Many such factors have been proposed, and the art is selectively to eliminate the less plausible. One popular view is that the key trigger to explosive metazoan diversification and skeletonisation was an increase in atmospheric oxygen above a critical theshold (Lipps *et al.* 1992). The protein collagen, found only in the Metazoa, is essential for providing a skeletal framework and is considered a prerequisite for the attainment of large body size and mobility in animals. Unfortunately there is no direct method of determining former atmospheric oxygen levels, but the stratigraphic record indicates clearly that the Early Cambrian was a time of significant global sea-level rise leading to widespread marine transgressions over the cratons. Brasier (1979, 1982) has indicated how the newly emergent animals could have exploited the opportunity thus afforded to colonise a succession of new, hitherto vacant ecological niches, with their diversity increasing correspondingly with expanded habitat area. This seems plausible enough, but it should be noted that the Lower Cambrian includes widespread black shales, with the associated high values of $\delta^{34}S$ supporting the notion of extensive anoxia in the ocean system, which does not seem to have been confined to temporally restricted anoxic events (Brasier 1995; Donnelly *et al.* 1990). Since extensive spreads of anoxic bottom waters can restrict habitat area as much as regression, there seems to be a need to analyse in much more detail than has been attempted hitherto the relationship between Lower Cambrian diversity and different types of sedimentary facies.

Sea-level rise was probably just one manifestation of extensive tectonic activity associated with the break-up of a late Proterozoic supercontinent. Another was a major phase of phosphogenesis and increased erosion rates on the continents manifested by high strontium isotope ratios. An increasing atmospheric CO_2 content, leading no doubt to global warming, is the likeliest reason for the change from predominantly aragonitic to predominantly calcitic carbonates. The change to predominantly mineralised skeletons may also relate in some degree to this factor (Donnelly *et al.* 1990; Kaufman *et al.* 1993).

With regard to the extinctions near the end of the Early Cambrian, they are most likely to be related to the extensive reduction of habitat area in epicontinental seas due to a global sea-level fall, as argued by Zhuravlev (1996). Zhuravlev terms this the Hawke Bay Event, of Early Toyonian age. This term was proposed by Palmer and James (1980) for an episode of shallowing recorded in Newfoundland and the Appalachians, but it was evidently more extensive and indeed probably global in scale (Brasier 1982; Hallam 1992). Nicholas (1994) argues that a sudden increase in $^{87}Sr/^{86}Sr$ ratios during and after the Early Toyonian may correlate with the beginning of global regression. The data of Brasier *et al.* (1994) demonstrate that the archaeocyath extinctions during the so-called Botomian crisis correlate with a time of mostly low $\delta^{13}C$ values (Fig. 2.3). The authors speculate that this could signify a time of reduced organic productivity associated with the extinction event (cf. Magaritz 1989).

The earlier extinction event of Zhuravlev (1996) and Zhuravlev and Wood (1996), the Sinsk, is dated as mid-Botomian. It is associated with an extensive spread of laminated black shale facies with accompanying positive excursion of $\delta^{13}C$. Since the black shales contain abundant monospecific assemblages of acritarchs, it is suggested that the extinctions were caused by an anoxic event perhaps related to a phytoplankton bloom and resultant eutrophication. Stratigraphic data are as yet, however, insufficient to determine whether or not the Sinsk Event was truly global.

Biomere boundaries: Late Cambrian mass extinctions?

By the Late Cambrian the trilobites had attained their acme of diversity and abundance, and they dominate nearly all fossil assemblages of this age. This is not to say that the trilobites had attained their utopia; on the contrary, the Late Cambrian trilobite record is punctuated by several major extinction and proliferation events: a boom-and-bust pattern rather reminiscent of the later evolutionary history of the ammonoids. Palmer (1965) was the first to recognise explicitly the iterative nature of Late Cambrian trilobite evolution in his studies of the fauna of the Great Basin of the United States. In this region he noted two abrupt trilobite mass extinctions separated by an interval of radiation of the pterocephalids from an oceanic stock that migrated into the area following the demise of the incumbent shelf taxa. This pterocephaliid clade forms the basis of the type biomere (Fig. 2.4) – Palmer's biostratigraphic concept defined as 'a regional biostratigraphic unit bounded by abrupt non-evolutionary changes [i.e. extinctions] in the dominant elements of a single phylum' (Palmer 1965: pp. 149–50).

Palmer's biomere concept has proved highly influential and the ensuing years have seen the discovery of several more examples. Stitt (1971, 1977) identified a Ptychaspid Biomere above and a Marjumiid Biomere beneath the original Pterocephaliid Biomere. A basal Ordovician Symphysurinid Biomere has subsequently been added and further biomeres may occur higher in the Ordovician although they have not been classified as such (Fortey 1989). Four biomeres are thus well established, along with three boundary-defining mass extinctions: two in the Late Cambrian and one at the Cambro-Ordovician (C-O) boundary.

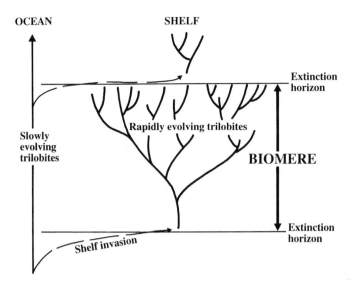

Fig. 2.4 Palmer's (1965) original biomere concept. *In situ* radiation of endemic shelf taxa occurs from an oceanic stock that migrated on to the shelf following the demise of the previous incumbents.

Stitt (1971, 1977) considerably developed Palmer's original concept and recognised four main stages within each biomere (Fig. 2.5):

(1) low-diversity trilobite assemblages consisting of short-lived, highly variable species;

(2) higher-diversity assemblages of longer-lived, less variable species;

(3) diverse assemblages of long-ranging species;

(4) paucispecific assemblages of short-lived species that can occur in huge numbers, including olenid trilobites which are 'interlopers from more peripheral [oceanic] sites' (Fortey 1989: p. 339).

The final, unnumbered stage witnesses a catastrophic extinction of endemic shelf trilobites often in 'only a few inches of rock' (Palmer 1965: p. 150, 1982). As outlined in Palmer's original model (Fig. 2.4), the trilobites of the subsequent Stage 1 are immigrants from continental slope and oceanic areas. Stitt attributed the repetitive evolutionary patterns of the biomeres to an initial adaptive radiation (Stage 1) followed by niche-partitioning and a 'filling-up' of the shelf ecosystem (Stages 2 and 3) and culminating in 'evolutionary senescence' (Stage 4) prior to the sudden mass extinction. The evolutionary senescence stage is more likely to reflect the onset of stressed conditions and the development of opportunistic faunas prior to the main extinction event.

The Palmer–Stitt model, particularly as originally depicted in Fig. 2.4, quickly became established as the shibboleth for interpreting Cambro-Ordovician trilobite evolution. However, in recent years the rigours of cladistic analyses have inevitably challenged this shrine. Thus, both Briggs *et al.* (1988) and Edgecombe (1992) demonstrated that most family-level extinctions at the C-O boundary are pseudo-extinctions of paraphyletic taxa; the Dikelocephalacea appear to be the only major monophyletic clade to disappear

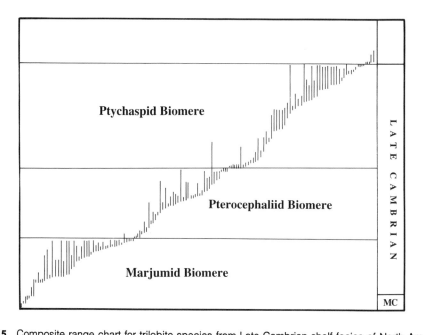

Fig. 2.5 Composite range chart for trilobite species from Late Cambrian shelf facies of North America. Note that short-lived species tend to characterise the initial and final stages of biomeres while longer-ranging species generally appear in the middle of biomeres. After Stitt (1977).

at this level (Fortey 1989). However, not all workers have downgraded biomere boundaries to mere species- and generic-level extinction events. Westrop (1989) considers the C–O mass extinction to have eliminated 50% of trilobite families and virtually all the endemic trilobite species on the North American craton. The original hypothesis of *in situ* evolution has also been challenged by cladistic examination. Many ptychaspid families are rooted in the Middle Cambrian, for example (Edgecombe 1992), and their appearance in Stages 2 and 3 of the Ptychaspid Biomere represents immigration not origination events (Westrop and Ludvigsen 1987). Thus, the timing of trilobite family cladogenesis and extinction is far from resolved but the mass extinction component of the biomere concept is reasonably well established.

In some regards the biomere boundaries do not qualify as mass extinction horizons because, as originally recognised, they are restricted to a single region, the shelf areas of the North American craton, and they involve only a single taxon, the trilobites. Similar patterns of evolution have been claimed for the trilobites of Australia and South-East Asia (Opik 1966), but the timing and details of such events have yet to be documented. Inarticulate brachiopods are one of the few other major components of Late Cambrian shelf assemblages, but they do not show extinction events (Rowell and Brady 1976). The cephalopods underwent a major radiation in the latest Cambrian of China roughly contemporaneous with the Ptychaspid Biomere (Chen and Teichert 1983). This entire fauna (36 genera) disappeared before the end of the Cambrian, suggesting, at face value, a major mass extinction. However, this could well be an artefact of poor preservation; earliest Ordovician cephalopods are very poorly known (Teichert 1986), and there is a likelihood of paraphyletic pseudo-extinction at this level early in the evolutionary history of the group (Fortey 1989).

Despite these reservations, it would be disingenuous to disregard the biomere boundaries as mere taxonomic artefacts within a single taxon in a single region as they represent the wholesale loss of a dominant and highly numerous group across a wide area of one continent. Furthermore, in his recent analysis Brasier (1995) identified contemporaneous extinction events on other continents, thus strenghtening their mass extinction designation. Investigations of facies changes across biomere boundaries provide further intriguing evidence of the reality of these events. In his original description, Palmer (1965) noted that biomere boundaries were sharp and often unrelated to any notable lithofacies change. Subsequent work has tended to confirm these observations (Stitt 1977; Palmer 1984; Westrop 1989) but with the important proviso that, on a regional scale, the boundaries are diachronous. The replacement of high-diversity, endemic shelf trilobites by low-diversity, pandemic, oceanic olenids appears to occur initially in more offshore shelf sites and move progressively shorewards (Ludvigsen and Westrop 1983), although Palmer (1984) doubts this diachrony. If it is accepted that biomere boundaries are diachronous then it suggests that the extinctions are associated with transgression and deepening. There is supporting lithofacies evidence for this conclusion because, in many boundary sections, palecoloured, trilobite packstones are often overlain by darker, thinner-bedded slope carbonates (Westrop and Ludvigsen 1987; Saltzman *et al.* 1995). Interestingly, though, this facies change commonly occurs a short distance (a few tens of centimetres) above the biomere boundary (Palmer 1984) and in some cratonic sections – for example, in central Texas and western Utah – shallow-water carbonate deposition continues uninterrupted across biomere boundaries (Westrop and Ludvigsen 1987; Westrop 1989).

Essentially the trilobite changes at biomere boundaries suggest deepening while the facies changes sometimes do not. A possible resolution of this enigma was hinted at by the work of Saltzman *et al.* (1995) on several Pterocephaliid–Ptychaspid boundaries. These consisted of a hardground overlain by a condensed basal transgressive lag (the trilobite packstones) in which the boundary occurs. The apparently abrupt nature of the extinction may relate to the slow sedimentation rates typical of transgressive phases. Independent evidence for transgression comes from Sr isotope ratios across this boundary which decline from 0.70925 to 0.70910, a trend explained by increased seafloor spreading which in turn provides a driving mechanism for eustatic sea-level rise (Saltzman *et al.* 1995).

A similar sequence of sea-level changes appears likely for the C-O boundary with a minor regression – major transgression couplet suggested by several authors (for example, Westrop and Ludvigsen 1987; Fortey 1989). It therefore appears that the trilobite extinctions are indeed connected with sea-level changes, although in many sections the trilobites are more subtle indicators of such changes than the lithofacies.

Biomere extinction mechanisms

The olenids are intimately associated with the biomere-terminating intervals, and an understanding of their palaeoecology may consequently provide clues as to the extinction mechanism. Throughout their history olenids characterised cold-water settings either in shallow water at high latitudes or in deeper water at low latitudes where they were restricted to continental margins. Olenids only occur in shallow-water, low-latitude sections around biomere boundaries, a fact that has led several authors to propose that the mass extinctions were triggered by the invasion of cold waters into shelf areas either due to transgression (Ludvigsen and Westrop 1983; Palmer 1984) or to a rise in the oceanic thermocline regardless of sea-level changes (Stitt 1977). Olenids also appear to have been specialist dysaerobic taxa as they are commonly encountered as the sole benthos in dark grey, pyritic shales (Fortey 1989). This raises the possibility that low oxygen levels were also implicated in the extinctions (Palmer 1984; Fortey 1989), although Westrop and Ludvigsen (1987) ruled this out due to a lack of any supporting evidence. However, at that time no attempt had been made to gather such evidence. More recently, Saltzman *et al.* (1995) documented a $+1‰$ swing of $\delta^{13}C$ values across the Pterocephaliid–Ptychaspid boundary which, they noted, was of similar magnitude to the excursion produced by organic carbon burial during the Cenomanian–Turonian oceanic anoxic event (cf. Chapter 8). By analogy with this well-documented example they therefore proposed an extinction caused by 'oceanographic changes from a generally stratified ocean to a mixed ocean driven by sea-level rise' in which regional upwelling produced cold, dysoxic conditions across the North American craton (Saltzman *et al.* 1995: p. 895).

Transgression and anoxia are thus the strongest contender for the biomere extinction mechanism, although supporting lithofacies evidence remains to be collected. The influence of regression prior to transgression has also to be fully explored. Both Briggs *et al.* (1988) and Fortey (1989) considered that regression immediately prior to the C-O boundary may have been responsible for many extinctions. However, this regression is a rather minor one and, as Palmer (1984) cogently noted, the biggest Late Cambrian regression on the North American craton occurs in the middle of the Pterocephaliid Biomere – a time of radiation not extinction.

The final potential extinction mechanism deserves only brief mention. Palmer (1982) mooted the possibility of bolide impact as a kill mechanism, but the failure to find iridium anomalies at any biomere boundaries has effectively quashed this alternative (Orth *et al.* 1984).

3 Latest Ordovician extinctions: one disaster after another

Between the end-Cambrian crisis and the next major mass extinction in the latest Ordovician, trilobites continued to flourish. However, their predominance in benthic communities was usurped by the increasingly diverse groups that belong to Sepkoski's (1981) 'Palaeozoic fauna'. These included articulate brachiopods, bryozoans, graptolites, conodonts, echinoderms, and the rugose and tabulate corals. Various mollusc orders (such as the nautiloids, bivalves, and gastropods) also diversified, although to a lesser degree than the Palaeozoic fauna. Peak Ordovician diversity coincided with the mid-Caradoc sea-level highstand which appears to have marked an all-time Phanerozoic high; both basinal black shale and platform carbonate deposition occurred over vast areas at this time (Thickpenny and Leggett 1987; Cocks and Fortey 1988; Brenchley 1989; Barnes *et al.* 1995). Ordovician palaeogeography was undoubtedly a contributory factor to this diversity maximum because the continents each contain their own endemic fauna (Fig. 3.1).

The long-term Ordovician diversity increase was punctuated by several minor and

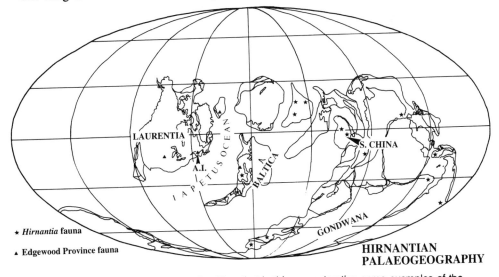

★ *Hirnantia* fauna

▲ Edgewood Province fauna

HIRNANTIAN PALAEOGEOGRAPHY

Fig. 3.1 Late Ordovician palaeogeography. Note that in this reconstruction some examples of the *Hirnantia* fauna occur in equatorial locations. After Scotese and McKerrow (1990).

little-known extinction events (see, for example, Barnes *et al.* 1995). Several appear to have been at least regionally significant. For example, a major end-Tremadoc regression in North America coincides with a significant conodont extinction event (Ji and Barnes 1993). Later in the Ordovician the migration of Baltica and Eastern Avalonia towards each other and, at the same time, their approach towards the south-eastern margin of Laurentia (cf. Fig. 3.1) produced a drastic decrease in provincialism and a concomitant decrease in total diversity (Brenchley 1989). However, this only resulted in generic-level losses and as a biotic crisis it is overshadowed by the calamity which struck at the end of the Ordovician. This was the first of the big five Phanerozoic extinction events, with 57% of genera and in excess of 25% of families going extinct (Sepkoski 1989; Benton 1995). No orders appear to have disappeared, although the Asaphina trilobite order nearly did and an entire class, the Graptolithina, escaped annihilation by a hair's breadth. Before turning to the details of the mass extinction, we examine briefly the biostratigraphic framework available for assessing this event.

Biostratigraphy

Both the conodonts and the graptolites have been utilised for biostratigraphic study in the Ordovician and Silurian; both have their advantages and disadvantages. Graptolites exhibit a high evolutionary turnover and consequently provide the most finely subdivided zonation scheme. However, they are only abundant in the fine-grained facies of oceanic, bathyal, and outer-shelf settings, while the more inner-shelf locations contain few if any graptolites. Conodonts, on the other hand, are common in all shelf facies (but less so in oceanic facies) where they are widely used in zonation. Unfortunately, conodonts illustrate a higher degree of endemism than the graptolites and it is often difficult to cross-correlate between the two zonation schemes and to achieve intercontinental correlation of shelf facies. These problems are manifest around the Ordovician–Silurian (O–S) boundary where the conodont-zoned carbonate shelf facies of Laurentia have proved difficult to correlate with the graptolite-zoned sections of Baltica. As a result two distinct stage

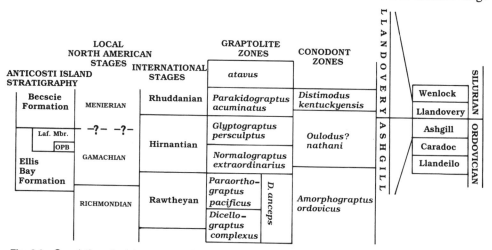

Fig. 3.2 Correlation chart for zones and stages around the Ordovician–Silurian boundary. OPB, Oncolite Platform Bed; Laf. Mbr, Laframboise Member. After Armstrong (1995).

nomenclatures have developed in the two areas (Fig. 3.2). These problems have been at least partly resolved by Armstrong (1995), who provided new conodont range data and applied graphic correlation analysis to Laurentian and Baltican sections. It is now possible to assign extinction levels to precise graptolite zones with some degree of confidence.

The base of the *acuminatus* graptolite Zone has been chosen as the O–S boundary with a section at Dob's Linn in the Southern Uplands of Scotland taken as the international stratotype (Cocks 1985). This level is slightly higher than the level previously taken as the O–S boundary, with the result that the end-Ordovician mass extinction of the older literature has now become a late Ashgill one. The main extinction events approximately coincide with the upper and lower boundaries of the *extraordinarius* Zone, the basal graptolite Zone of the terminal Ordovician Hirnantian Stage. Conodonts are few and poorly preserved at Dob's Linn, but it is probable that the base of the *acuminatus* Zone lies within *Oulodus? nathani* conodont Zone (Melchin *et al.* 1991; Armstrong 1995).

In North America the precise timing of the extinctions has proved difficult to determine because the latest Ordovician Gamachian Stage is absent over broad areas of the former Laurentian continent. One of the few complete O–S boundary sections occurs on Anticosti Island, near Quebec, where the main extinction occurs low in the predominantly Silurian Becscie Formation. Zonally significant graptolites are absent from the O–S sections of Anticosti, and it has long proved difficult to compare these extinctions with those elsewhere in the world. Graphic correlation of conodont ranges from Anticosti, plotted against the standard graptolite zonation, has recently shown that the main extinction level is once again around the *extraordinarius–persculptus* boundary (Armstrong 1995).

What went extinct?

Graptolites

For the graptolites, the late Ashgill crisis was the worst in their history and, but for a few species, the entire group would have been eliminated. However, only a short time before the extinction graptolites were flourishing. For example, several species and the new genus *Paraorthograptus* appeared in the late Rawtheyan (Barnes *et al.* 1995). The end-Rawtheyan extinction was therefore both sudden and unexpected: it eliminated all the dicellograptids and retiolitimorphs and left a highly impoverished fauna devoid of the 'standard' Ordovician biserial graptolite morphology (Fortey 1989). At Dob's Linn, the extinction occurred at the end of the *pacificus* Zone (Wilde *et al.* 1986), and graptolites are absent from the *extraordinarius* Zone apart from a thin band containing the zonal species (Fig. 3.3). In most areas of the world the graptolite fauna of the *extraordinarius* Zone consist of a few surviving species of glyptograptids and normalograptids. In western Laurentia the appropriately named *Climacograptus miserabilis* is the only common species in this immediate post-extinction interval (Berry *et al.* 1990). Even this meagre diversity is not found in the high southern latitudes of Gondwana where graptolites are absent in the late Hirnantian (Berry *et al.* 1990). Only in the equatorial latitudes of China did graptolites temporarily escape the end-*pacificus* disaster: the group remained diverse in this region before their extinction at the end of the *extraordinarius* Zone (Chen and Zhang 1995).

The miserable time for graptolites essentially coincided with the *extraordinarius* Zone and the first signs of recovery are already seen in the succeeding *persculptus* Zone where

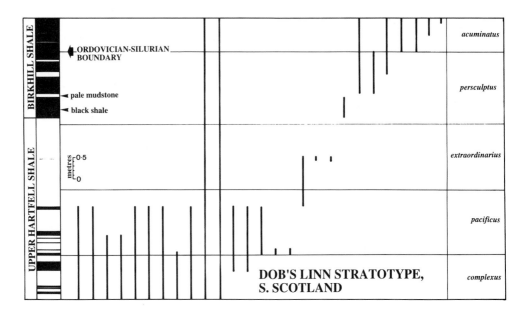

Fig. 3.3 Sedimentary log of the Dob's Linn stratotype from the Southern Uplands of Scotland, showing the alternations of black shales and mudstones and ranges of graptolite species. Graptolite extinctions are concentrated at the end of the *pacificus* Zone and new species begin to appear in the *persculptus* Zone. After Williams (1988).

the first monograptid *Atavograptus ceryx* and the diplograptids make their rare appearance (Fortey 1989; Berry 1996). The radiation continued into the *acuminatus* Zone with the appearance of *Parakidograptus* and *Cystograptus*, although normalograptids remained numerically dominant at this level. Monograptids subsequently underwent a spectacular radiation in the post-Rhuddanian interval of the Llandovery Series: more than 200 graptolite species are known by the end of the Early Silurian (Berry and Boucot 1973; Berry 1996).

Conodonts

The late Ashgill marked a crisis for conodonts that was only slightly less severe than that for graptolites. In the sections of Anticosti, 33 out of 38 species disappear in the interval from the base of the *persculptus* Zone to the earliest Silurian (Barnes and Bergström 1988), while in the higher latitude sections of Gondwana, conodonts entirely disappeared in the Hirnantian. Globally, of the approximately 100 species known from the Ashgill only 20 appear to have survived into the Silurian (Barnes *et al.* 1995) – an extinction that translates to the loss of 18 of 25 genera and 7 of 14 families (Goodfellow *et al.* 1992).

Despite the magnitude of the conodont extinction, the recovery was rapid, at least in lower latitudes. On Anticosti Island new species appear during the extinction interval, thereby producing a 'transition zone' containing both Ordovician and Silurian forms (McCracken and Barnes 1981; Barnes 1988; Melchin *et al.* 1991). By the base of the Silurian 16 new species had appeared that were not present in the early Hirnantian

(Barnes and Bergström 1988). The first Silurian conodonts to appear in higher latitutudes were derived from the surviving warm-water stocks (Barnes *et al.* 1995). Overall, in the earliest Silurian, the seven surviving conodont genera were joined by eight new ones and familial diversity increased from seven to ten (Goodfellow *et al.* 1992). Armstrong's (1996) detailed investigation of the origins of the progenitor taxa in the transition zone has shed an interesting light on the nature of the extinction event. The newly arriving species appear to be derived from deep-water or bathyal stocks, while the extinctions are concentrated among the shelf taxa. The radiation of the bathyal taxa in their new shelf habitats began in the *persculptus* Zone and continued into the Silurian.

Nautiloids

It would be interesting to compare the distinctive extinction/radiation patterns of the graptolites and conodonts with the other major pelagic group of the Ordovician, the nautiloids. Unfortunately, the details of nautiloid ranges across the O–S boundary are poorly known and the available evidence contradictory. Thus, Chen and Rong (1991) noted the disappearance of 87 out of 109 genera in the Ashgill – a major mass extinction – while Teichert (1986) considered that the O–S crisis was a minor event for the group and only noted the loss of one monogeneric order. Both observations may be correct, but it is a little surprising that an 80% generic extinction had virtually no effect at higher taxonomic levels.

Plankton

The record of the plankton during the late Ashgill crisis is available for both the acritarchs and chitinozoans, although at a less than desirable stratigraphic resolution. It is none the less clear that a serious crisis afflicted the base of the Late Ordovician food chain. Grahn (1988) recorded the loss of 9 out of 11 chitinozoan species around the Rawtheyan–Hirnantian boundary – approximately the same level as the graptolite extinctions. A similar crisis for acritarchs is less precisely dated but the *acuminatus* Zone appears to be the diversity low point, perhaps pointing to a preceding Hirnantian extinction event, and it was almost immediately followed by a prolonged radiation throughout the remainder of the Llandovery (Kaljo 1996).

Brachiopods

The O–S interval witnessed a major turnover in brachiopod communities, with more than 150 of the 180 pre-Hirnantian genera failing to survive into the Llandovery. At higher taxonomic levels this translates into the loss of 13 out of 27 families (Sheehan 1982). For superfamilies the extinction is inevitably less marked, with only the Gonambonitacea going extinct, although the Lazarus effect is considerable: 5 super-families (and 30 genera) are not known from the Hirnantian to Rhuddanian fossil record (Cocks 1988; Fortey 1989). When viewed in more detail, the brachiopod extinction can be resolved into two distinct crises separated by a short interval characterised by a highly distinctive, low-diversity, near-cosmopolitan brachiopod community: the *Hirnantia* fauna (Fig. 3.4; Rong and Harper 1988; Harper and Rong 1995). The first crisis came at the end of the Rawtheyan when 60% of genera went extinct. The effects were particularly devastating among shallow-water tropical and temperate assemblages (Robertson *et al.* 1991) although some deeper-water communities did not escape.

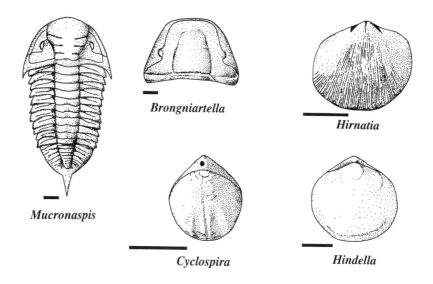

Brongniartella

Hirnatia

Mucronaspis

Cyclospira

Hindella

Fig. 3.4   Examples of brachiopods and trilobites from the early Hirnantian. All scale bars are 1 cm, except for *Cyclospira* which is 5 mm.

For example, the dwarf *Foliomena* brachiopod community that lived in the deep-water, dysoxic environments of Baltica was entirely eliminated (Harper and Rong 1995).

After the first extinction, most *extraordinarius* Zone brachiopod communities consist of the *Hirnantia* fauna; only in a narrow equatorial zone are higher diversity communities of the Edgewood Province encountered. In some palaeogeographic reconstructions the *Hirnantia* faunas of south China and Australia are placed in equatorial latitudes (Scotese and McKerrow 1990; Fig. 3.1) while Rong and Harper (1988) place these occurrences at 20°S. However, all reconstructions are in agreement that the occurrences of *Hirnantia* faunas in Baltica were found at 20°S. The *Hirnantia* fauna consisted of taxa that had not been previously encountered together; they came from a range of facies and several different provinces but, on the whole, high-latitude Gondwanan forms tended to predominate (Rong and Harper 1988). The remarkable spread of the *Hirnantia* fauna to all latitudes except for a narrow equatorial belt can thus be viewed as a spread of cold-water taxa in the early Hirnantian (Rong and Harper 1988; Cocks and Fortey 1988; Fortey 1989; Harper and Rong 1995). In comparison with the diverse, highly provincial brachiopod communities of the Rawtheyan, the pandemism of the *Hirnantia* fauna is truly remarkable. A modest amount of latitudinal variation is seen within *Hirnantia* assemblages: in high latitudes there are rarely more than ten genera present, while in the subtropics (for example, in Baltica and south China) there is at least double this diversity (Rong and Harper 1988; Owen and Robertson 1995).

The high-diversity Edgewood faunas are confined to the carbonates of Laurentia, Baltica and Kolyma in north-east Russia (Fig. 3.1) where brachiopods, rugose corals, bryozoans and several other taxa continued to thrive into the early Hirnantian. The Edgewood fauna is known from Anticosti Island but, even at this low-palaeolatitude location, a *Hirnantia* fauna has been recorded from the *extraordinarius* Zone. This suggests that cold-water brachiopods even reached the equator for a time (Cocks and

Copper 1981). However, several authors have questioned the validity of the Anticosti *Hirnantia* fauna because the identifications are not based on internal moulds which are, apparently, essential for brachiopod identification (Harper and Rong 1995; Lespérance 1985).

The second brachiopod mass extinction struck at the end of the *extraordinarius* Zone and eliminated the *Hirnantia* fauna and many of the brachiopods of the Edgewood Province. In a few areas, notably the Lake District of England, the *Hirnantia* fauna may have temporarily lingered on into the *persculptus* Zone (Cocks and Fortey 1988). In comparison to the earlier extinction, shallow-water brachiopods fared somewhat better than their deeper-water cousins during this crisis (Harper and Rong 1995). By the Rhuddanian brachiopod fortunes were at their low point; only 57 genera are known from this time, of which over half were Lazarus taxa and many others were extremely rare (Cocks 1988). Despite this apparent malaise, the brachiopod radiation may already have begun by this point; Copper (1995) has recorded the first spire-bearing atrypids from the late Hirnantian of Anticosti Island. Further collecting may reveal the presence of further progenitor taxa in this interval. However, the major detectable radiation began in the late Rhuddanian with the diversification of the athyrids, pentamerids and spiriferids (Harper and Rong 1995) and, for the remainder of the Llandovery, a series of increasingly more diverse brachiopod communities were developed. Initially the *Cryptothyrella* community, consisting of widespread, opportunistic species, appeared in many shallow-water settings. This was replaced in the mid-Llandovery by a *Virgiana* community which in turn gave way to the *Pentamerus* community in the late Llandovery (Sheehan 1975). It is noteworthy that both the *Cryptothyrella* and *Virgiana* communities were derived from Ordovician stocks restricted to Baltica; *Cryptothyrella*, in particular, was found in Ashgill reefs of Sweden. All of the Llandovery communities were cosmopolitan and it was only with the establishment of the *Pentamerus* community that both the degree of endemism and within-community diversity began to reach pre-extinction levels (Sheehan 1975; Watkins 1994).

Trilobites

The trilobite record is closely comparable to that of brachiopods; both show two phases of extinction punctuated by an interval of low-diversity, cosmopolitan, cold-water faunas. In brief, 113 Rawtheyan genera were reduced to 71 in the Hirnantian and then to 45 in the early Llandovery (Briggs *et al.* 1988). At the family level the second, late Hirnantian event was the more significant (Fortey 1989). Thus, only three or four families were lost at the end of the Rawtheyan while ten or eleven families (33% of the total) disappeared in the late Hirnantian (Brenchley 1989). The extinctions were selective in the sense that they eliminated all pelagic trilobites (the agnostids and cyclopygids) and many deep-shelf taxa, while leaving shallow-water taxa relatively unscathed (Briggs *et al.* 1988). Only trilobites encountered in carbonate mudmounds escaped any generic-level extinctions (Brenchley 1989). However, those trilobites that went extinct had a bewildering array of morphologies and presumably life-styles. The Asaphina, one of the most diverse and numerous of Ordovician trilobite orders, disappeared in the late Ashgill together with all its constituent families: the Trinucleidae, Nileidae, Asaphidae, Cyclopygidae, Remopleuridae, and the Raphiophoridae (Fortey 1989). The only exception was *Raphiophorus*, which persisted into the Late Silurian; a solitary reminder of a great Ordovician clade. Examination of the protaspis, the larval stage of trilobites, has provided a possible clue to the demise of the Asaphina (Chatterton and Speyer 1989; Fortey 1989). Some protaspes appear to have been adapted to a planktonic existence, and it is significant that all taxa possessing this larval morphology, together with all trilobites that pursed a pelagic existence in the adult stage, became extinct. Therefore, for trilobites, the crisis was probably caused by deleterious changes in the water column. This interesting observation helps to explain why trilobites break one of the 'rules' of mass extinction during the late Ashgill crisis – widespread forms were wiped out while more endemic taxa (such as phacopids, encrinurids, odontopleurids, and aulacopleurids) survived with only generic-level extinctions (Chatterton and Speyer 1989). The majority of extinct taxa had a planktonic larval stage that ensured widespread dispersal, while the survivors had a benthic protaspis with limited dispersal ability.

In the interval between the end-Rawtheyan and late Hirnantian extinctions most surviving trilobites were extremely rare (Fortey 1989). The only common trilobites consisted of a few genera of Gondwanan affinities that expanded their range to become virtually cosmopolitan in the Hirnantian (Briggs *et al.* 1988; Lespérance 1988). This trilobite equivalent of the *Hirnantia* fauna is dominated by the closely related, or possibly congeneric, *Dalmanitina* and *Mucronaspis* and the homalonotid *Brongniartella* (Fig. 3.4). Like the brachiopods, these trilobites are found in a broad range of facies, indicating an extreme eurytopy.

The late Ashgill was clearly a major crisis for the trilobites and they took a considerable period of time to re-establish something of their former diversity, although they never again produced pelagic adult life-styles. Monitoring the diversification is difficult due to the dearth of taxonomic studies on Llandovery trilobites (Lespérance 1988). However, only a single genus, *Acernaspis*, appears to have originated in the Rhuddanian, pointing to a very slow initial rate of diversity increase. The radiation in the later Llandovery was derived, at least partly, from trilobites that lived in Ordovician reef habitats or shallow inshore clastic settings (Fortey 1989). Like the Llandovery brachiopods, the trilobite communities were extremely widespread at this time: only a single Remopleurid Province can be recognised (Chatterton and Speyer 1989).

Ostracods

It has been known for some time that the ostracods experienced a diversity low point around the end of the Ordovician (Copeland 1973) but the generic-level turnover has yet to be elucidated. At the family level, the compilation of Whatley *et al.* (1993) indicates a substantial extinction and origination event: of 32 Ashgill families, 10 go extinct during this interval (31% extinction) while 8 new families appear sometime in the Llandovery. The precise timing of these events is not fully documented but, in the sections of Anticosti Island and the Mackenzie Platform of Canada, the extinction appears somewhat later than the conodont and trilobite extinctions (Copeland 1973, 1989; Wang *et al.* 1993a).

Bryozoans

Bryozoans were a diverse and successful group in many Late Ordovician shallow-shelf settings. Most studies have been undertaken on faunas from Laurentia and Baltica, and consequently our knowledge of the late Ashgill bryozoan crisis is mostly derived from these regions. It would appear that the faunas of the two continents responded differently to the crisis. For the North American bryozoans the picture is complicated by the widespread absence of Hirnantian strata and thus the truncation of the fossil record at the end of the Rawtheyan. Thus, 21% of genera disappeared at the end of the Rawtheyan, primarily from mixed carbonate/siliclastic shelves, but this may be a back-smeared Hirnantian extinction event. In Baltica, where there are ample Hirnantian strata, bryozoans lost 23% of their genera near the end of the Hirnantian, this time from carbonate environments (Tuckey and Anstey 1992). Overall though, the extinction was only a modest one when compared to trilobites and brachiopods, and no families or orders were lost. Llandovery faunas are, however, rather impoverished, and most genera of the cystoporate and cryptostome orders are absent from the fossil record at this time. These Lazarus taxa reappear in the late Llandovery and early Wenlock (Taylor and Larwood 1988).

Echinoderms

Due to its reliance on rather exceptional preservational conditions, the fossil record of echinoderms is rather patchy (cf. Donovan 1996). The details of their fortunes during the late Ashgill mass extinction are therefore not as precise as those for other groups. It is none the less clear that, for many groups, this interval marks a serious crisis. The crinoid record is reasonably well known from North America and Britain; this shows 70% generic extinction in the late Rawtheyan followed by a 30% extinction near the end of the Hirnantian (Eckert 1988). For edrioasteroids the extinction appears to have been even worse, as very few genera appear to have survived the crisis interval (Eckert 1988). Cystoids suffered similarly, with only five genera known from the Llandovery (Paul 1988), although the Lazarus effect is considerable at this time: eight cystoid families crossed the O–S boundary but only one is found in the Llandovery fossil record, five reappear in the Wenlock and the remaining two later in the Silurian after an outage of nearly 20 m.y. (Paul 1982).

Bivalves

Bivalves were in the early stages of their diversification in the Ordovician, and it was only in nearshore level-bottom communities that they had become common. However, the relative youth of the bivalve clade was no proof against the vicissitudes of the late Ashgill crisis. Hallam and Miller (1988) noted the extinction of 16 out of 25 genera of pteriomorphs and 8 of 15 genera of nuculoids in the latest Ordovician. Mytilaceans appear to have discovered a rock-boring habit in the Ordovician, but they were eliminated at the end of the period, and this specialised mode of life was not successfully rediscovered by the bivalves until the mid-Jurassic, 270 m.y. later (Pojeta and Palmer 1976). In a similar resetting of the evolutionary clock, Frey (1986) noted that communities in offshore muddy environments in the Late Ordovician of Ohio and Indiana consisted of bivalves and trilobites. 'These first intrusions into offshore shelf environments by pelecypods were truncated by the Late Ordovician extinction event. Filter-feeding pelecypods did not successfully re-establish themselves in these environments until the Late Devonian or Carboniferous' (Frey 1986: p. 263): an 80 m.y. legacy of the late Ashgill mass extinction.

Rugose corals

The O–S crisis did not have a significant effect on rugosan standing diversity (Scrutton 1988) despite the 70% generic extinction (62 of 90 genera) claimed by Kaljo (1996). However, the details of the faunal changes have yet to be evaluated. The history of North American solitary Rugosa is among the best known (Elias 1989). Extinctions began in the late Richmondian Stage, contemporaneous with the withdrawal of epicontinental seas in the region, and were generally preceded by declines in abundance and range restrictions prior to final disappearance. The crisis persisted into the early Gamachian Stage and thus coincides with the *extraordinarius* Zone extinctions of elsewhere. Taxa from the continental margin survived preferentially, perhaps due to their ability to withstand relatively cool waters (Elias 1989), although such corals also had broad geographic ranges – a factor that may have been of greater importance in their survival. Globally, the survivors of the crisis were tiny and the subsequent radiation did not begin in earnest until the late Llandovery.

Reef taxa

Reefs have proved highly prone to extinction events throughout their history and, after most mass extinction events, they are usually absent from the fossil record for several million years (Fagerstrom 1987). The partial exception to these 'rules' occurred during the late Ashgill crisis when reefs were not as badly affected as level-bottom communities. Reefs are absent for much of the Llandovery, reappearing again at the end of this interval (Brunton and Copper 1994), but, unlike some other mass extinctions, the late Ashgill crisis did not cause a wholesale change in the composition of the main reef-forming taxa. Both Ordovician and Silurian reefs are primarily constructed of tabulate corals and stromatoporoids. At the generic level there were clearly losses for both groups; these were perhaps as high as 70% for tabulate genera, but this left 20 or so genera to provide the stock for the widespread reefs of the Wenlock (Kaljo and Klaaman 1973).

Discussion

The above review has not discussed every known fossil group because not all have been studied in the requisite detail. However, enough is known about a diverse range of groups to draw some general conclusions about the crisis (Fig. 3.5).

The extinction phase began abruptly at the end of the Rawtheyan and nearly eliminated all the graptolites. In interesting contrast to the single extinction phase of this pelagic groups, most benthic taxa experienced two extinction phases, an end-Rawtheyan and a late Hirnantian one (Fig. 3.5). The earlier phase was worst in low latitudes, particularly in Laurentia where trilobites, brachiopods, bryozoans, and echinoderms all suffered major extinctions. However, it is important to note that these disappearances may have more to do with the the widespread regression and consequent absence of Hirnantian strata in Laurentian sections. In a similar vein, Fortey (1989) has argued that the extreme rarity of deep-water Hirnantian sections in Laurentia suggests the apparent loss of deep-water taxa, particularly trilobites, may be an artefact. In conclusion, it is possible that the very limited amount of strata available for collecting Hirnantian fossils may have exaggerated the end-Rawtheyan extinction and back-smeared the early *persculptus* extinction.

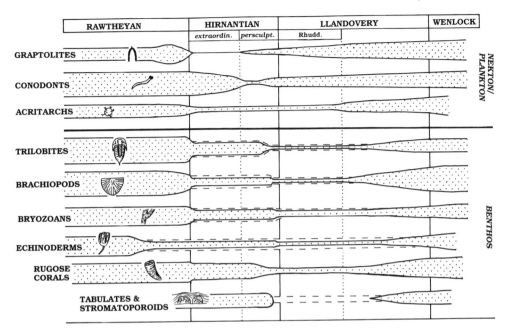

Fig. 3.5 Summary of diversity changes among selected taxa during the Late Ordovician crises. Bar width depicts qualitative diversity changes within groups. Dashed lines indicate known diversity with the addition of Lazarus taxa.

The short interval between the two extinction events, which was perhaps as little as half a million years, was characterised by the widespread occurrence of a low-diversity, cold-water fauna of brachiopods and trilobites. This distinctive *Hirnantia-Mucronaspis* fauna was wiped out by the second extinction event in the basal *persculptus* Zone. This crisis was particularly severe for mid- and deep-shelf taxa (the victims included many trilobites and some rather advanced bivalve communities) and, for reasons outlined above, was probably the greater of the two late Ashgill extinction events. Among the brachiopods and trilobites shallow inshore areas, especially reefs, were a refuge and many Silurian clades originated from Ordovician taxa encountered in these environments.

The early Llandovery shows all the hallmarks of a post-mass extinction interval and many of the features are seen again following the end-Permian event (cf. Chapter 5). Thus, reefs are absent, level-bottom communities consist of depauperate but cosmopolitan assemblages, and many survivors temporarily disappear from the fossil record (that is, Lazarus taxa are common). Also comparable is the slow pace of radiation in the initial 5 m.y.; it was not until the Wenlock, with the reappearance of Lazarus taxa and reefs and the re-establishment of faunal provinces, that the marine ecosystem could be said to have fully recovered from the mass extinction.

Boundary sections

Dob's Linn

The global stratotype section for the O–S boundary was chosen in the mid-1980s at Dob's Linn in the Southern Uplands of Scotland (Cocks 1985), a choice that raised the ire of many geologists interested in this interval. Lespérance *et al.* (1987: p. 221) considered it 'a hasty and illogical decision, which must be quickly reappraised', while Barnes (1988: p. 195) similarly thought it 'a serious error of judgement'. The main criticism against the choice of Dob's Linn lies in the near absence of either a benthic fauna or conodonts in the section, which therefore renders correlation with many sections difficult, particularly Anticosti Island, the preferred choice of the protoganists noted above. In its favour Dob's Linn contains the most complete succession of graptolites known from across the boundary interval (graptolites are absent from the Anticosti section) which allows correlation with many similarly deep-water sections, for example in China and Arctic Canada (Fig. 3.3; Wilde *et al.* 1986; Wang *et al.* 1993b).

The Southern Uplands are a tectonically complicated region that is widely considered to be an accretionary prism complex formed on the Laurentian margin of the Iapetus Ocean (Fig. 3.1). The accumulation rates in the Dob's Linn section, 97 m in 25 m.y. or 4 m/m.y., are very slow and strongly suggest deposition in a distal, deep-water setting (Lespérance *et al.* 1987). The interesting succession of facies changes recorded in the Dob's Linn strata are therefore all the more important as they probably record the fluctuating palaeoceanographic conditions of Iapetus at a time of crisis. Up to a level high in the *pacificus* Subzone, deposition alternated between organic-rich black shales with a common graptolite fauna and unfossiliferous, bioturbated pale mudstones (Fig. 3.3). There then followed a prolonged interval of mudstone deposition punctuated by only a few millimetres of black shale accumulation containing *Normalograptus extra-ordinarius*. Black shales reappear suddenly at the base of the *persculptus* Zone and dominate the succession up into the Silurian. The graptolite mass extinction coincides with the base of the thick, barren mudstone (Fig. 3.3; Williams 1988) and is clearly related to the loss of black shale facies.

Anticosti Island

In complete contrast to the Dob's Linn section, the numerous sections displayed on Anticosti Island near Quebec contain no black shales and only the occasional graptolite but do contain an abundant benthic fauna preserved in a range of mixed clastic and carbonate strata. These are the only known, ostensibly complete, shallow-marine O–S boundary sections.

The regional setting of the Anticosti sections has been variously ascribed to 'a craton-margin carbonate sequence' (Goodfellow *et al.* 1992: p. 2) consisting of a carbonate ramp (Long and Copper 1987b) or platform (Lespérance 1985) that formed in a 'residual foreland basin' (Long 1993: p. 51). This was a legacy of the Taconic orogeny that affected a large portion of the south-east margin of Laurentia in the Caradoc. Barnes (1988) noted that Anticosti subsidence rates were considerable in the later Ordovician, and it is this factor that probably ensured the continuity of deposition during the O–S boundary interval.

The major faunal turnover on Anticosti occurs around the contact between the Ellis Bay Formation and the overlying Becscie Formation. The precise age of the base of the lower formation is uncertain but there is a general consensus that, in its upper part, it is of Hirnantian age. The shales, siltstones, sandstones, and limestones of the 75 m thick Ellis Bay Formation record a whole series of minor transgressive–regressive cycles, in which the influence of both tidal and storm processes can be detected (Long and Copper 1987a). Detailed lateral correlation is difficult due to the lack of useful zone fossils and, more especially, the rapid lateral facies changes; for example, thick, localised sandbodies are prevalent in the north-east but not the west of the island (Long and Copper 1987b). However, a few metres from the top of the formation is a distinct, 30 cm to 1.2 m thick bed of limestone containing oncolites, large intraclasts and coarse calcarenitic material (Fig. 3.6). This useful marker bed, informally known as the Oncolite Platform Bed (OPB), can be traced through all the sections in Anticosti. The top surface of the OPB may be a hardground (Cocks and Copper 1981; Wang *et al.* 1995), in which case it records one of the few breaks in deposition in the O–S interval, probably at a level within the *extraordinarius* Zone (Melchin *et al.* 1991). Small bioherms up to 4 m high are developed upon the OPB (Fig. 3.7) and these contain a diverse Ordovician fauna of calcareous algae, tabulate corals, phaceloid rugose corals, stromatoporoids (especially *Aulacera* and *Eoclimadictyon*) and bryozoans (Long and Copper 1987a). The OPB and the bioherms constitute the Laframboise Member, the highest unit of the Ellis Bay Formation, and they are sharply overlain by the 'semi-lithographic limestones' of the Becscie Formation (Long and Copper 1987a). These thinly bedded, fine-grained, light grey limestones show a remarkably uniform development throughout Anticosti Island, in contrast to the rapid facies changes of the underlying formation, and they seem to have accumulated in much deeper water.

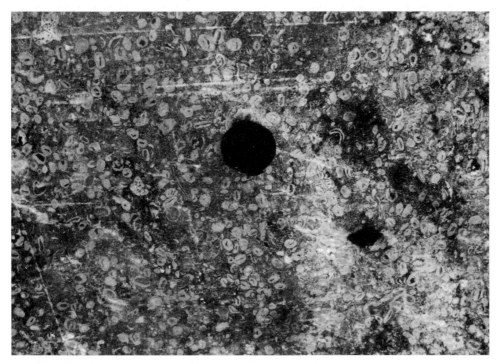

Fig. 3.6 Bedding plane view of the Oncolite Platform Bed at the Lac Wickenden road section on Anticosti Island. The parallel scratches are glacial striations formed during the Pleistocene (not the Hirnantian!) glaciation. Lens cap is 5 cm in diameter.

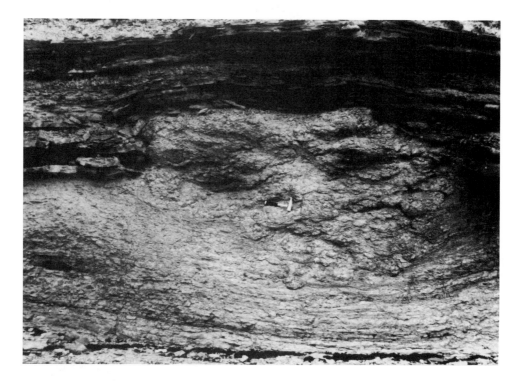

Fig. 3.7 Patch reef in the Laframboise Member seen at Point Laframboise, Anticosti Island. It is overlain by the thinly bedded carbonates of the Becscie Formation.

The mass extinction on Anticosti Island is spread through the topmost metres of the Ellis Bay Formation and the basal metres of the Becscie Formation (Fig. 3.8). Trilobites are the first group to display losses, with six genera disappearing several metres beneath the OPB at the same level that Cocks and Copper (1981) noted the brief appearance of a *Hirnantia* fauna. Many conodont species disappear within the bioherms but the major crisis for most groups is at the formational boundary. Only the ostracods appear to be out of kilter with the other groups; their crisis occurred between 35 and 45 m above the base of the Becscie Formation (Barnes 1988) at a time when other groups were in the initial stages of radiation.

Arctic Canada

In addition to Anticosti Island, Canada is also blessed with many other O–S boundary sections, from both basinal and platform carbonate settings (see, for example, Good-fellow *et al.* 1992; Wang *et al.* 1993a). These are scattered over a large area of the Yukon Territory and Northwest Territories. The platform locales invariably display a major Upper Ordovician hiatus overlain by Llandovery carbonates but most basinal sections are complete. On Cornwallis and Truro islands in the Arctic, the basinal Ashgill sediments consist of laminated black shale and burrowed limestone alternations up to top of the *pacificus* Zone. At this point a distinctive bed of laminated siltstone is developed. Many graptolite species became extinct at the base of the siltstone, which

Fig. 3.8 Basal Becscie Formation at Point Laframboise. The Ordovician–Silurian boundary occurs around the level of the hammer.

contains only rare examples of *Neodiplograptus bohemicus* indicating an *extraordinarius* age. Graptolites only become common again at the base of the *persculptus* Zone coincident with the re-establishment of black shale deposition (Melchin *et al.* 1991). The geological history of the Arctic Canadian islands clearly bears close comparison with the Dobs Linn section, although the laminated siltstone of the *extraordinarius* Zone is not such convincing evidence of an oxygenation event at this time.

South China

The numerous O–S boundary sections of the Yangtze Basin of South China provide an important additional perspective on the crisis. This large basin was centred on central and southern Sichuan and the Anhui Province of south-west China and lay to the north of the principal land area of the continent. During the course of the Ordovician, the Yangtze Basin became progressively more isolated as an archipelago developed around its western and northern margin until, in the Ashgill, the only marine connection with the world ocean lay through a shallow seaway to the north-east of the basin (Xu 1984). The semi-isolated nature of the basin is reflected by the presence of several endemic graptolite genera and the prevalence of black shale accumulation.

In the Upper Yangtze Basin black shales constitute virtually the entire Ashgill to

Llandovery record with the exception of a thin mudstone of probable *extraordinarius* age (Wang *et al.* 1992, 1993b), although the diagnostic graptolite species is absent. In the shallower Lower Yangtze Basin alternations of black shales and shelly limestones with a diverse benthic fauna predominate. At Yichang in the Upper Yangtze Basin, the top of the *extraordinarius?* mudstone witnessed a major graptolite extinction with 84 out of 88 species and 16 out of 19 genera disappearing (Wang *et al.* 1993b). In the Jinxian section of the Lower Yangtze Basin, the mass extinction occurred at the same stratigraphic level within a thin mudstone bed; numerous species of trilobites, brachiopods, nautiloids, ostracods, bivalves, and gastropods were eliminated.

In marked contrast to elsewhere in the world, the South Chinese sections therefore reveal a single mass extinction event. The end-Rawtheyan crisis had little effect in this area although, interestingly, a *Hirnantia* fauna became established in the shallow marine sandstones of the 'north-east passage' during the early Hirnantian (Xu 1984). Cold-water upwelling in this area may have facilitated the colonization of this cold-adapted fauna at an exceptionally low-latitude site (Fig. 3.1). Notably, *Hirnantia* and its associates were not able to invade the interior of the Yangtze Basin.

Glaciation and sea-level changes

Direct evidence for glaciation in the Late Ordovician initially came from the Sahara where diamictites and striated pavements provide the tell-tale clues (Beuf *et al.* 1966). Much additional evidence has since come from Algeria, Spain, southern France, and some South American countries (Berry and Boucot 1973). All these areas together constituted the continent of Gondwana which lay at high southern palaeolatitudes at this time (Fig. 3.9). It has become clear that Gondwana was covered by a substantial ice

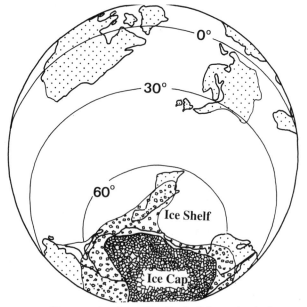

Fig. 3.9 Palaeogeographic reconstruction of the South Polar region in the early Hirnantian showing the extent of the Gondwanan glaciation. After Brenchley *et al.* (1991).

sheet and glaciomarine facies and dropstones indicate that floating ice extended up to 45°S (Brenchley 1984, 1989; Brenchley *et al.* 1991).

The duration of the glacial episode has been variously attributed within a mid-Ordovician to mid-Silurian interval. Berry and Boucot (1973) assigned an Ashgill to early Llandovery age, while Frakes *et al.* (1992) suggested an even longer cool climatic episode lasting 35 m.y. from the Caradoc to the Wenlock. However, most recent authors have opted for a much shorter, more intense glacial episode concentrated in the million or so years of the Hirnantian Stage (see, for example, Brenchley 1989; Brenchley *et al.* 1991). The cause of the uncertainty stems from the rather inaccurate age assignments of some of the evidence. Much recent redating has tended to vindicate the shorter estimates. For example, the supposedly Wenlock tillites of Argentina are undoubtedly older as they are overlain by strata with a *Hirnantia* fauna (Brenchley *et al.* 1994). Oxygen isotope evidence discussed below further points to a glaciation of extremely short duration.

Indirect but telling evidence of a Hirnantian glaciation comes from sea-level changes at this time. As already related, complete O–S boundary sections are very rare and, with the exception of the Anticosti sections, tend to be encountered in deeper, basinal settings. In most areas of the world the Hirnantian is marked by a hiatus. In Laurentia in particular vast areas of carbonate platform became emergent late in the Ashgill (Eckert 1988) although evidence for sea-level retreat is also seen in Baltica and Siberia (Berry and Boucot 1973; Brenchley 1984; Brenchley and Newall 1984). The build-up of the Gondwanan ice sheet lowered sea levels by an amount that has been variously estimated at 40 m (Long 1993), 50 m (Eckert 1988), 70 m (Brenchley and Newall 1984) and up to 100 m (Brenchley 1989). Deep karstic features which formed in carbonate mudmounds in Sweden at this time suggest a minimum of 50 m of sea-level fall (Marshall and Middleton 1990; Brenchley *et al.* 1995).

The Upper Ordovician of the Prague Basin in the Czech Republic provides one of the more detailed and complete records of Hirnantian sea-level changes (Brenchley and Storch 1989). This shows that regression was initiated at the base of the stage but was punctuated by a short transgression immediately prior to the main lowstand in the mid-Hirnantian (Fig. 3.10). Sea level began to rise at the end of the *extraordinarius* Zone and continued uninterrupted into the Silurian, probably a reflection of the rapid melting of the Gondwanan ice cap. A Llandovery transgression is widely recognised and, in many sections throughout the world, strata of this age rest on an Upper Ordovician unconformity (McKerrow 1979; Brenchley and Newall 1984; Brenchley *et al.* 1995).

Stable isotopes

Recent analyses of both carbon and oxygen isotopes from O–S boundary sections have provided important and intriguing evidence of global changes. In the earliest study Orth *et al.* (1986) measured the $\delta^{13}C$ values of carbonates from Anticosti Island and noted a major negative shift within the Laframboise Member. However, subsequent studies on Anticosti have a shown a positive shift from 0‰ to 4‰ at the same level followed by a return to values of 0–1‰ in the Becscie Formation (Long 1993; Brenchley *et al.* 1994). The inconsistency may derive from the use of whole-rock samples by Orth *et al.* (1986). The most reliable results appear to come from unaltered bioclasts which indicate a positive excursion in the early Hirnantian (Fig. 3.10; Marshall and Middleton 1990;

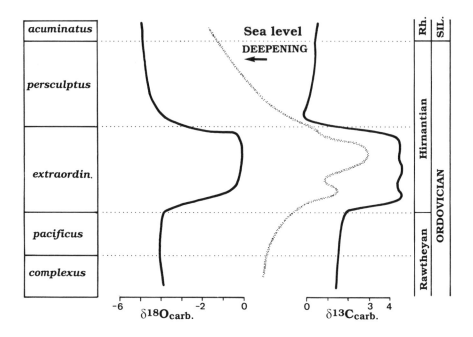

Fig. 3.10 Comparison of eustatic sea-level fluctuations in the O–S interval (Brenchley 1989) with carbon and oxygen isotope variations in unaltered carbonates (Brenchley *et al.* 1994).

Brenchley *et al.* 1994) – a trend that is seen in both low and high palaeolatitudes (Marshall *et al.* in press). Kerogen analyses show a similar pattern of $\delta^{13}C$ fluctuations (Wang *et al.* 1993b; Marshall *et al.* in press). However, measurements from several Arctic Canadian sections have either shown a negative $\delta^{13}C_{carb}$ swing (Goodfellow *et al.* 1992) or both positive and negative fluctuations with no clear trend (Wang *et al.* 1993a).

If the positive excursion records the most truthful trend then it poses problems for interpretation. A positive inflection of $\delta^{13}C_{carb}$ suggests that large volumes of (isotopically light) organic carbon are being buried and removed from the exogenic carbon cycle and yet, as we have seen in most sections (for example, Dob's Linn), the early Hirnantian is characterised by the cessation of black shale deposition. This should have had the opposite effect on carbon isotope ratios. Brenchley *et al.* (1994, 1995) have argued, rather unconvincingly, that the positive shift of $\delta^{13}C_{carb}$ values reflects increased 'productivity and/or sedimentation of carbon' and consequently 'Lower levels of . . . dissolved pCO_2' (Brenchley *et al.* 1994: p. 298). However, elevated productivity is incapable of producing a 4–6‰ shift of $\delta^{13}C$ values and, in any case, the sedimentary record indicates a decrease of organic carbon burial. The $\delta^{13}C$ record currently remains one of the more enigmatic aspects of the Late Ordovician crisis.

Interpretations of the $\delta^{18}O_{carb}$ record are in greater accord with the geological evidence but again they raise several problems. Bioclast analyses from a range of Baltican and Laurentian localities indicate a positive shift of oxygen isotope ratios in the early Hirnantian following a prolonged period of Late Ordovician stability (Fig. 3.10; Marshall and Middleton 1990; Long 1993; Brenchley *et al.* 1994). Such shifts have been observed for the Pleistocene during intervals of ice advance when the volume of isotopically light water trapped in the ice caps increases. The Hirnantian excursion clearly suggests that the glaciation was a brief affair rather than one spread throughout the Late Ordovician and Early Silurian as proposed by some. However, the great magnitude of the $\delta^{18}O$ excursion (+ 2‰ on Anticosti and 4‰ in the Baltic States) is disquieting. Even the lower value is considerably in excess of any fluctuations observed during the Quaternary glaciation when $\delta^{18}O$ values do not vary by more than 1.2‰ (Brenchley *et al.* 1995). For the Hirnantian $\delta^{18}O$ changes to be entirely due to the glaciation would require an inconceivably thick Gondwanan ice cap and a glacioeustatic regression measured in hundreds of metres. A positive shift of $\delta^{18}O$ values can also be achieved by cooling in the tropics, although here again the magnitude would have to be large. A combination of global cooling and glacioeustatic regression is of course possible, but Brenchley *et al.* (1994) calculate that a sea-level fall of at least 100 m combined with a 10°C drop of tropical ocean temperatures would be required to achieve a 4‰ swing of $\delta^{18}O$ values. Both estimates are only just within the realms of possibility. Clearly the oxygen isotope record suggests almost inconceivably severe climatic changes in the late Ashgill.

Trace metals

There has been little success in the search for an iridium anomaly in O–S boundary sections. The Dob's Linn section failed to yield any Ir-enriched levels (Wilde *et al.* 1986) nor did several sections in the Canadian Arctic (Goodfellow *et al.* 1992). A very weak anomaly detected on Anticosti Island (Orth *et al.* 1986) was confirmed by Wang *et al.* (1995) on top of the OPB but the value, 0.058 ppb, borders on the insignificant. Furthermore, as noted above, this level is a hardground and the extremely slow sedimentation rates may account for the slight elevation of Ir values. Slightly greater success in the hunt for an Ir anomaly has been had with the Yangtze Basin sections. Wang *et al.* (1992) recorded Ir values averaging 0.092 ppb (the peak value was 0.23 ppb) at the base of the *persculptus* Zone, the level of mass extinction. However, the Co/Ir ratios were not consistent with a meteoritic origin (Wang *et al.* 1992).

The sporadic enrichment of Ir may be explained by the widely reported elevated concentrations of other trace metals at the base of the *persculptus* Zone. Goodfellow *et al.* (1992) noted that Ni, V, As, Hg, Cd, and Mo were all enriched at this level in the Yukon, a signature characteristic of intense anoxia and sulphide formation. In the adjacent Northwest Territories, a uranium peak at the same level again suggests anoxic deposition (Wang *et al.* 1993a). In the Yangtze Basin sections As, Mo, Sb and U peaks coincide with the Ir anomaly in the basal *persculptus* Zone; a combination of sediment starvation and anoxia during initial transgression are the favoured mechanism of enrichment for all the trace metals (Wang *et al.* 1992, 1993b).

Evidence for anoxia is not so obvious in the shallower sections of Anticosti but Mn enrichment at the base of the Becscie Formation is probably indicative of a brief interlude of dysoxic deposition (Wang *et al.* 1995).

Extinction mechanisms

The late Ashgill provides a fascinating interval for those interested in studying complex global crises. More than any other mass extinction, the event is clearly separable into two distinct phases divided by an extraordinarily intense but short-lived glaciation. Further-more, any extinction mechanism must account for the fact that, while the extinction was 'only' a severe crisis for benthic forms (reef taxa in particular survived surprisingly well), it was a near-complete catastrophe for pelagic forms (phytoplankton, graptolites, and trilobites with a planktonic larval and/or adult stage).

Cooling

Berry and Boucot (1973) were among the first to note that global cooling associated with the Gondwanan glaciation could have lead to the Ashgill diversity decline, an idea fervently propounded by S.M. Stanley (1984, 1988). Clearly this mechanism cannot account for the demise of many high-latitude, cool-water taxa (especially graptolites, conodonts and trilobites) but it may still be applicable to the demise of lower-latititude forms. Generally, at times of glaciation, the warm tropical belt contracts but does not disappear; the survival of the Edgewood fauna in a narrow tropical belt of carbonates stretching from Laurentia to Baltica indicates that this was the case during the Hirnantian glaciation. Cooling is therefore also unlikely to account for the loss of the numerous tropical taxa unless, of course, the $\delta^{18}O$ values are a true proxy for major equatorial cooling (p. 57).

Regression

Glacioeustatic regression constitutes the other obvious manifestation of the Gondwanan glaciation and, in this case, equatorial locations were severely affected, with vast areas of carbonate platform becoming emergent (Brenchley and Newall 1984; Sheehan 1988). Several authors have therefore proposed that the loss of shallow-water habitat area was a major cause of extinction (Berry and Boucot 1973; Sheehan 1988; Owen and Robertson 1995; Armstrong 1996). Armstrong (1996) has developed this idea and postulated that shallow-shelf taxa may not have been able to migrate down the bathyal slope as sea level fell because they could not migrate through the cold water beneath the oceanic thermocline. In effect, taxa from shallow, warm waters may have run out of habitat space due to the combined effects of cooling and regression.

The regression and habitat loss mechanism has not gained universal acceptance primarily because the detailed timing of extinction and regression is not a close one. The end-Rawtheyan extinctions occurred before the peak of regression in the mid-Hirnantian, while the second extinction phase developed during the ensuing transgression (Eckert 1988; Owen and Robertson 1995). The demise of numerous pelagic forms is also unlikely to be directly related to sea-level fall (Fortey 1989). In an interesting variation on the anti-regression expositions, Wyatt (1995) argued that, because global sea-level stood so high in the Late Ordovician (cf. Hallam 1984a) many continents were probably deeply flooded. As a result, a 50–100 m regression would have *increased* the area of shallow seas (assuming that continental hypsometry was comparable to that of today). However, despite Wyatt's cogent argument, one is left wondering where all the supposed shallow-marine shelf areas were in the early Hirnantian. The geological evidence emphatically indicates that vast areas of shallow marine deposition were lost at this time. In a fair summary, Brenchley (1989: p. 121) concluded: 'The relationship of the extinctions to sea-level change in the mid Hirnantian is far from clear.'

Oceanographic changes and oceanic anoxia

The double mass extinction event of the Ashgill is most clearly related to first the sudden onset and then the sudden termination of Gondwanan glaciation. This strongly implies that sudden changes in climate and oceanography were important components in the extinctions.

A substantial body of evidence indicates that, for a prolonged period prior to the Hirnantian, the deeper waters of the world's oceans were anoxic. The relative abundances of Ce and other rare earth elements in shales provide one such set of supporting data (Wilde *et al.* 1996). This situation changed with the onset of glaciation, whereupon cold, dense, oxygen-rich waters, generated around the shores of Gondwana, sank and oxygenated the deep oceans in a manner analogous to the Antarctic Deep Water of the modern oceans. The loss of black shales in the *extraordinarius* Zone at Dob's Linn is probably a result of this oceanic ventilation. Supplying oxygen to the deeper levels of the ocean may not immediately suggest a cause of mass extinction, but the change of nutrient circulation dynamics provides the key. In particular, phosphorus, the main biolimiting nutrient, is much more efficiently removed from oxic than anoxic oceans due, essentially, to its removal by iron oxyhydroxides (Van Cappellen and Ingall 1994, 1996). Consequently, on a global scale, oxic oceans are considerably more oligotrophic than euxinic oceans. Therefore, the crisis among pelagic forms was probably triggered by a drastic productivity decline. The loss of an anoxic lower water column was particularly disastrous for graptolites because they were apparently adapted to the high-nutrient, dysoxic conditions around the oceanic redoxcline (Berry *et al.* 1990). Benthic dysaerobic assemblages, such as the *Foliomena* fauna of Baltica, similarly succumbed to the loss of their dysoxic habitats.

Wilde and Berry (1984, 1986) proposed, as an additional kill mechanism, the upwelling of poisonous deeper waters triggered by the onset of deep, cold-water circulation. Assuming that this process occurred sufficiently fast, the upward displacement of anoxic deep water – euphemistically called 'biologically unconditioned water' by Wilde and Berry – would have poisoned many shallow-water taxa, presumably because of its lethal H_2S content. However, H_2S oxidises rapidly in the presence of oxygen and, unless upwelling was exceptionally rapid, it is unlikely that the deep anoxic waters posed much of a threat to shallow marine taxa.

It is a sad irony for those fond of dysaerobic taxa that, for many groups that went extinct at the end of the Rawtheyan, their favoured dysoxic habitats were re-established during the rapid *persculptus* Zone transgression a mere half million years later. Despite the frequency with which such facies occur in the Palaeozoic, oxygen-poor conditions are clearly implicated in the second and probably greater of the two Ashgill mass extinctions (Briggs *et al.* 1988; Rong and Harper 1988; Fortey 1989). Perhaps the half-million-year respite from dysoxia ensured that dysaerobic taxa perished, leaving survivors that were ill adapted to low oxygen levels. The second phase of extinctions were concentrated among the mid and outer-shelf faunas where the intensity of anoxia was probably great: a conclusion supported by the trace metal enrichments at the base of the *persculptus* Zone.

Whether the anoxia kill mechanism can account for all base-*persculptus* extinctions is debatable. In south China, for example, the graptolites curiously became extinct during

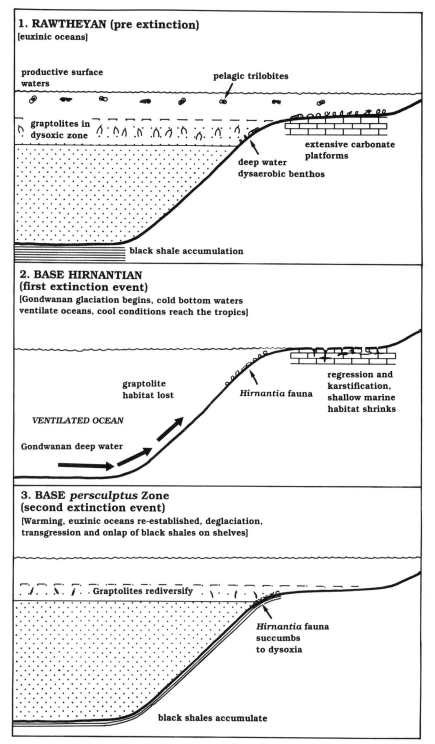

Fig. 3.11 Schematic three-stage model of the late Ashgill crisis.

this second phase of extinction when their preferred, dysoxic habitats become widely re-established. Brenchley *et al.* (1995) also highlighted the fact that the shallow-marine extinctions on Anticosti Island are apparently not related to the development of oxygen-restricted facies. In fact the Mn enrichment at the base of the Becscie Formation indicates the temporary development of dysoxic conditions (Wang *et al.* 1995) although whether this was sufficient to cause the extinctions is a moot point. Further study of the geochemistry and facies of the lower metres of the Bescsie Formation (cf. Fig. 3.8) would be highly worthwhile.

Combined causes

As should by now be clear, the late Ashgill crisis was the product of more than one cause and most authors favour a combination of kill mechanisms (Robertson *et al.* 1991; Brenchley *et al.* 1995; Harper and Rong 1995; Owen and Robertson 1995; Armstrong 1996). Although differing in detail, all these sequential models illustrate a considerable consensus (Fig. 3.11). The initial phase of extinctions clearly relates to a fundamental change in oceanic circulation and chemistry that caused the demise of the graptolites and probably the other pelagic groups. Cooling and regression may have had a part to play at this stage, but the tropical Edgewood fauna continued to thrive at this time and most extinctions occurred considerably before the main regression. The second phase of the extinction again relates to a rapid change in the oceanic state, this time back to euxinic conditions. The onlap and spread of bottom-water anoxia during the Llandovery transgression effectively eliminated mid- and outer-shelf habitats and may even have affected inner-shelf faunas (Fig. 3.11).

4 Crises of the Late Devonian: the Kellwasser and Hangenberg events

Between the Late Ordovician and the next great mass extinction in the Late Devonian the world's marine fauna was afflicted by numerous lesser extinction events. Between them, Kaljo *et al.* (1996) and Walliser (1996) document over 24 bio-events in this interval. These minor extinctions are beyond the scope of this book but, by the Late Devonian the events were increasing in magnitude to such an extent that the crisis at the Frasnian–Famennian (F–F) boundary is widely regarded as one of the big five mass extinctions of the Phanerozoic. The subsequent end-Famennian or Devonian–Carbo-niferous (D–C) event has also been termed a mass extinction. The Devonian crises have been named by House (1985); the F–F and D–C examples were named the Kellwasser and Hangenberg events respectively, after characteristic black shale horizons in German sections. There are also grounds for considering an earlier mid-Givetian Taghanic event as a mass extinction.

The recognition of major Late Devonian mass extinctions is a relatively recent event in the palaeontological literature. Newell (1967) and Bretsky (1968) were the first to identify the F–F and D–C crises respectively, but only a few years later Pitrat (1970: p. 55) considered that 'There is no evidence of an important crisis of extinction of marine invertebrates in the Late Devonian'. However, in the previous year Digby McLaren, in his presidential address to the Paleontological Society, had proposed an abrupt F–F mass extinction caused by a meteorite impact (McLaren 1970), a suggestion which, in the days before Ir anomalies and the Alvarez team, met with embarrassed silence. The first detailed documentation of the timing and victims of the Devonian crises dates from a series of papers by Michael House (see, for example, House 1971, 1975). Unlike McLaren, House envisaged a protracted crisis that began 'towards the close of the Givetian' (House 1971: p. 90). The debate between sudden and gradual extinction proponents continues to this day.

Raup and Sepkoski's (1982) initial compilation of Phanerozoic extinction rates provided somewhat equivocal evidence for a Devonian mass extinction. Unlike the other four major extinctions, which stand out as distinct peaks above background extinction rates, the later Devonian is characterised by a less distinct 'hump' composed of the last four stages of the Devonian (Fig. 1.1). Initially it was unclear if this hump merited the title 'mass extinction' but subsequent analyses, particularly of Sepkoski's generic data set, have sharpened a Late Devonian peak (Sepkoski 1996).

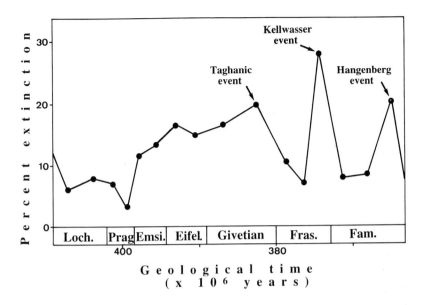

Fig 4.1 Percentage extinction rates for well-preserved marine genera through the 7 stages of the Devonian. After Sepkoski (1996).

The mid-Givetian still appears as an interval of elevated extinction rates separated from the F–F mass extinction by a significant interval of reduced rates, although whether this constitutes a discrete mass extinction or a prolonged intra-Devonian interval of elevated extinction rate is a moot point. There is, however, no doubt that the F–F event constitutes one of the big five mass extinctions, comparable in magnitude to the celebrated end-Cretaceous event (Sepkoski 1986).

Correlation and the Late Devonian world

The complex diversity changes of the Devonian can only be documented within a high-resolution biostratigraphic framework. Fortunately, the period is blessed with an unusually high diversity of pelagic groups with biostratigraphic utility. These include the enigmatic cricoconarids, the entomozoacean (finger-print) ostracods and the ammonoids. However, conodonts provide the best biostratigraphic control because of their widespread distribution in a broad range of facies. Conodont zones are therefore used to define the F–F and D–C boundaries (Fig. 4.2; Ziegler and Sandberg 1990; Klapper *et al.* 1993). Important recent changes to the conodont zonation in the crucial F–F interval include the replacement of the uppermost *Palmatolepis gigas* Zone by the *P. linguiformis* Zone in the topmost Frasnian (Sandberg *et al.* 1988). It has also been suggested that the *P. gigas* Zone be replaced by a *P. rhenana* Zone. For the D–C boundary, the transition between *Siphonodella praesulcata* and *S. sulcata* provides the marker although, in many boundary sections, rapid facies changes commonly make the precise placement of the zonal boundary difficult. However, compared to the contentious state of affairs for the Upper Permian (cf. Chapter 5), Late Devonian biostratigraphy is reasonably well established.

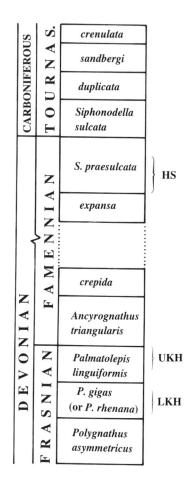

Fig 4.2 Conodont zones from the late Frasnian to the early Tournasian. Note that several mid-Famennian zones have been omitted. The levels of the Lower Kellwasser Horizon (LKH), Upper Kellwasser Horizon (UKH) and Hangenberg Shale (HS) are indicated.

A bigger issue in Devonian studies concerns the plate-tectonic reconstructions (cf. the recent reviews of Hallam 1994b; and McGhee 1996). The main uncertainty hinges on whether the large continents of Gondwana and Euramerica were united in the Late Devonian or whether they were separated by a large southern ocean. These debates have more than passing relevance to the discussion of Late Devonian crises because the latitudinal bias to extinction, reported in particular by Copper (1977, 1986), can only be evaluated if the palaeogeography is reasonably well constrained. Some palaeomagnetic evidence favours the placement of Gondwana at high southern latitudes with an ocean 3000 km wide separating it from Euramerica. Other palaeomagnetic data and much faunal evidence place Gondwana at considerably lower latitudes with only a narrow seaway separating it from Euramerica (Fig. 4.1; Scotese and McKerrow 1990). Despite the discrepancies, all models generally place Euramerica in an equatorial position, while the South Pole is usually found somewhere between central Africa and central South America. The location of the easterly regions of Gondwana, in contrast, is poorly constrained. Thus, Australia and South China have been placed anywhere between 70°S and 10°N. Faunal data suggest a relatively warm setting (see, for example, Becker and House 1994; Ormiston and Oglesby 1995) but there is as yet no consensus emerging from these important debates.

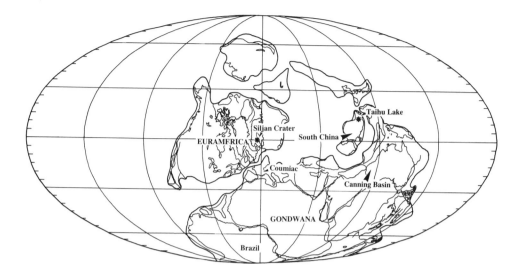

Fig 4.3   World palaeogeography during the Frasnian–Famennian interval. From Scotese and McKerrow (1990).

Extinctions of the Late Devonian

Brachiopods

Brachiopods were diverse and abundant members of Devonian benthic communities and their fortunes during the later Devonian crises provide one of the best data sets. This reveals that the entire Givetian to Frasnian interval was a time of elevated extinction rates. The crisis began during the mid-Givetian Taghanic event when six families were lost along with the common, widespread genus *Stringocephalus* (House 1985). The tropical to subtropical atrypoids were reduced from an early Givetian high of 12 genera to 5 genera by the end of the stage (Copper 1986). The subsequent appearance of five new atrypoid genera at the base of the Frasnian confirms the general picture, seen in Sepkoski's data (Fig. 4.3), that the Taghanic crisis was distinct from the Late Devonian crises and was separated by an interval of radiation and recovery. However, the recovery was brief and the remainder of the Frasnian saw the progressive elimination of numerous brachiopod families. A total of 33 families were lost of which, notably, 30 were restricted to tropical settings (Copper 1977). Mid- and outer-shelf brachiopods also appear to have suffered disproportionately (Bratton 1996). The Pentamerida and Atrypida orders were entirely eliminated at this time, while the Orthida and Strophomenida suffered severely (Johnson 1979). Global-level species data are not available for these extinctions but, in an interesting study, McGhee (1988) looked at species turnover rates in New York state and the southern Urals. This revealed that, whereas extinction rates were high throughout the Frasnian, they were balanced by high origination rates for much of the interval. Only in the late Frasnian did origination rates fall, precipitating a rapid decline of species diversity.

The early Famennian is marked by impoverished brachiopod diversity and the presence of several holdover taxa; *Peratos*, the last atrypoid, is seen in the Upper *crepida* Zone of Morocco, for example (Schindler 1990). The later Famennian witnessed the rapid recovery of brachiopod diversity thanks to radiations in the spiriferids and rhynchonellids, the two orders least affected by the Kellwasser crisis, and the productids (Johnson 1979; Ji 1989). By the Famennian the prolonged ordeal of the brachiopods appears to have been over. The strong similarity of Famennian and Tournasian brachiopod communities suggests that they were little affected by the Hangenberg crisis, although Simakov (1993) suggested that there was some turnover at the time.

Bryozoa

Bryozoans were also highly successful in the Devonian; Cuffey and McKinney (1979) counted over 1000 known species in this interval. Curiously, their diversity variations are out of kilter with other groups at this time. Thus, the major crisis occurs at the Givetian–Frasnian boundary (not at the slightly early Taghanic crisis) when 42% of families, 64% of genera and 69% of species went extinct (Horowitz and Pachut 1993). This constitutes the second greatest extinction in their history, after the Late Permian mass extinction, and corresponds to a loss of over 500 species. The bryozoans were little affected by the later Devonian crises and they maintained a diversity in excess of 200 species throughout the interval. Diversity began to rise once again in the Tournasian when the fenestrates in particular began to radiate into a variety of 'reef' environments (Fagerstrom 1994).

Foraminifera

Foraminifera underwent a rapid diversification throughout the Middle to Late Devonian, the first in their history, and none of the crises of this interval had any effect on generic diversity (Tappan and Loeblich 1988). However the larger, more complex taxa appear to have been vulnerable to extinction and several genera were lost during the F–F crisis, probably because of the disappearance of their preferred reef habitats (Brasier 1988). Complex, multiseptate foraminifera (for example, *Quasiendothyra*) again evolved in the Famennian in tropical locations, only to decline in importance during the Hangenberg crisis (Kalvoda 1990). Agglutinating foraminifera, on the other hand, became exceptionally abundant in D–C boundary sediments along with the simple, conical genus *Earlandia* (Hance *et al.* 1994). For the foraminifera as a whole, none of these crises is deserving of the 'mass extinction' epithet but the repeated loss of multiseptate, tropical taxa is a clear signal.

Bivalves

Throughout their history the bivalves have shown themselves to be relatively immune to mass extinction events. This was the case in the Devonian; only two families were lost during the Kellwasser crisis and none at all during the Hangenberg event (Hallam and Miller 1988). In fact, one of the most significant features of bivalve history in the Devonian was the explosive radiation of several epibyssate groups in offshore environments (Kriz 1979). Despite these successes, there is some suggestion that bivalve extinction rates may have been unusually high in the Devonian. Thus, Bretsky (1973) noted a 31% loss of Late Devonian bivalve genera, a figure that compares with only a 3% loss in the succeeding Tournasian. Precisely when these losses occurred is not known but it is clear that, even for the bivalves, the Late Devonian was not without its stresses.

Crinoids

The fossil record of most echinoderm groups is too patchy to provide much useful data on extinction events. However, the crinoids lost 32% of their familial diversity some time in the Late Devonian (McKinney 1985). The rarity of Famennian crinoid records (McIntosh and Macurda 1979) may point to an end-Frasnian age for this crisis.

Corals

In contrast to the echinoderms, the fortunes of the Rugosa during the later Devonian are well known. The interval marked one of their most serious crises, second only to their complete extinction at the end of the Permian. The initial losses are associated with the Taghanic event when eight families suffered heavy generic-level losses (House 1985; Scrutton 1988). This crisis was particularly severe among the endemic taxa of Euramerica (Oliver 1990). Despite this setback, the Rugosa recovered quickly and diversified in the Frasnian when they were noticeably more cosmopolitan than their forebears (Scrutton 1988; Oliver and Pedder 1994). The radiation came to an abrupt halt during the Kellwasser crisis (*linguiformis* and Lower *triangularis* Zones) when extinction swept away nearly all the rugose corals. There is an interesting selectivity to this crisis; of 148 shallow-water species only four survived into the Famennian (97% extinction), while of the ten deep-water rugose species, four survived, making a 60% extinction rate (Pedder 1982). More recent investigations suggest that the differential survival rate that Pedder originally discovered was even more marked. Thus, Sorauf and Pedder (1986) could trace only two or three genera and no species of shallow-water Rugosa across the F–F boundary, while all 12 deepwater genera were found to cross the boundary.

The Rugosa recovered rapidly in the Famennian and in the early stages many species show an unusually high degree of intraspecific variability (Poty 1996), presumably a reflection of reduced constraints on morphology in post-crisis shallow-marine habitats. Most of the new Famennian taxa are derived from deep-water Frasnian species that invaded shallow-shelf habitats following the demise of the previous incumbents. By the late Famennian the Rugosa had recovered their pre-Kellwasser diversity values and colonial taxa had re-evolved (Pedder 1982). The subsequent fortunes of the group during the Hangenberg crisis are problematic. Some authors suggest that there were few losses at this time and that Tournasian Rugosa are closely related to Famennian ones (Sorauf and Pedder 1986; Oliver and Pedder 1994). However, other specialists consider the Hangenberg event to be a major crisis interval (Scrutton 1988; Fagerstrom 1994; Poty 1996). Poty, for example, notes that no rugose species are known to cross the D–C boundary in Belgian sections.

Tabulates were important components of Devonian reefs along with stromatoporoids (see below) and their fate is clearly linked to the demise of reefs during the F–F interval. Their record is less well resolved than that of the Rugosa but it is sufficient to indicate that the Kellwasser event marked a major crisis for the group – McGhee (1996) records 80% generic extinction at this time. The Famennian represents a nadir in tabulate fortunes; only six or seven genera are known, but these survivors appear to have been rather resilient as they all crossed the D–C boundary (Scrutton 1988; Fagerstrom 1994).

Stromatoporoids

As noted above, stromatoporoids were the principal constructors of reefs in the mid-Palaeozoic and the fate of many reef-dwelling taxa (such as large foraminifera, harpid trilobites, and some brachiopods) was closely tied to the fortunes of this group. The initial phase in the decline of Devonian reefs began in the early Frasnian of Europe when a major transgression drowned numerous examples (House 1985; Narkiewicz and Hoffman 1989). However, reefs locally survived in Europe and, in the Frasnian of the Canning Basin of Australia, stromatoporoids constructed some of the largest reefs of all time. The global decline of reefs and stromatoporoids occurred rapidly in the late Lower *gigas* Zone and, by the end of the Frasnian, the stromatoporoids had lost 50% of their familial and generic diversity (Stearn 1987) and reefs had disappeared from most regions (Eder and Franke 1982; Fagerstrom 1994). A few stromatoporoid–tabulate reefs persisted into the *crepida* Zone in Poland and Moravia (Buggisch 1991) and small stromatoporoid patch reefs are encountered in the mid-Fammenian of Alberta (Stearn *et al.* 1987). Clearly, stromatoporoids had not forgotten how to make reefs in the Famennian but for the most part they are encountered in level-bottom communities at this time, while binding calcareous algae took over the role of reef formation (Fagerstrom 1994). The only stromatoporoids to illustrate significant recovery from the Kellwasser crisis belong to the long-ranging, primitive Labechiida order; the genus *Stylostroma* is particularly common and widespread in the Famennian. The success of this group has been ascribed to its cold-water preference (Stearn *et al.* 1987). However, following their partial Famennian recovery, the stromatoporoids, labechiids included, received their *coup de grâce* in the late Famennian when the six surviving families went extinct (Fagerstrom 1994). A few records of Lower Carboniferous stromatoporoids may represent holdover taxa, but all younger Palaeozoic occurrences are attributable to misidentifications of chaetetids (Bogoyavlenskaya 1982).

The late Frasnian decimation of stromatoporoids effectively marked the end of true reef formation for a prolonged interval. Bridges *et al.* (1995) suggest that it was not until the Permian that large reefs, with a biogenic framework comparable to the Frasnian reefs, reappeared. Cyanobacterial or stromatolite reefs are quite common in the Famennian and reefs of a kind occur as curious micritic mudmounds, commonly called Waulsortian mounds. These are common in the Tournasian but lack any obvious constructing or binding taxa. Crinoids and bryozoans are common in the mounds but they do not occur in sufficient density to constitute a framework (Bridges *et al.* 1995).

In an important recent discovery, Sano and Kanmera (1996) have described reefs of mid-Carboniferous age from a seamount terrane in Japan. These are constructed of rugose corals, bryozoans, and chaetetids and indicate that the reappearance of true framework reefs first occurred on oceanic islands, tens of millions of years before they reappeared in continental shelf locations.

Calcareous algae

Various groups of calcareous algae were common components of both reef and shallow-marine level-bottom communities in the Devonian. Generally, all the groups appear to have weathered the Kellwasser crisis well. The following Famennian proliferation of reef algae is likely a consequence of the loss of the stromatoporoid competition, but this was a short-lived success: all calcareous, reef-dwelling algae went extinct in the mid-late Famennian. Only the single holdover taxon *Renalcis* survived for a short interval into the Tournasian but the pre-Hangenberg crisis marks one of the most important extinctions in the history of calcareous algae (Chuvashov and Riding 1984; Fagerstrom 1994).

Sponges

The stratigraphic ranges of sponges are poorly known and strongly biased by the few monographic works. However, an interesting facet to their history concerns the proliferation of hexactinellids (glass sponges) during the F–F boundary interval (McGhee 1982). These have been recorded from the Appalachians, Poland and western Canada (McGhee 1996). As all these sites occur close to the Devonian equator (Fig. 4.3), it points to an unusually low-latitude occurrence of a group normally encountered in cool, deep waters.

Trilobites

The Devonian was not a pleasant time for trilobites, and the group was progressively whittled down by both background and mass extinctions. The end result was that the 8 orders, 13 families and 23 subfamilies known from the Givetian were reduced to a single subfamily and only eight genera at the end of the Famennian (Alberti 1979; Feist 1991). Extinction rates were high throughout the Givetian (Briggs *et al.* 1988) although there does not appear to have been a discrete Taghanic event. Losses remained high during the Frasnian but were partially offset by elevated origination rates, spurred primarily by the invasion of deep-water, cephalopod limestone environments for the first time (Feist 1991). This produced the tropidocoryphines, small trilobites with reduced eyes (Feist and Clarkson 1989). This modest radiation was terminated by an abrupt extinction within the *linguiformis* Zone which affected trilobites in all their habitats. Thus, the deep-water tropidocoryphines were lost, together with the asteropygids, odontopleurids, and the reef-dwelling harpids (Briggs *et al.* 1988). At the family level this was the most severe of the Devonian crises with only three families – the proetids, aulacopleurids, and phacopids – surviving (Feist 1991). There is little selectivity in these extinctions except perhaps a slight hint that cool-water taxa were less severely affected (Briggs *et al.* 1988).

Famennian trilobites recovered quickly from the Kellwasser crisis and, by the Upper *triangularis* Zone, they had recontinued their expansion into deep-water sites (Feist 1991). This again produced reduced-eyed forms (phacopids) and this time also blind forms (proetids and phacopids). No comparable radiation among shallow-water taxa is evident, although the phillipsids, a family of proetids with normal eyes, radiated rapidly in the late Famennian (Brauckmann *et al.* 1992). All these radiations were abruptly terminated by the Hangenberg event when all but one subfamily of the phillipsids disappeared (Brauckmann *et al.* 1992). The loss of the long-ranging Phacopida order is one of the more notable losses of this trilobite catastrophe.

The surviving phillipsids were entirely nearshore taxa and, in the Upper *praesulcata* Zone, one of these, *Pudoproteus*, became prolifically abundant for a short time. The last trilobite radiation of their history began immediately after the Hangenberg event and, by the mid-Carboniferous, they had re-established a presence in deep-water dysaerobic settings. However, the trilobites never again achieved their Devonian diversity and abundance and the Hangenberg event in effect marks the start of their swan-song.

Ostracods

Three major ecological groups of ostracods can be defined in the Late Devonian: benthic filter feeders, benthic deposit feeders, and the planktonic entomozoacean (finger-print) ostracods. All groups show distinct extinction patterns. Some species of entomozoacean are prolifically abundant in the anoxic Kellwasser facies but, for the group as a whole, the F–F boundary is a major crisis, with many species disappearing in the *linguiformis* Zone (Groos-Uffenorde and Schindler 1990). In contrast, the appearance of several new species during the Hangenberg event indicates that this was not a crisis interval for the entomozoaceans (Bless *et al.* 1992) and all Famennian genera survived into the Tournasian (Wilkinson and Riley 1990).

The benthic ostracods illustrate a distinctly different record than their planktonic cousins. Only 3 of 33 families were lost in the Frasnian but, in the F–F boundary section at Coumiac, France, Lethiers and Feist (1991) recorded a 70% species extinction, with deposit feeders being preferentially lost. Overall, warm-water, dysoxia-tolerant ostracod species appear to have survived the Kellwasser crisis best. The subsequent Famennian recovery of deposit feeders is therefore attributed to improved oxygen levels and colder conditions in many shelf seas (Lethiers and Whatley 1994).

The fortunes of benthic ostracods during the Hangenberg crisis are similar to those during the Kellwasser event. Thus, modest family-level losses (5 of 30) contrast with up to 70% species extinction in some sections (Blumenstengel 1992). That the D–C event was a mass extinction for the group is suggested by the rapid radiation shown by Tournasian ostracods, which may have been reoccupying the niches emptied by a preceding extinction.

Eurypterids

The major decline in eurypterid diversity occurred during the Early Devonian and, by the Late Devonian, they had become rather rare members of the marine biota. Briggs *et al.* (1988) could find no discrete mass extinction events in this Devonian-long decline but, in the more recent compilation of Selden (1993), elevated extinction rates were seen in both the Frasnian and Famennian stages, when four and five families went extinct, respectively. The three surviving Carboniferous families were entirely freshwater groups.

Ammonoids and nautiloids

The first ammonoids appeared in the Lower Devonian (Emsian Stage) and immediately embarked on the roller-coaster ride of extinction and radiation that was to characterise their 330 m.y. history. House (1985) has identified numerous extinction events in the Devonian but only three of these correspond to crises among other groups. The Taghanic event represents the first of these when, according to House (1985), only two genera are known to have survived, although presumably others also survived because four families outlived the crisis. The succeeding Kellwasser crisis eliminated the important Gephuroceratidae and Beloceratidae families within the *linguiformis* Zone and left only a single anarcestid genus and several tornoceratids (House 1989). The tornoceratids were a deep-water group that could presumably survive in cool waters as they are the only ammonoids known from the high-latitude Frasnian sections of South America (House 1989): their ability to live in deep and/or cool waters may have contributed to their survival.

Ammonoids recovered rapidly in the Famennian and, towards the close of the stage, several bizarre paedomorphic taxa evolved, including forms with triangular coiling. Not until the heteromorphs of the Cretaceous were ammonoids again to produce such evolutionary novelties. However, the Famennian 'experiments' were short-lived as the Hangenberg crisis eliminated both the paedomorphs and the vast majority of other ammonoid lineages in a mass extinction that was one of the worst in the group's history (House 1985, 1989; Becker 1992; Korn 1992). A total of 26 families disappeared, including the long-lived tornoceratids, while only a single family, the Prionoceratidae (and perhaps only a single genus), survived into the Carboniferous.

Despite the evolutionary bottle-neck, this single taxon was sufficient to found a rapidly diversifying dynasty of Carboniferous lineages (House 1989, 1992; Becker 1992). New genera appeared within the earliest *sulcata* Zone (for example, *Gattendorfia*) and by the mid-Tournasian the Prolecanitina and Goniatitina superfamilies were well established. Early Tournasian ammonoids show some degree of evolutionary novelty in their early whorl stages (House 1992), perhaps reflecting a relaxation of morphological constraints in the early recovery interval.

The nautiloids were somewhat eclipsed by the ammonoids in the Late Devonian and they are rather less well known. House (1985) observed that the Kellwasser crisis was a relatively minor event for the group while the Hangenberg crisis was more serious, with the diverse Discosorida order disappearing and the Oncocerida nearly so. At the family level King (1993) noted that 11 of 19 taxa failed to cross the D–C boundary. Therefore, as for the ammonoids, the Hangenberg event marks one of the worst extinctions in nautiloid history. However, the Carboniferous recovery of nautiloids was a more sedate affair; only one new family appeared in the Tournasian followed by seven more in the Visean (King 1993).

Fishes

The passing of the marine eurypterids marked the loss of the first large predators of the seas but, by the Devonian, they had been largely surpassed in this role by a plethora of fish groups; it is with good reason that the Devonian has long been known as 'the age of fish'. Throughout their history fish have proved virtually immune to mass extinction events; only the insects surpass them in invulnerability (and even this group was decimated during the end-Permian event). This makes the mass extinctions suffered by fish in the Late Devonian all the more remarkable. With the exception of the sharks, all the marine chordates were affected by the Kellwasser and Hangenberg crises. For the jawless agnathans the first event marked their virtual extermination, particularly of armoured forms; all nine families failed to cross the F–F boundary (Halstead 1993). No Famennian to Lower Carboniferous agnathans are known, although they reappeared in the Upper Carboniferous. The naked lampreys and hagfishes of today are a mere shadow of the former agnathan diversity.

The more 'advanced' jawed fishes fared equally badly in the Late Devonian extinctions. The placoderms were the first fishes to acquire jaws and, by the Devonian, they had become the most diverse and abundant of the fishes. The arthrodires were the most diverse of the placoderms and their well-documented fate mirrors that of the group as a whole. Frasnian arthrodires underwent a spectacular radiation, with 61 new genera appearing during the stage; but these proved to be short-lived, as 63 of 70 genera failed to cross the F–F boundary (Gardiner 1990). Interestingly, McGhee (1982) has shown that this crisis was much more severe for marine than for freshwater placoderms. The arthrodires partially recovered from this débâcle and 13 new genera appeared in the Famennian. These included the fearsome-looking eastmanosteids and titanichthyids of the late Famennian: giant predators with lengths in excess of 8 m. This diversification was abruptly truncated by the Hangenberg crisis and all placoderms, large and small, marine and freshwater, went extinct at the end of the Devonian (Gardiner 1990).

The Tournasian witnessed the spectacular radiation of the actinopterygians or ray-finned fishes (Gardiner 1993), a radiation that ultimately produced the teleosts. The actinopterygians had previously been a rather small and insignificant component of fish populations, and it is tempting to view their post-Devonian success as a filling of empty ecospace in much the same way as the mammals followed the dinosaurs.

Conodonts

The history of the conodonts during the Kellwasser crisis has been amply documented and, despite McLaren's (1982) curious observation that they were 'unaffected' at this time, it was a major mass extinction for the group. Conodont losses occurred in three closely spaced steps within the *linguiformis* Zone (Morrow and Sandberg 1996). The first step, at the base of the Upper Kellwasser Horizon (UKH), reduced species diversity by half from 70 to 35. Several more species were lost at the second step within the UKH and, at the top of the UKH, the final step removed a further 20 species. The losses were partially offset by the appearance of six progenitor species within the UKH. Conodonts from a range of environments were eliminated, and it is difficult to see any selectivity in the extinctions. Thus, all shallow-water conodonts went extinct with the exception of a single species, *Icriodus cornutus*, which, perversely, became prolifically abundant in the late *linguiformis* and early *triangularis* Zone (Sandberg *et al.* 1988; Buggisch 1991). *Palmatolepis*, a genus that may either have occupied a pelagic (Buggisch 1991) or deep basinal life-site (Sandberg *et al.* 1988), was common up to the extinction interval and then became rare.

The conodont record during the Hangenberg crisis has received less detailed scrutiny and diversity changes are consequently less well known. Several genera disappeared during the *praesulcata* Zone, including *Icriodus* and *Palmatolepis*, but generally it appears to have been a less severe crisis than the Kellwasser event (Bless *et al.* 1992). Curiously though, the family-level record of the conodonts reveals a different picture: no families were lost in the Frasnian but four of ten disappeared in the Famennian (Aldridge and Smith 1993).

Cricoconarids

The tiny, cone-like shells of the cricoconarids are common in Frasnian strata and occasionally reach rock-forming densities. The affinities of the group are a matter of conjecture (molluscs?) but their broad facies distribution points to a pelagic life-style possibly analogous to modern pteropods. Four cricoconarid families are recognised: the styliolinids, nowakiids, tentaculitids, and homoctenids. All went extinct during the

Kellwasser crisis and their downfall, which represents one of the major losses of the mass extinction event, has been recorded in detail. The styliolinids perished first along with many tentaculitids in the earliest Lower *gigas* Zone, slightly below the Lower Kellwasser Horizon, although a few styliolinids may have lingered on into the *linguiformis* Zone in China (Schindler 1990, 1993). Nowakiids were the next to go in the *linguiformis* Zone, quickly followed by the remaining tentaculitids at the end of the Zone (Narkiewicz and Hoffman 1989; Schindler 1990). The homoctenids survived into the Famennian, but only for a brief interval as they disappeared rapidly in the Lower *triangularis* Zone (Schindler 1990). It is a pity that little is known about cricoconarid ecology because the stepped extinction pattern may relate to a progressive deterioration in the water column analogous to that recorded by planktonic foraminifera during the Cenomanian – Turonian crisis (Chapter 8). It must suffice to say that the Kellwasser event was a serious crisis for the world's zooplankton (as also testified by the entomozoacean extinctions).

Microplankton

A clue to the demise of the cricoconarids may lie in diversity reductions among the microplankton. McGhee (1996) notes the extinction of 60% of prasinophyte genera and 81% of acritarch species during the Late Devonian. These are substantial losses, but it is unclear if they relate to the Kellwasser or Hangenberg crises or both. More clear is the extinction of the chitinozoa at the end of the Famennian (Tyson 1995). At the F–F boundary section at Coumiac, France, Paris *et al.* (1996) recorded abundant tasmana-cean prasinophytes from the extinction interval followed by exceptional abundances (19000 specimens per gram) of a single species of chitinozoan in the basal bed of the Famennian. In a detailed review of the occurrence of tasmanaceans, Tyson (1995) noted that they are usually associated with anoxic and/or cool waters; both may have been present at this low-latitude site in the late Frasnian. The chitinozoan bloom has been variously ascribed to cool waters, elevated nutrient supply or a reduction of zooplankton (cricoconarid?) grazing (Paris *et al.* 1996).

The Hangenberg crisis, in addition to eliminating the chitinozoa, also caused marked reductions among the acritarchs and prasinophytes (Wicander 1975), although further studies are required to elucidate the timing and global nature of these extinctions.

Plants

Plants experienced the first major crisis of their history when their diversity declined to a prolonged low point during the early mid-Frasnian to mid-Famennian period. This interval was, however, marked by the evolution of the first tree, *Archaeopteris*, a gymnosperm that reached heights in excess of 30 m. This was able to colonize flood-plain environments, thereby producing the first forests that spread over large areas of the world. The lack of provinciality or, in the words of Raymond and Metz (1995), 'phytogeographic differentiation' may go some way to explain the low values of floral diversity at this time. The hegemony of *Archaeopteris* came to a rapid conclusion in the late Famennian following the rise of seed plants (sphenopsids) and early ferns (Algeo *et al.* 1995).

The changes in the plant macrofossil record are partly recorded in the miospore record. Thus, the Frasnian–Famennian interval of low diversity is marked by a reduction of miospore species from greater than 50 in the early Frasnian to substantially less than 30 around the F–F boundary (Boulter *et al.* 1988). The detailed changes during the Kellwasser crisis are not known, but the miospore record of the Hangenberg crisis shows major extinctions and floral turnover, suggesting a major terrestrial event coincident with the marine one (Bless *et al.* 1992).

How selective and how sudden were the extinctions?

The foregoing review reveals that the later Devonian was one of the more complex periods of diversity change and extinction in the world's history (Figs 4.4 and 4.5). The wide range of variations between different groups is one of the unique features of this interval. Probably the most perplexing is the bryozoan mass extinction at the end of the Givetian – a crisis that was all their own – although the disappearance of the reef-forming calcareous algae in the mid-late Famennian is also curious as this was a phase of radiation in most groups. However, the majority of extinctions occurred during the three great Taghanic, Kellwasser and Hangenberg crises.

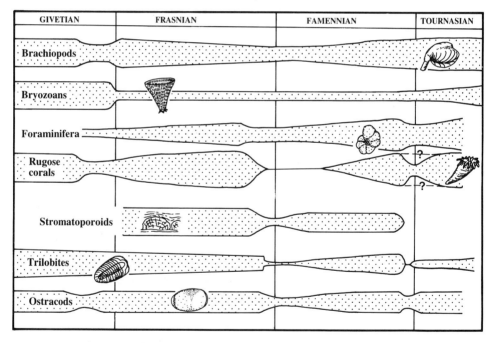

Fig 4.4   Summary of later Devonian to basal Carboniferous diversity changes for benthic groups.

The Taghanic crisis was the least severe but it is also the least known of the three events. The principal victims – brachiopods, rugose corals, and ammonoids – come from low-latitude and shallow-water environments. The contemporaneous radiation of planktonic ostracods and palmatolepid conodonts (House 1985) suggests that the crisis was confined to the benthic environment.

The selectivity of the Kellwasser extinctions is better documented; one of the clearer signals is the preferential loss of warm-water taxa (for example, atrypid and pentamerid brachiopods) recorded by Copper (1977, 1986), although benthic ostracods indicate the opposite selectivity (Lethiers and Whatley 1994). The proliferation of glass sponges at low-latitude sites suggests the importance of cooling during the crisis (McGhee 1996) and/or the preferential success of deeper-water taxa – a phenomenon that is also well illustrated by the rugose corals. The most noticeable victims of the crisis were the reefs and their resident taxa of stromatoporoids, tabulates, large foraminifera, and trilobites (Fig. 4.4). The extinction also caused substantial losses among all the main pelagic groups. Thus, cricoconarids were wiped out and the agnathans nearly so, whereas the conodonts, ammonoids, and placoderms 'escaped' with heavy losses (Fig. 4.5).

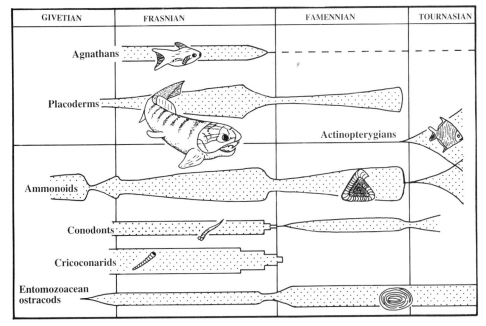

Fig 4.5 Summary of later Devonian to basal Carboniferous diversity changes among pelagic groups.

The Kellwasser crisis eliminated taxa from a broad ecological spectrum, and it is clear that the event deserves its 'mass extinction' designation. It is by no means so certain that the Hangenberg crisis belongs in the same category. Extinctions occurred among the trilobites and the stromatoporoids during this later event but many benthic taxa, notably brachiopods, tabulates and bryozoans, were unperturbed. Better documentation of the rugose corals and ostracods across the D–C boundary may establish the significance of the crisis for the benthos as a whole. The total extinction of the placoderms and chitinozoans and the near-eradication of the ammonoids and nautiloids indicates that the Hangenberg crisis was at its most severe in the water column, although notably it did not affect the entomozoaceans. Interestingly, this 'worse in the water column better on the seafloor' conclusion is the opposite to that observed for the Taghanic event. Perhaps not surprisingly, the most long-lasting changes of the Hangenberg event are seen in the water column where new ammonoid and fish (actinopterygians) lineages rapidly arose.

Following McLaren's (1970) proposal of a sudden mass extinction at the F–F boundary, there has been much discussion of the rapidity of this event. Improvements in knowledge over the ensuing decades have generally strengthened the arguments in favour of a more gradual, protracted crisis (Copper 1986; Farson 1986; Schindler 1990, 1993; Becker *et al.* 1991) with the vast majority of losses occurring between the Lower *gigas* and top *linguiformis* Zones. Others, notably those in favour of bolide impact extinction scenarios, continue to support a catastrophic component to the extinctions at the end of the *linguiformis* Zone (McLaren 1982; Sandberg *et al.* 1988; McGhee 1988, 1996; Goodfellow *et al.* 1989; McLaren and Goodfellow 1990; Wang *et al.* 1991). Arguments can be found for both viewpoints and it depends primarily on which group is under scrutiny. Brachiopods and probably trilobites were declining throughout the Frasnian but other groups radiated during this stage (ammonoids, placoderms, and rugose corals) and for them at least the F–F crisis probably came as something of a surprise. The rapid versus gradual debate also depends on the time-scale. Thus, both the conodonts and cricoconarids underwent rapid extinction in the space of a million years or less but, viewed in detail, this can be resolved into a series of steps (Fig. 4.5).

Boundary sections

Further clues to the cause of the F–F and D–C extinctions can be had from the sedimentary record of boundary sections and fortunately, unlike other crises (such as the end-Triassic event), there is a wealth of sites available for study.

Germany

The Late Devonian sections in Germany provide the eponymous horizons for both the Kellwasser and Hangenberg crises (House 1985). These are thin horizons of dark, organic-rich, laminated shales or limestones interbedded with shallow marine carbonates. The Kellwasser Horizons contain a distinctive fauna of *Buchiola* (a supposedly pseudo-planktonic bivalve), tornoceratids, orthocone nautiloids, homoctenids and arthrodires (Schindler 1990). This facies and fauna is typical of Devonian euxinic basinal settings and their occurrence in shelf settings records the brief expansion of such conditions into shallow-water sites (Schindler 1993). The Kellwasser Horizon is divisible into lower and upper horizons (Fig. 4.2) separated by a bioclastic limestone that apparently marks the extinction level for many pelagic forms – among them homoctenids, entomozoaceans, and ammonoids. The majority of benthic extinctions occur either at the base of the Lower Kellwasser Horizon or the base of the Upper Kellwasser Horizon – most trilobites in particular survive to the higher level (Schindler 1993). Pelagic extinctions have been variously reported to occur either between the Kellwasser Horizons (Sandberg *et al.* 1988; Groos-Uffenorde and Schindler 1990) or within the Upper Kellwasser Horizon (Lottman *et al.* 1986).

Germany also provides many excellent D–C sections, particularly in the Rhenish Slate Mountains, where the mid-*praesulcata* Hangenberg Shale provides a widespread marker even though only a few centimetres thick (Becker 1992; Kürschner *et al.* 1992).

France

The international stratotypes for both the F–F and D–C boundaries are found in France, at Coumiac and La Serre, respectively (Paproth *et al.* 1991; Klapper *et al.* 1993). The section at Coumiac, in the Montagne Noire region, is comparable with the shallow-marine German sections and displays equivalents of the Lower and Upper Kellwasser Horizons although, in France, they consist of dark, laminated dolomites interbedded with pink biomicrites (Becker *et al.* 1989). The extinctions also show a similar pattern to those seen in Germany.

The La Serre section, also in the Montagne Noire, is developed in a deep-water cephalopod 'griotte' facies but, at the boundary, a thin, sharp-based, bioclastic oolite is developed. This contains large intraclasts, abundant and diverse brachiopods and of course the all-important, defining conodont *S. sulcata* (Flajs and Feist 1988; Paproth *et al.* 1991). The presence of estherians at this level led Flajs and Feist (1988) to suggest reduced salinity conditions, but this is contradicted by the bulk of the fauna which unequivocally indicates fully marine conditions. Walliser (1996) lamented the choice of the La Serre stratotype which probably has a sequence boundary or omission surface as the D–C boundary. The boundary may have been better chosen in China.

China

South China is replete with numerous F–F and D–C boundary sections which record a broad range of marine palaeoenvironments (Bai and Ning 1988; Ji 1989). The shallow-marine Frasnian to Famennian, carbonate sections from around the city of Guilin, Guangxi Province, record an abrupt mass extinction of benthos (stromatoporoids, pentamerids, and atrypoids) coincident with the brief development of Kellwasser-like facies in the *linguiformis* Zone (Ji 1989).

Guangxi also contains one of the best D–C boundary sections at Nanbiancun (Ziegler *et al.* 1988). The slope facies at this site contain both deep-water and transported shallow-marine fossils, thus providing a rare opportunity to examine changes in the benthic fauna from a range of water depths. The *praesulcata–sulcata* boundary falls within a bed of bioclastic limestone that contains an abundant and diverse assemblage of brachiopods (Fig. 4.6). Unlike many European sections, there is no evidence for a break in deposition nor a black shale. The absence of any faunal turnover among the Nanbiancun benthos lends further weight to the suggestion that the Hangenberg crisis was at its worst in the water column.

Australia

No discussion of F–F boundary sections is complete without mention of the Canning Basin of north-west Australia. The extensive stromatoporoid–coral reefs of the area are one of the marvels of the fossil record. Reef growth in the region ceased in the late Frasnian and the overlying strata consist of thinly bedded, ammonoid-bearing limestones interbedded with breccia horizons (Becker *et al.* 1991). The transition from reefs to this deeper-water facies association is contemporaneous with the onset of deposition of the Kellwasser Horizons in Germany, although Becker *et al.* (1991) did not consider the thin-bedded limestones of Australia to be the product of oxygen-poor deposition. The highest Frasnian beds are generally devoid of benthic fossils but, at one locality, an

Fig 4.6 Devonian–Carboniferous boundary section at Nanbiancun, near Guilin, Guangxi Province. The section, from a carbonate ramp, shows interbedded limestones with a diverse allochthonous fauna, and thinner shales. The boundary occurs within a limestone at the level of the hammer head.

immense diversity of benthic taxa, particularly gastropods, is found immediately below the F–F boundary (Becker *et al.* 1991). After this extraordinary occurrence benthos became exceptionally rare in the Canning Basin and the cyanobacterial reefs that developed in the Famennian are virtually unfossiliferous.

Sea-level changes and glaciation

The Late Devonian was a time of a generally rising global sea level that climaxed in the Frasnian; this was followed by a significant fall in the Famennian Stage (Hallam 1992). This much is reasonably well established, but the details of the changes during the crisis intervals are more debatable, particularly for the Kellwasser event. Copper (1986) proposed sea-level fall across the F–F boundary, whereas Eder and Franke (1982) proposed a rise, but most authors are in general agreement that this was a period of high-frequency fluctuations (Buggisch 1972, 1991; Johnson *et al.* 1985; Sandberg *et al.* 1988; Muchez *et al.* 1996). The most widely accepted sea-level curve was produced by Sandberg and colleagues (Fig. 4.7) based on a combination of facies analyses and conodont biofacies evidence. This recognises two deepening events coincident with the Kellwasser Horizons bracketed by regressive intervals. Sequence stratigraphic analysis of Chinese and Belgian sections has confirmed the general outline of this curve although not the abrupt nature of some of the inflexions (Muchez *et al.* 1996). A notable feature of the curves is the major regression in the basal Famennian *triangularis* Zone. Evidence for

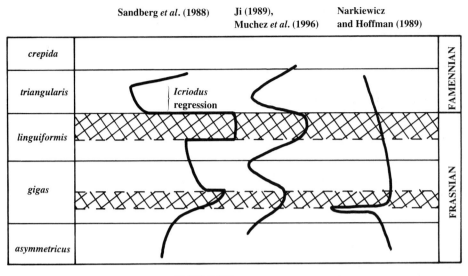

Sandberg *et al.* (1988)   Ji (1989),   Narkiewicz
                           Muchez *et al.* (1996)   and Hoffman (1989)

DEEPENING →

Fig 4.7 Summary of three proposed eustatic sea-level curves for the Frasnian to Famennian interval; sea-level rise to the right. The cross-hatched areas depict the intervals marked by the Lower and Upper Kellwasser Horizons.

this includes a frequent early *triangularis* age hiatus (see, for example, Narkiewicz and Hoffman 1989; Geldsetzer *et al.* 1993), the sharp top contact of the Upper Kellwasser Horizon and the abundance of *Icriodus*, a shallow-water conodont, in nearly all *triangularis* sections (Sandberg *et al.* 1988; House 1989; Ji 1989; Buggisch 1991; Muchez *et al.* 1996).

A substantially different series of sea-level changes was proposed by Narkiewicz and Hoffman (1989), based primarily on facies analysis of Polish sections (Fig. 4.7). They observed that the reefs of the region are capped by a karstic surface and overlain by the Lower Kellwasser Horizon – an identical succession occurs in the Anti-Atlas Mountains of Morocco (Buggisch 1991) – and they therefore proposed a short, sharp regression–transgression in the Lower *gigas* Zone. No further major sea-level changes were proposed for the F–F interval and an early *triangularis* regression was significantly absent from their interpretation. Rather, they suggested that the absence of strata of this age was due to the virtual shutdown of carbonate production in the immediate post-mass extinction interval (Narkiewicz and Hoffman 1989).

At this point it is worth considering further the evidence for the earliest Famennian regression because it forms the key aspect of many mass extinction models. In most sections there is rarely a facies change at this level, beyond the condensation or omission surface already noted. In the Belgian sections the base Famennian is notable for the spread of shale deposition over the entire region (Muchez *et al.* 1996: Fig. 4), hardly the signature of a major sea-level fall. Some of the best evidence for regression comes from shallow-water F–F sections seen at Trout River in the Northwest Territories of Canada. Here the late Frasnian Kakisa Formation consists of carbonates with small stromatoporoid biostromes (Geldsetzer *et al.* 1993). This formation, which is entirely of Upper *gigas* age on the basis of conodont evidence, is capped by a karstic surface containing neptunean dykes infilled with mid-*triangularis* Zone sediments. The sandstones and siltstones of the overlying Trout River Formation are of late *triangularis* to *crepida* age.

There is thus a *linguiformis* to mid-*triangularis* hiatus in the section. However, Geldsetzer and colleagues argued that the *linguiformis* Zone must be present in the uppermost Kakisa Formation because the formation is too thick to have accumulated entirely in Upper *gigas* Zone. This deduction was based on calculations of expected thickness derived from typical platform carbonate accumulation rates and the estimated 0.5 m.y. duration for the conodont subzone. Having apparently side-stepped the problem of having no *linguiformis* conodonts, Geldsetzer *et al.* (1993) were thus able to claim a major end-*linguiformis* Zone regression in accord with the Sandberg *et al.* sea-level curve (Fig. 4.7). The fallacy of their arguments is immediately apparent, however, when it is realised that the 0.5 m.y. subzone duration is merely a guess – cf. McGhee (1996) for the wide range of estimates – and the carbonate accumulation rates are simply an average of a large range of rates. A parsimonious interpretation of the Trout River section would attribute the onset of karstification to the earlier sea-level fall seen in the *gigas* Zone – an event that is recorded in numerous sections throughout the world (see, for example, Narkiewicz and Hoffman 1989; Buggisch 1991).

The remaining evidence for a major end-Frasnian regression comes from the sudden increase in *Icriodus* abundance but, in this interval immediately following a mass extinction, should we expect the normal conodont depth distributions to prevail? As Bottjer *et al.* (1995: p. 19) have noted, 'palaeoecological models for determining palaeoenvironments may be less useful for mass extinction aftermaths . . . when normal ecological conditions have experienced breakdown'. In particular, the rapid evolutionary changes among conodonts during the F–F crisis severely disrupted the normal biofacies distributions (Belka and Wendt 1992). As *Icriodus* was one of the few conodonts to survive the mass extinction, we interpret its earliest Famennian proliferation as an opportunistic expansion into a much broader range of environments than are normally typical for the genus. This argument does not rule out the possibility of major basal Famennian regression but the currently available evidence suggests little sea-level change across the F–F boundary.

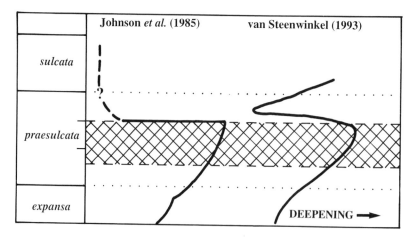

Fig 4.8 Summary of two proposed eustatic sea-level curves for the D–C boundary interval. The cross-hatched areas depict the interval marked by the widespread deposition of the black shale associated with the Hangenberg crisis.

Devonian–Carboniferous sea-level changes have also been the subject of detailed scrutiny, and this has also revealed some spectacular changes (Fig. 4.8). A major transgressive phase in the early mid-*praesulcata* Zone produced vast areas of black shale deposition in both basinal and shelf locations throughout the world (see, for example, Savoy and Harris 1993; Robison 1995) of which the Hangenberg Shale is the best known (House 1985; Johnson *et al.* 1985). Only in China do black shales seem to have been absent in some shallow-shelf and ramp settings (Ziegler *et al.* 1988), but even here the black Changshun Shale is widely developed in the *praesulcata* Zone (Bai and Ning 1988; Chai *et al.* 1989). In some areas black shale deposition continued uninterrupted into the Carboniferous, notably in North America, but elsewhere black shale formation ceased at the end of the mid-*praesulcata* Zone. Johnson *et al.* (1985) placed a dramatic sea-level fall at this level and, thanks to the work of van Steenwinkel (1992, 1993), this interpretation is strongly supported by sequence stratigraphic evidence. Detailed analysis of the sections in the Rhenish Slate Mountains of Germany revealed a major type 1 sequence boundary associated with up to 95 m of incision and valley generation (Fig. 4.9). These incised valleys and their fills are only encountered in basinal sections while the surrounding shelf areas display a mid-late *praesulcata* hiatus (van Steenwinkel 1992). The impressive 95 m incision depth in basinal settings implies that the eustatic fall was considerably in excess of 100 m. Further evidence for this spectacular regression comes from the sudden appearance of oolitic facies in the deep-water La Serre stratotype (p. 77). It therefore appears that the *praesulcata* Zone records a transgressive–regressive couplet of exceptional magnitude.

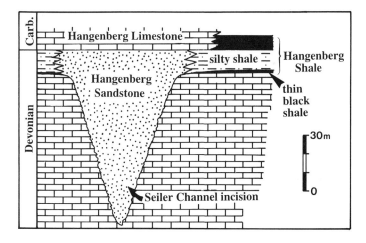

Fig 4.9 Stratigraphic relations in the D–C boundary sections of the Rhenish Mountains, Germany. The Hangenberg Shale is a composite unit encompassing a thin, widespread black shale at its base. The Hangenberg Sandstone infills a major incised valley that has cut down from a level within the Hangenberg Shale. After van Steenwinkel (1992).

Strontium isotope data from conodont apatite further supports the proposed *praesulcata* sea-level changes. Ratios of $^{87}Sr/^{86}Sr$ show a sudden swing from values of 0.7084 to 0.7086 at the mid–late *praesulcata* boundary followed by a rapid return to lighter values at the base of the *sulcata* Zone (Kürschner *et al.* 1992). The short-term enrichment in [87]Sr coincides with, and strongly reinforces the existence of, van Steenwinkel's major regression.

The rapidity and magnitude of some Late Devonian sea-level changes, particularly the late *praesulcata* regression, have led to suggestions of ice-cap formation on Gondwana (Kürschner *et al.* 1992), for which there is substantial evidence from South America. Diamictites, striated pebbles, dropstones, and erratics are all known from the Famennian basins of northern Brazil and point to at least localised glaciation of Gondwana (Caputo 1985). Only palynostratigraphic dating is available and this indicates a mid-Famennian age (Caputo and Crowell 1985). Therefore the glaciation may have occurred between the Kellwasser and Hangenberg crises. Oxygen isotope data are not quite in accord with the Brazilian data because they indicate a long-term cooling beginning in the mid-Famennian and culminating within the Tournasian (Brand 1989, 1992). This suggests that glaciation was more likely to have been a factor in the D–C than in F–F sea-level changes.

Geochemistry

Stable isotopes

Carbon isotopes

Changes in carbon isotope ratios in limestones can provide valuable evidence in the analysis of global environmental events, so the contradictory nature of much of the published Late Devonian $\delta^{13}C$ data is unfortunate. Initially, McGhee *et al.* (1986) reported a positive shift of values across the F–F boundary, but Goodfellow *et al.* (1989) found a negative shift at the same level. The discrepancy may be explained by the measurement of differing stratigraphic levels because, as Buggisch and Joachimski have subsequently shown, there were several positive and negative shifts during the Kellwasser crisis (Buggisch 1991; Joachimski and Buggisch 1993). These authors have obtained consistent results from numerous European sections which reveal two positive

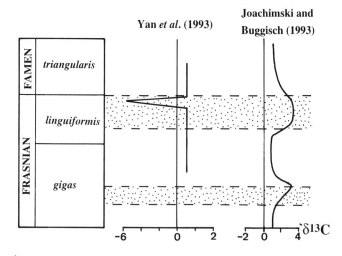

Fig 4.10 Reported carbon isotope trends across the F–F boundary. The stippled areas correspond to the depositional intervals of the Lower and Upper Kellwasser Horizons.

excursions coincident with the Lower and Upper Kellwasser Horizons (Fig. 4.10). The same curve has been obtained from both micritic carbonate and organic C (Joachimski and Buggisch 1994) and appears to be the most reliable one available for this interval. The inevitable interpretation of the twin positive peaks is that they record the burial of large amounts of isotopically light organic C during the Kellwasser events. That the δ^{13}C returns to lighter values after both the events further suggests that much of the organic-rich strata was rapidly oxidised after the events.

The veracity of the Joachimski and Buggisch curve is indicated by its close correlation with sedimentary events, but recent analyses of Chinese sections have revealed an entirely different curve. Wang et al. (1991) found a sharp negative spike of 3.5‰ in the late linguiformis Zone but no positive excursion. Yan et al. (1993) found the same trend but with a −7.0‰ spike. In a subsequent analysis of a Canadian section, Wang and colleagues found a positive anomaly from across the F–F boundary that almost certainly corresponds to the upper excursion of the Joachimski and Buggisch curve, but they failed to find the sharp negative spike (Wang et al. 1996). The absence of the spike was attributed to the very short duration of the event which was attributed by Wang et al. – and also by Yan et al. – to a brief interval of zero-productivity (Strangelove) oceans. A more prosaic explanation of the negative spike may come from a consideration of the strata in which it is found. The Chinese anomaly derives from a thin, organic-rich, black shale interbedded within limestones. The shift to negative values in the shale could therefore record the enrichment of carbonate with light CO_2 derived from anaerobic degradation of organic matter. Yan et al. (1993) discount this alternative because 'the erratic behaviour of δ^{13}C values in the boundary layer [black shale] indicates that the carbon isotope record reflects environmental changes'. Curious logic indeed.

No such complexities exist for D–C carbon isotope trends. All studies indicate that there were no excursions at this level (Xu et al. 1986; Bai and Ning 1988; Schönlaub et al. 1988) although Xu et al. (1986) documented a huge −8.0‰ shift within the Tournasian. Both this event, which has yet to be duplicated in other sections, and the lack of excursions in the D–C interval are rather puzzling.

Oxygen Isotopes

Reported oxygen isotope trends from across the F–F interval are as bewildering as those seen for carbon isotopes. Brand's (1989) analyses of brachiopod calcite revealed a prolonged Late Devonian negative shift continuing uninterrupted across the F–F boundary and reaching a low point in the mid-Famennian, but details across the Kellwasser extinction interval are lacking. Whole-rock analysis of carbonate boundary sections suggests little change of δ^{18}O at this time (McGhee et al. 1986; Buggisch 1991). In contrast, Yan et al. (1993) discovered a +2.0‰ spike in the same Chinese black shale that produced the δ^{13}C anomaly and Goodfellow et al. (1989) found a latest Frasnian positive excursion followed by a Famennian negative excursion in both the Canning Basin and Coumiac. This evidence at least suggests a short interval of cooling in the linguiformis Zone followed by Famennian warming.

Few analyses of δ^{18}O values have been undertaken from D–C boundary sections but those that have reveal, like the δ^{13}C results, little change (Bai and Ning 1988; Schönlaub et al. 1988). Only Xu et al. (1986) have found any significant excursions, a −3.0‰ spike in the mid-praesulcata Zone carbonates of China.

Sulphur isotopes

Devonian evaporites record a rapid increase in $\delta^{34}S_{sulphate}$ values during the Frasnian, peaking at $+25.0‰$ around the F–F boundary, followed by a fall to $+14.0‰$ at the D–C boundary (Holser 1984). Details of the fluctuations in proximity to the crisis intervals are not available from evaporite data but pyrite isotope trends during the F–F crisis reveal that the $\delta^{34}S$ peak may have been sharply focused at the F–F boundary. Values of $\delta^{34}S_{pyr.}$ from Alberta, Canada, show a rapid rise to $+20.8‰$ in a pyritic siltstone straddling the F–F boundary (Geldsetzer *et al.* 1987). This is a very heavy value for pyrite and may indicate isolation of the basinal waters (Geldsetzer *et al.* 1987). This explanation is entirely feasible, but the discovery of equally heavy values in boundary sediments at Coumiac, France suggests a global phenomenon (Goodfellow *et al.* 1989), such as the burial of large amounts of pyrite during the Kellwasser events.

Iridium and trace metals

Given the widespread sedimentary evidence for anoxia during the accumulation of the Kellwasser and Hangenberg horizons, it is not surprising that their trace-metal concentrations also suggest reducing depositional conditions. This is particularly the case for the Hangenberg Shale and its correlatives which are enriched in Ni, V, and numerous chalcophile elements. This is due to their insolubility in reduced form and, in the case of Ni and V, their affinity for organic compounds in reducing conditions (Bai and Ning 1988; Chai *et al.* 1989). Iridium is also enriched in these shales, particularly at the base of the Changshun Shale in China (Wang *et al.* 1993c). Most authors have attributed this Ir spike to the ready precipitation of platinum-group elements at redox boundaries (Bai and Ning 1988; Wang *et al.* 1993c).

The most celebrated Ir anomaly from the Late Devonian interval comes from the stromatolitic reefs of the Canning Basin. Here, Playford *et al.* (1984) discovered an Ir peak of 0.3 ppb, 20 times above background values, in a bed containing the cyanobacterium *Frutexites*. At the time of publication the precise age of this horizon was not well known, but it was thought to be close to the F–F boundary. Subsequent work has shown that the level actually occurs in the early *crepida* Zone, long after the Kellwasser crisis (Nicoll and Playford 1993). Furthermore, the association of Ir and *Frutexites* is probably not coincidental because the siderophile-metabolising properties of this cyanobacterium may also have caused the Ir enrichment.

The search for an Ir anomaly more directly associated with the F–F boundary has generally proved fruitless, with a single exception. Wang *et al.* (1991) investigated the rather notorious thin black shale between the *linguiformis* and *triangularis* Zones of south China (from whence the negative $\delta^{13}C$ spike came) and found an Ir peak of 35 ppb. In an extraordinary split decision, Orth and Attrep suggested that there 'may be a causal relationship between Ir enrichment and a reducing environment' (Wang *et al.* 1991: p. 779), whereas the remaining four authors, with their eyes on the heavens, opted for a bolide origin.

Causes of the extinctions

Bolide impact

As already noted, the suggestion that the F–F crisis was caused by a bolide impact has long standing (McLaren 1970) and predates the end-Cretaceous scenario by a decade. McLaren suggested that a large meteorite landed in an ocean in the Late Devonian and generated a giant tsunami, exceeding a kilometre in height. This flooded large areas of the world's land surface and the resultant runoff then made the seas fatally turbid and hence caused mass extinction. At the time of this proposal the only supposed supporting evidence came from the apparently abrupt nature of the F–F extinction. More direct evidence, in the form of an Ir anomaly, has not been forthcoming, the dubious Chinese example excepted, but Ir is not the only 'smoking gun' of bolide impact.

Microtektites are tiny, glassy particles produced during bolide impact. They have characteristic splash and tear-drop morphologies. Diligent searching has produced microtektitie-like spherules from sections in Belgium and China but the former examples occur at a level just beneath the early–mid-*triangularis* Zone boundary (Claeys *et al.* 1992) whereas the latter are of *crepida* age (Wang 1992). The Chinese examples may be coincident with the Canning Basin Ir anomaly, and the possibility that a bolide struck at this time is further strengthened by the presence of an impact crater – Taihu Lake in China – of the correct age (Wang 1992). The Belgian microspherules may, in their turn, be derived from the Siljan Ring, a Swedish impact structure (Claeys and Casier 1994).

Microtektites have also been reported from the D–C interval of China, specifically in the early *praesulcata* Zone (beneath the basal Changshun Shale Ir anomaly) and in the *duplicata* Zone of the Tournasian (Bai and Ning 1988; Fig. 4.12). However, probably the most conclusive evidence for bolide impact comes from the early Frasnian *punctata* Zone of Nevada and adjacent states (Leroux *et al.* 1995b). Shock-metamorphosed quartz and a modest Ir anomaly both occur, but the most spectacular evidence is the Alamo breccia. This 70 m thick pile of large blocks of limestone covers 4000 km^2 of a shallow marine platform and is probably rubble caused by bolide impact.

With such supporting data, McLaren's (1970) original hypothesis has gained numerous adherents, although with modifications to the kill mechanism (Geldsetzer *et al.* 1987; Sandberg *et al.* 1988; Goodfellow *et al.* 1989; and some authors of Wang *et al.* 1991). Thus it is claimed that the most lethal effect of oceanic impact would be the overturn of the water column and the poisoning of surface waters with hydrogen sulphide – the enrichment of pyrite with ^{34}S in the boundary beds thereby becomes a signature of such an overturn (Geldsetzer *et al.* 1987; Goodfellow *et al.* 1989).

Whether such an effect would extend much beyond the impact site is debatable, but a more serious criticism of the impact hypothesis is the failure to find evidence for impact during the extinction intervals despite careful searching (see, for example, McGhee *et al.* 1986). Good evidence for impact comes from the *punctata* Zone, and it is reasonable to conclude that bolides may have struck in the *triangularis* and *crepida* Zones (Fig. 4.11). These are all intervals of radiation not extinction, and it could therefore be concluded, somewhat tongue in cheek, that bolide impact produces beneficial effects for life. Alternatively, McGhee (1996: p. 244) concluded that, although the 'search for iridium anomalies [etc.] has been frustrating, ... multi-impacts [remain] the most viable

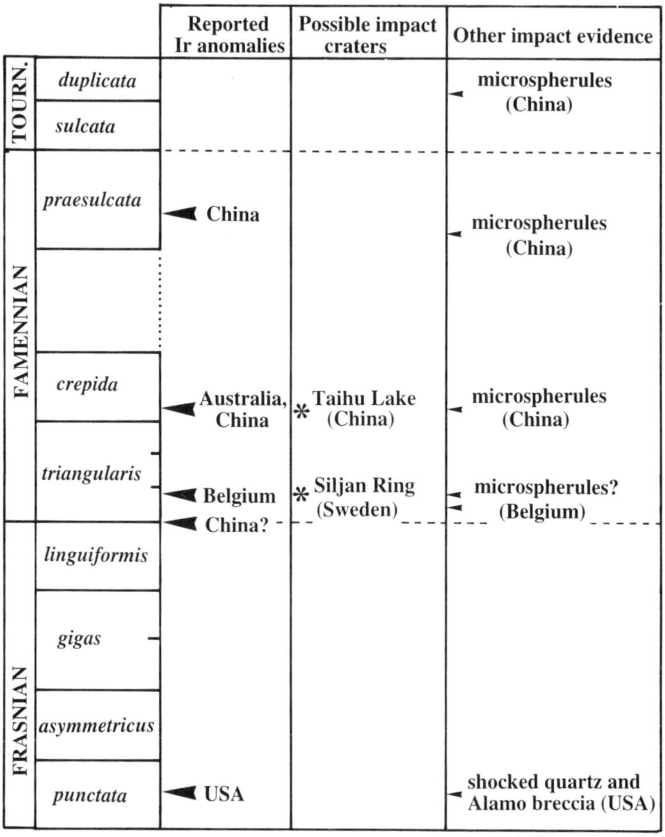

		Reported Ir anomalies	Possible impact craters	Other impact evidence
TOURN.	duplicata			microspherules (China)
	sulcata			
FAMENNIAN	praesulcata	◄ China		microspherules (China)
	crepida	◄ Australia, China	* Taihu Lake (China)	microspherules (China)
	triangularis	◄ Belgium ◄ China?	* Siljan Ring (Sweden)	microspherules? (Belgium)
FRASNIAN	linguiformis			
	gigas			
	asymmetricus			
	punctata	◄ USA		shocked quartz and Alamo breccia (USA)

Fig 4.11 Stratigraphic location of the various lines of evidence for bolide impact in the Late Devonian to Early Carboniferous interval.

explanation for the ecological and temporal patterns of [Kellwasser] events'. On the contrary, there are many more and better extinction scenarios.

Anoxia

Anoxia has long been one of the most plausible candidates for the Kellwasser extinctions (Buggisch 1972). The near-conclusive evidence includes:

(1) the widespread development of anoxic and dysoxic facies;

(2) the enrichment of trace metals in boundary sediments;

(3) the heavy $\delta^{34}S_{pyr.}$ values, indicating burial of large amounts of pyrite;

(4) the two positive $\delta^{13}C$ excursions, indicating the burial of large amounts of organic C;

(5) the close correspondence between the development of dysoxic/anoxic facies and the extinctions;

(6) the preferential survival of dysoxia-tolerant benthos (ostracods, bivalves) and many deeper-water taxa among the sponges, rugose corals, and ammonoids, which were probably also able to withstand low oxygen levels.

Not surprisingly, anoxia figures wholly or partly in many extinction scenarios (Schlager 1981; House 1985; Buggisch 1991; Joachimski and Buggisch 1993; Becker and House 1994; Algeo *et al.* 1995). The lethality of the extinction may have been exacerbated by the rapid sea-level changes in the latest Frasnian. To have killed reef taxa, the oxygen-deficient conditions would need to have spread into very shallow waters. Narkiewicz and Hoffman (1989) envisaged precisely this: they suggested that anoxic conditions initially became established in basinal areas but only affected the shallowest-water benthos during the subsequent regression when they were unable to migrate downslope due to the presence of basinal anoxia – a case of 'trapped between the devil and the deep blue sea'.

However shallow the depths reached by oxygen-poor waters during the Kellwasser (and Hangenberg) events, dysoxic conditions are unlikely to have been a direct cause of the demise of the surface-dwelling chitinozoans and cricoconarids. Rather, their extinction may have hinged on changes in nutrient dynamics caused by changes in oceanic redox chemistry; such changes have also been implicated in the end-Ordovician (p. 59) and end-Permian extinctions (p. 140).

Anoxia is also the leading candidate for the Hangenberg extinctions (Wang *et al.* 1993b) and the evidence is essentially the same: global development of black shale facies, trace-metal signature of intense anoxia, and the correspondence between the extinctions and the onset of anoxic deposition. Sea-level fluctuations were equally dramatic during this crisis and may also have contributed to the extinctions as taxa failed to keep track with the rapidly shifting environments.

The lack of a positive $\delta^{13}C$ excursion associated with the Hangenberg event could relate to inadequate sampling or possibly because the event was of too short duration to leave a signature. More disquieting is the selectivity of the extinctions: pelagic groups suffered disproportionate losses and some benthic taxa were devastated (trilobites, stromatoporoids) while other benthic groups, notably brachiopods and foraminifera, were unscathed. Anoxia may not have spread into every shallow-water habitat, thereby providing refugia for some shallow-marine taxa; several of the ramp and shelf boundary sections in China are notably free of black shale (see, for example, Fig. 4.6).

The hegemony of the anoxia–extinction nexus is not complete, and many authors have suggested that there is no link between the two (Copper 1986; Claeys *et al.* 1994; McGhee 1996). The principal argument hinges on the fact that black shale horizons are numerous in the Late Devonian (cf. Algeo *et al.* 1995) but the majority are not associated with extinctions. Black shales are indeed frequent at this time, but this reasoning misses the crucial point that only the Kellwasser and Hangenberg black shales are truly global in extent. Goodfellow *et al.* (1989) alternatively argued that anoxia was not the culprit because the mass extinction was instantaneous whereas the anoxic transgression took several million years and was therefore too slow. Both points are highly dubious; the mass extinction spans at least two conodont zones whereas the spread of black shales occurred very rapidly indeed, well within the duration of the conodont zones.

Cooling

Copper (1977, 1986) is the chief proponent of a model involving Frasnian cooling and extinction. Supporting evidence is diverse:

(1)   the preferential elimination of reef taxa and other inhabitants of the low latitudes;

(2)   the proliferation of cold-water groups during and after the crisis interval (for example, hexactinellid sponges, labechiid stromatoporoids, tornoceratid ammonoids, and chitinozoans);

(3)   the presence of glacial sediments in northern Brazil;

(4)   the rapidity of sea-level changes during the crisis interval, suggesting glacially driven eustatic changes.

Copper envisages the cooling to be a consequence of the closure of an equatorial seaway as Gondwana and Euramerica docked in the Frasnian (Fig. 4.3). The resultant plate configuration is suggested to have restricted the circulation of equatorial currents and deflected cool, high-latitude currents into the tropics. Of all the extinction mechanisms this is the most slow-acting because it is a product of continental drift. Copper (1986) therefore emphasises the gradual, Frasnian-long duration of extinctions.

Despite the substantial evidence in favour of the cooling hypothesis, its tenability is rather doubtful. Thus, the glacial evidence from Brazil is undoubtedly too young; it is in fact more likely to be a factor in the Hangenberg crisis. The faunal data are also open to alternative interpretation because the most successful surviving taxa have not only a cool-water tolerance but also a deep-water habitat. Therefore, their proliferation during the crisis interval could equally well reflect the spread of dysaerobic, deep-water facies in association with the Kellwasser transgressions. Oxygen isotope data lend no support to the cooling hypothesis because, apart from tentative evidence of cooling in the *linguiformis* Zone, they indicate that cooling did not begin until the mid-Famennian (Brand 1992).

Warming

Rather than cooling, Brand's (1989) $\delta^{18}O$ data indicate exceptional warming in the late Frasnian and early Famennian, with equatorial ocean temperatures apparently reaching 40°C. As most marine organisms perish at temperatures in excess of 30°C, both Thompson and Newton (1989) and Ormiston and Oglesby (1995) therefore suggested 'death by cooking' as the Kellwasser extinction mechanism. Low oxygen levels are also likely to be a factor in such an extinction scenario due to the low solubility of this gas at such temperatures.

This extinction scenario must rank as one of the least likely contenders. Firstly, it is highly unlikely that the calculated 40°C temperatures are in any way realistic because they would have caused 100‰ extinction; even a more modest 30°C should have produced wholesale defaunation at equatorial latitudes. Oxygen isotopes are highly susceptible to diagenetic distortion and this undoubtedly has happened to Brand's samples. None the less, global warming may be implicated in the F–F extinctions because certain aspects of the extinction selectivity, notably the preferential survival of taxa from deep, cold-water environments, support this contention. However, other evidence argues against this extinction mechanism. Global warming should preferentially eliminate habitats at the highest latitudes, but the polar faunas of South America are among the least affected by the Kellwasser crisis (McGhee 1996). Finally, the peak of warming post-dates the mass extinction in Brand's data.

Brackish oceans

In many locations the D–C boundary beds contain large numbers of small, concentrically ornamented valves variously identified as the bivalve *Posidonia* or the conchostracan '*Estheria*' (House 1985; Flajs and Feist 1988). On the basis of shell structure, Flajs and Feist (1988) suggested that many of these records, if not all, belong in the latter category. This raises the question why an exclusively freshwater taxon should become abundant in marine facies. Flajs and Feist suggested the widespread development of a stratified water column with a brackish or freshwater upper layer in which the estheriids lived. This radical interpretation is partially supported by Brand's (1992) $\delta^{18}O$ data from Europe, but not elsewhere in the world.

The widespread development of brackish oceanic surface waters constitutes a potential extinction mechanism for the Hangenberg crisis, but the evidence is rather tenuous. The estheriids might more plausibly be interpreted as disaster taxa temporarily invading the marine realm or as *Posidonia* bivalves with an estheriid-like shell structure.

Carbon cycling models

Both the warming and anoxia scenarios provide proximate explanations for the extinctions but they do not address the ultimate causes. Several teams of workers have, however, modelled the Late Devonian world with a view to interpreting global environmental changes and thus the primary cause of extinction. Their results share numerous facets in common, particularly their involvement of the carbon cycle and the selection of anoxia as the main agent of marine extinction, and yet they all differ in detail.

The Buggisch model

Buggisch's (1991) model involves a close link between sea level, climate and oceanic productivity (Fig. 4.12). In essence, sea-level rise is envisaged to trigger black shale formation due to the high productivity of the flooded shelves and ultimately to cause mass extinction as anoxic waters spread far and wide. The model was later fine-tuned by Joachimski and Buggisch (1993) who borrowed the concept of warm saline bottom water (WSBW) from models of Cretaceous oceans (Brass *et al.* 1982). The flooding of low-latitude continental shelves may have generated WSBW which, due to its density, would have sunk to intermediate or abyssal depths of the Late Devonian oceans. The resultant stratification thereby exacerbated the oxygen deficiency of the world's oceans.

Buggisch's model is self-limiting because, as organic C was buried beneath the anoxic waters, atmospheric CO_2 levels were drawn down, leading to global cooling. Ultimately ice caps are suggested to have formed leading to regression, exposure of continental shelves, oxidation of black shales, and a gradual increase in atmospheric CO_2. The whole scheme is therefore cyclic, and Buggisch (1991) envisaged the Late Devonian world lurching repeatedly between icehouse and greenhouse conditions (Fig. 4.12). Herein lies the major weak link in the chain of events: only one minor glacial event is known (from the Famennian), while the most severe glaciations should have bracketed the Upper Kellwasser event. There is no evidence for these excepting perhaps the rapidity of sea-level change at this time.

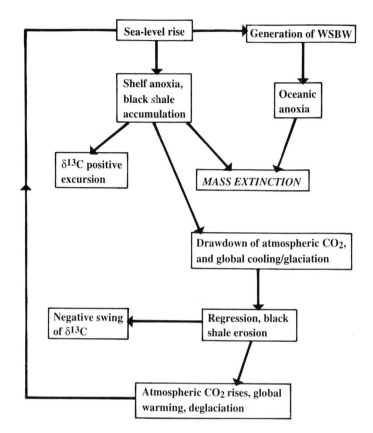

Fig 4.12 Buggisch's cyclical model for the Late Devonian climatic oscillations and extinctions.

The Becker–House model

In a similar model to Buggisch's, Becker and House (1994) discarded the glaciation link and introduced oceanic volcanism as the ultimate cause of the Kellwasser crisis. They proposed that increased volcanic activity in the oceans caused both eustatic sea-level rise and increased atmospheric CO_2 levels – both factors that would have favoured the generation of WSBW. Rather than stratifying the oceans, Becker and House suggested that the increased flux of water into the ocean depths would have caused the upwelling of cold, deep water to the ocean's surface. As in the modern ocean, sites of upwelling are highly fertile and the increased productivity is therefore suggested to have generated large areas of oxygen-poor waters and thus mass extinction.

The Becker–House model neatly combines the evidence for both cooling and anoxia, but the efficacy of upwelling as a mechanism for the widespread generation of oxygen-poor shelf waters is open to doubt. Modern upwelling is restricted to a few specific locations, notably the western margins of mid-latitude continents, but its effects in the flooded interiors of continents, far removed from the ocean, are likely to be negligible (Ormiston and Oglesby 1995).

The Algeo model

The final model under consideration points the finger of blame at plants and the development of the world's first forests (Algeo *et al.* 1995). Several workers have previously noted the propinquity between the rise of land plants and the marine extinctions of the Late Devonian. Tappan (1970) considered that plants would trap nutrients, thereby decreasing the nutrient flux to the oceans, while Calef and Bambach (1973) suggested that an increased flux of terrigenous organic matter debris into the oceans would have increased Late Devonian productivity. Algeo *et al.* (1995) highlight the increased chemical weathering of rocks as the most salient aspect of floral colonization and thus favour an increased nutrient flux to the oceans. This, they suggest, is the underlying reason behind the frequency of black shale events in the Late Devonian (Fig. 4.13). A side-effect of the increased burial of organic C on both land and in the sea was the drawdown of atmospheric CO_2 and thus global cooling. Thus, once again, both anoxia and cooling feature in this model although only the former is held as a cause of the Kellwasser and, in this case, the Hangenberg extinctions as well.

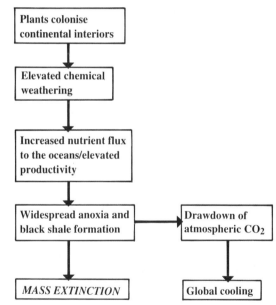

Fig 4.13 Summary of the Algeo model for the Late Devonian extinctions.

Perhaps the only question arising from the Algeo model lies in the degree to which chemical weathering increased in the Late Devonian. The *Archaeopteris* forests were restricted to floodplain environments, whereas more upland areas may not have been colonised until later in the Famennian, with the appearance of seed plants. Increased chemical weathering may not therefore have become significant until the very end of the Devonian.

A mid-Carboniferous mass extinction?

Following the 80 m.y.-long vicissitudes of Silurian and Devonian time, the dawning of the Carboniferous appears to have marked the start of a 120 m.y. respite from major biotic crises. Many groups diversified in the Lower Carboniferous, notably the actinopterygians, gastropods, crinoids, rugose corals, productid brachiopods, and foraminiferans (see, for example, Batten 1973; Tappan and Loeblich 1988; Gardiner 1993; Ausich *et al.* 1994). The only crisis to interrupt this halcyon interval has been reported from the mid-Carboniferous boundary. This level falls between the Arnsbergian and Chokierian Stages of the European Namurian Series and corresponds to the Mississippian–Pennsylvanian boundary of North America and to a level in the upper part of the Russian Serpukhovian Series. The importance ascribed to this event varies from author to author: Saunders and Ramsbottom (1986) called it a major extinction, and others (Ziegler and Lane 1987; Weems 1992), with undoubted exaggeration, ranked it as a mass extinction comparable to the big five. Sepkoski's (1986) original analysis of generic extinction rates revealed a pre-Arnsbergian peak composed primarily of foraminiferan, brachiopod, and coral extinctions and an end-Arnsbergian peak of cephalopod, crinoid, and bryozoan extinctions. In his latest analysis (Sepkoski 1996), the older extinction event has curiously disappeared but there is still a distinct Arnsbergian (intra-Serpukhovian) extinction peak. Very little detailed work has been done on this mid-Carboniferous extinction but it is worth briefly reviewing what little is known.

Many major ammonoid lineages disappeared at the mid-Carboniferous boundary and there is little doubt that, for this group at least, the interval was a major crisis (Ramsbottom and Saunders 1985; Saunders and Ramsbottom 1986). However, ammonoid extinction events are frequent (House 1989) and do not merit a 'mass extinction' designation on their own. Conodonts were also undergoing a crisis and reached a diversity low point at the mid-Carboniferous boundary (Ziegler and Lane 1987; Nemirovskaya and Nigmadganov 1994). Significantly, crinoids reportedly lost 42% of their generic diversity at the same time (Saunders and Ramsbottom 1986). This claim was not supported by the detailed investigation of Ausich *et al.* (1994) who noted that, although the Carboniferous was a time of rapid turnover among crinoid taxa, no distinct intra-Carboniferous crisis could be discerned. Brachiopods are one of the few groups for which biogeographic and range data have been compiled for the mid-Carboniferous (Raymond *et al.* 1989, 1990). The gigantoproductids appear to have been the only family to succumb, while the only significant generic extinctions occurred in Gondwana. Thus, the late Visean brachiopod faunas of New South Wales, Australia, were reduced from 51 to 5 genera in the Namurian. Raymond *et al.* (1990) concluded that the mid-Carboniferous boundary only marks a minor brachiopod crisis but with the proviso that better knowledge of southern hemisphere faunas might considerably increase the apparent importance of the event.

The status of the mid-Carboniferous crisis is therefore ambivalent; Sepkoski's data suggest it was a major event, while the few detailed studies suggest otherwise. The terrestrial record suggests an equally modest extinction event; Benton (1989) records the loss of only four low-diversity families of amphibians in the Serpukhovian. However, Weems (1992) emphasised that the Serpukhovian extinctions eliminated cosmopolitan taxa, thereby implying their global significance.

The apparent absence of a major mid-Carboniferous extinction is somewhat surprising as the interval is characterised by many rapid, global environmental changes. A major phase of Gondwanan glaciation began at this time, triggering faunal migrations as climatic zones narrowed (Kelley and Raymond 1991). The major sea-level fall at this time is also no doubt a consequence of ice-cap growth (Saunders and Ramsbottom 1986). Only the demise of many ammonoids and some Gondwanan brachiopods appears closely related to these changes. Perhaps the significant message here is that episodes of major glaciation are not necessarily the cause of mass extinctions.

5 Palaeozoic nemesis

The mass extinction at the end of the Permian has long been recognized as the greatest in the history of life (see, for example, Phillips 1841); the enormity of the faunal turnover made it an obvious choice for the Palaeozoic–Mesozoic erathem boundary. Estimates of the species-level extinction (extrapolated from generic and familial data) range from 96% (Raup 1979) to a no less impressive, conservative estimate of 75% (Hoffman 1986). More recent estimates have tended towards the higher end of this range – for example 90% (McKinney 1995). The effect was to reduce an all-time Palaeozoic high of species diversity to the lowest post-Ordovician value of the entire Phanerozoic; Permian diversity levels were not attained again until the Cretaceous (Allison and Briggs 1993; Benton 1995). Raup (1979) suggested that a quarter of a million species – a figure comparable with modern-day marine species diversity (Nicol 1979) – were reduced to less than 10 000 while family diversity was reduced from 650 to 420. With such wholesale wipe-out, survival must have been a matter of pure chance for many groups.

Until recently the end-Permian mass extinction was viewed as one of the less amenable events to study due to a (perceived) rarity of complete boundary sections: this has been attributed to a Phanerozic sea-level low (Erwin 1993, 1994). Extinction mechanisms for marine benthic invertebrates have therefore traditionally tended to focus on the limited shelf sea area as a causal mechanism – the species–area effect. Diversity is considered to have gradually declined as the seas gradually retreated from the continental shelves during the last 10 m.y. of the Permian (Schopf 1974; Forney 1975). However, this model is driven more by a lack of data than by any conclusive evidence, and the picture has become rapidly more complicated as data-gathering has accelerated since the mid-1980s.

Recent major advances in Permo-Triassic studies have included the demonstration that the eruption of the huge West Siberian flood basalt province almost exactly coincided with the mass extinction; the resultant desire to implicate the former as a cause of the latter is proving irresistible (see, for example, Renne *et al.* 1995). For the first time deep-water, oceanic sediments have been discovered in numerous boundary sections in the accretionary prisms and allochthonous terranes of Japan (Isozaki *et al.* 1990). The faunal data have also improved and better resolution of diversity trends has shown that the supposedly prolonged Late Permian diversity decline actually consists of two distinct extinction events separated by perhaps as much as 16 m.y. of recovery and radiation (Jin *et al.* 1994b; Stanley and Yang 1994) according to the latest time-scale of Menning (1995). Undoubtedly the single most important development in the study of the end-Permian mass extinction has been the detailed documentation of the numerous excellent P-Tr boundary sections in South China. Before 1980 virtually nothing was known in the West about these sections but, after a decade and a half of intensive study, the situation has reversed and the Chinese sections are now the best known in the world. Ironically, Erwin (1994) considered that the story from South China may not be representative of the world as a whole. It may be fairer to say that the South China story is only atypical in the sense that it is better documented than elsewhere. However, before exploring these new discoveries it is necessary to review the current state of the chronostratigraphic scheme for the P-Tr interval.

Stratigraphy

Permian chronostratigraphy has long been in a mess, and interregional correlation can barely be achieved even at stage level. The nature of the problem derives from the endemism of ammonoids at this time and the palaeogeography (Fig. 5.1). Marine Permian strata are encountered in three widely separated regions; mid- to high-latitude Boreal shelf sea sections, a major equatorial embayment in the United States and, by far the most important, the numerous Tethyan sections which stretch from equatorial to high southern palaeolatitudes (Fig. 5.1). Correlating between these three broad regions has proved extremely difficult, particularly for the Boreal sections where ammonoids can be extremely rare. This is exemplified by the difficulties of dating the Kapp Starostin Formation of Spitsbergen, the youngest Permian unit in the region where, due to the absence of conodonts and ammonoids, Nakamura *et al.* (1987) were forced to use

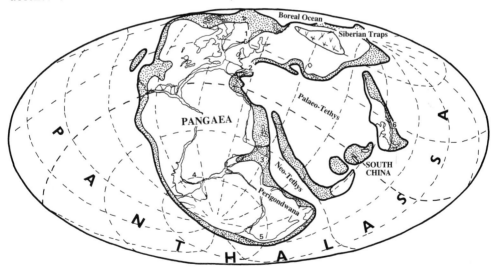

Fig 5.1 Global palaeogeography during during the Permo-Triassic boundary interval, showing the supercontinent of Pangaea stretching from one pole to another. The world's two oceans, Panthalassa and Tethys, were somewhat isolated from one another at this time due to the presence of several small continents at the eastern end of Tethys. Shallow-marine shelves (stippled) were extensive on the smaller continents but were restricted to a narrow fringe around Pangaea. The exceptions were three large embayments in the western USA, down the north-east coast of Greenland and between Africa and India where a seaway extended across Pakistan to Madagascar. Projection: Lambert equal area full globe, from Scotese *et al.* (1979).

brachiopod biostratigraphy. This led them to conclude an age ranging from the Ufimian to the Dzhulfian (cf. Fig. 5.2) but, as pointed out by Stemmerik (1988), the over-whelmingly ecological control on the distribution of brachiopod taxa makes even such a broad age assignment highly suspect. The carbon isotope stratigraphy of the Kapp Starostin Formation in fact suggests a much younger Dzhulfian to Dorashamian age (Malkowski *et al.* 1989; Gruszczynski *et al.* 1989). The Zechstein Basin of north-west Europe, a southerly extension of the Boreal Ocean, has proved equally hard to date and has been leaping up and down the Permian chronostratigraphic column throughout its history of research (for a review, see Kozur 1994a).

In the older literature the long-established Permian series of the Urals are frequently used as global standards (Fig. 5.2) even though in post-Artinskian times they are 'virtually useless in intercontinental correlation' (Jin *et al.* 1994a: p. 1); this is particularly so for the Tatarian Series which is entirely developed in terrestrial facies. Good marine sections are developed in the mid-Permian of the southern United States for which a regional stage nomenclature is also available (and the Guadalupian Stage is often applied elsewhere), but the later Permian is poorly developed and correlation with Tethyan regions is problematical. Younger Permian marine sections occur in the border region between Iran and Armenia where the Dzhulfian and Dorashamian Stages have their type localities. The question of the age of the uppermost Dorashamian is, however, unresolved (Erwin 1993: p. 77; Kozur 1994a). Undoubtedly the best and most continuous marine Permian sections occur in South China, and several stages have been devised for the region (Fig. 5.2). Of these, the latest Permian Changxingian Stage has become widely adopted, but there is little reason why the other post-Artinskian Chinese stages should not also become the global standards, particularly as the complete conodont zonation devised in the region at last offers the potential for global correlation (Ding 1992; Jin *et al.* 1994a).

Early Triassic stage nomenclature is thankfully less complicated than that of the Permian. This is primarily a reflection of the much more cosmopolitan distribution of the ammonoids at this time which allows for easier interregional correlations. A fourfold stage division – Griesbachian, Dienerian, Smithian, and Spathian – is the most widely utilised scheme (Fig. 5.2). The stages were named after type sections in a series of creeks in Axel Heiberg and Ellesmere islands in Arctic Canada which in their turn were named in commemoration of some of the distinguished ammonoid workers of the late nine-teenth and early twentieth centuries (Tozer 1965, 1967, 1994). Moves are afoot to reduce the Early Triassic to only two stages, the Induan and Olenekian, but Tozer's fourfold division still has utility.

All the Canadian stages can be correlated globally, with the unfortunate exception of the lower part of the Griesbachian Stage; that is, the crucial interval around the P-Tr boundary. The basal Griesbachian is defined by the appearance of *Otoceras concavum* which is first found about 15 m above a major regional unconformity in its type section. The age of this lowermost stratum is presumably Late Permian although until recently it had yielded no zonally significant fossils. The overlying zone is defined by the total range of *Ot. boreale* while the overlying *Ophiceras commune* Zone is an acme zone, the zonal species making its first rare appearance in the upper part of the *Ot. boreale* Zone. The *Oph. commune* Zone marks the first development of a truly cosmopolitan ammonoid fauna and the zonal species (or closely related species) can be found in virtually all the

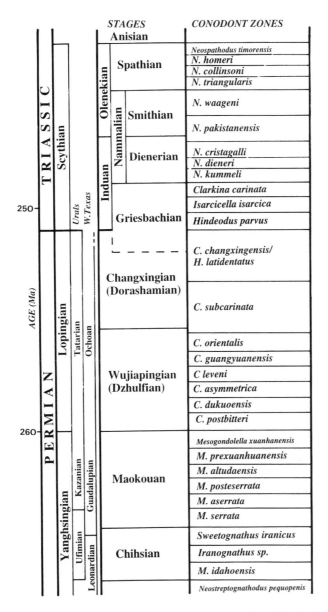

Fig 5.2 Series, stages and conodont zones for the mid-Permian to Middle Triassic interval. Absolute ages are based on Claoué-Long *et al.* (1991) age for the youngest Permian and Brack *et al.*'s (1995) 241 Ma age for the late Anisian.

regions of the world (the appearance of abundant *Ophiceras* also coincides with the first swarm abundances of the bivalve *Claraia*). The *Otoceras* faunas, in contrast, are only found in moderate to high latitudes while, with rare exceptions, they are frustratingly absent from the low-latitude Tethyan sections such as those of South China. The problem of defining a synchronous P–Tr boundary could be avoided by placing it at the base of the *Ophiceras* Zones, but this would only side-step the fundamental issue of the timing of the end-Permian mass extinction because, in most regions, the event pre-dates the *Ophiceras* Zone.

Conodonts have proved of great value in correlating Tethyan boundary sections, and their recent discovery from higher latitudes may soon resolve the correlation problems around the P-Tr boundary. In Tethyan sections *Hindeodus latidentatus* has been shown to evolve via a continuous morphocline into *H. parvus* (Wignall *et al.* 1996: Fig. 1); such lineages make ideal choices for the definition of zonal boundaries provided a consistent choice is made along the morphocline. The development of the first true *H. parvus* also coincides with the demise of *Clarkina changxingensis* (Wang 1995), or it might slightly pre-date this extinction as a slight overlap in ranges is seen in a few sections (Yin 1985a; Orchard 1994). The first *H. parvus* therefore marks a significant and easily recognized event in conodont history, and it is widely favoured as the biostratigraphic marker for the base of the Triassic (Yin *et al.* 1992, 1994; Kozur 1993; Paull and Paull 1994a). Sweet (1977) has previously regarded *H. parvus* as conspecific with *Isarcicella isarcica* of the subsequent zone, although he now recognizes it as a separate species of *Isarcica* (Sweet 1992). *H. parvus* appears consistently earlier than *I. isarcica* in all sections, and it is therefore reasonable to retain a distinct *H. parvus* Zone.

Conodonts from *Otoceras*-bearing sections are poorly known with the exception of the intensively studied Guryul Ravine of Kashmir, a high-latitude Perigondwanan section (Fig. 5.1, location 3). Recent re-examination of conodonts collected from this site in the 1970s by Heinz Kozur has revealed that the *H. latidentatus* to *H. parvus* transition occurs in the middle of the *Otoceras woodwardi* Zone at a level where *Ophiceras* also appears (Kozur 1994b). The *Ot. woodwardi* Zone is thought to equate with the *Ot. boreale* Zone of Boreal sections, although the presence of *Ophiceras* suggests that its upper boundary lies within the *Oph. commune* Zone. If the *H. parvus* Zone is chosen as the base of the Triassic then the Griesbachian Stage in its type area is substantially equivalent to the Changxingian Stage of Tethys. This is a conclusion long suspected by conodont workers (Kozur 1979; Sweet 1979) although it has been hotly contested by, for example, Tozer (1988).

The accurate correlation between Boreal and Tethyan areas is fundamentally important to the understanding of P-Tr events, particularly sea-level changes. The absence of the *Otoceras* zones in Tethyan sections has often been used as evidence of a major hiatus and regression at this level (see, for example, Nakazawa *et al.* 1980; Dagys and Dagys 1988; Baud *et al.* 1989) but never with supporting sedimentological evidence. The conodont stratigraphy indicates that, on the contrary, there are numerous, complete P-Tr boundary sections throughout Tethys (Wignall *et al.* 1996). In Boreal sections the unconformity beneath the Griesbachian was also held as evidence of a major P-Tr regression. In fact this break is basal Changxingian (or even older) and considerably pre-dates the P-Tr boundary.

What went extinct?

The sheer magnitude of the cataclysm that devastated life in the Late Permian was clearly demonstrated by Raup's (1979) figure quoted above. These approximations were derived from the inadequate data base available in the mid-1970s but the ensuing decades of research have done nothing to suggest that Raup's calculations were in any way exaggerated. This is exemplified by Labandeira and Sepkoski's (1993) examination of the insect fossil record. Insects have been extremely resilient to the numerous biotic crises of the Phanerozoic – therein lies the reason for their high diversity today. New groups have continued to appear throughout their history but extinctions rarely seem to happen above the family level. That is with the exception of the Late Permian, when seven orders disappeared, clearing the way for the radiation of the more familiar groups in the Triassic (for example, beetles began their explosive radiation in the Middle Triassic). Thus, even the insects were unable to weather the Late Permian holocaust – clear testimony to the severity of the extinction mechanism (of which more later). However, to understand the timing and potential selectivity of the extinction we must turn to the stratigraphically better-resolved marine record.

Reefs

Reefs are well known throughout the Permian; they include the large and famous Guadalupian reefs of Texas (Fagerstrom 1987). In recent years numerous latest Permian (Lopingian) reefs have also been discovered, and they have been revealed as a repository for a hitherto unappreciated diversity of marine invertebrates. Lopingian reefs are recorded from throughout Tethys where they show a tendency towards increasing size and diversity of their component taxa towards the end of the Changxingian (Flügel and Reinhardt 1989). The sphinctozoans, in particular, show a rapid diversity increase at this time (Rigby and Senowbari-Daryan 1995). The largest reef occurs in the *changxingensis* Zone at Laolongdong in south-east Sichuan, southern China; it is 140 m wide and 70 m high (Reinhardt 1988). There has been some debate as to whether the reefs are stratigraphic reefs, produced by differential compaction (Reinhardt 1988), or patch and barrier reefs (Fan *et al.* 1990) but all authors are agreed that they are built of abundant, baffling calcareous sponges (sphinctozoans and inozoans, with minor tabulozoans and the hydrozoan *Disjectopora*). Lamellar, blue-green algae and the alga(?) '*Tubiphytes*' helped bind the reefs together while crinoids, bryozoans (six families), non-fusuline foraminifers and brachiopods were important contributors to the overrall reef diversity (Fan *et al.* 1982, 1990). Reef termination occurred abruptly in the late Changxingian and in Sichuan the reefs are sharply overlain by deeper-water wackestones a few metres beneath the base of the *parvus* Zone (Wignall and Hallam 1996). Of the 30 genera of sphinctozoan encountered in the reefs, 21 failed to cross the P-Tr boundary. The survivors were long-ranging, conservative forms (Rigby and Senowbari-Daryan 1995).

Following their demise, reefs are absent from the fossil record for 7–8 m.y., one of the longest 'reef gaps' of the Phanerozoic (Fagerstrom 1987). Reefs reappeared in the mid-Anisian as modest-sized patch reefs in the mid- and deep-ramp carbonate facies of the Dolomites of northern Italy. Superficially at least, these oldest Triassic reefs resemble the Late Permian sponge–algal reefs, as the dominating reef-forming taxa are the calcisponge *Girtyocoelia* (known from the Permian), '*Tubiphytes*', *Archaeolithoporella* mats, and several genera of trepostome bryozoans (Flügel and Stanley 1984; Fois and Gaetani 1984). However, recent reappraisals of the Anisian reef faunas indicate that Elvis rather than Lazarus taxa may have been more important (Senowbari-Daryan *et al.* 1993; Flügel 1994). Thus all the sponge genera are new, '*Girtyocoelia*' is probably a simple homeomorph (Elvis impersonator) of a Permian genus, while the microstructure of Triassic '*Tubiphytes*' is different from the Permian examples, suggesting a distant relationship.

The most surprising members of the Anisian reefs are not the Palaeozoic home-omorphs but the highly diverse, rare scleractinian corals (Flügel and Stanley 1984; Fois and Gaetani 1984). A total of five families, seven genera and ten species are known and the typical Mesozoic forms of *Isastrea* and '*Thecosmilia*' are already present. Thus, even at this initial moment in scleractinian history they had rapidly attained a high diversity.

The reef story during the P-Tr crisis is therefore one of abrupt and total extinction in the latest Changxingian, a prolonged absence from the fossil record, followed by an equally abrupt radiation in the Anisian. Several lineages of calcisponge must have survived the crisis in the Scythian but the number of Lazarus taxa has been overstated in the past and this has tended to obscure a significant calcisponge radiation of Elvis taxa in the Anisian.

Corals

For the rugose and tabulate corals the Late Permian was the final chapter in their long history. Traditionally the demise of the Rugosa has been viewed as a progressive diversity decline, with the initial loss of the more complex, larger colonial corals such that 'only solitary, probably deep water corals [existed] at the end of the Permian' (Flügel and Stanley 1984: p. 178). This view is still substantially correct (Fedorowski 1989) but it has been considerably modified by the detailed work of Yoichi Ezaki on Lopingian Rugosa.

Most Permian tabulates did not survive the late Maokouan and only a few michelinids persisted to the end of the Changxingian (Fedorowski 1989). The Rugosa also under-went a diversity decline at the end of the Maokouan but remained diverse in the Wujiapingian and even underwent a modest radiation – ten new genera appeared at this time. Fedorwoski considered the subsequent history of the Rugosa to be one of gradual diversity decline and disappearance, first from higher-latitude regions of the world and then western Tethys until eventually they were only found in South China. The decline first eliminated massive colonial corals, followed by fasciculate colonies, then solitary corals with dissepiments, and finally simple, non-dissepimentate, solitary corals (Fedorowski 1989). This sequential loss is best seen in the sections of Transcaucasia and Iran where only the solitary *Pentaphyllum* lasts to the end of the Changxingian (Ezaki 1993a,b). The diversity decline is intimately associated with the development of deep-water facies (unfavourable for most Rugosa) throughout the region in the Changxingian. It is noteworthy that in South China, where shallow-water facies can be found right up to the end of the *changxingensis* Zone, the disappearance is considerably more abrupt than elsewhere. Changxingian Rugosa of this region consist of both solitary plerophyllids and lophophyllids and colonial waagenophyllids (Ezaki 1994). In many sections Rugosa are common to within a metre or so of the P-Tr boundary, and in several sections it is the colonial taxa which are the last to disappear (Ezaki 1994: p. 173). Clearly, given the right facies, corals were able to thrive until almost the end of the Permian.

The supposed latitudinal contraction of the Rugosa also does not hold up to closer scrutiny because it ignores the occurrence of corals in the topmost Permian of Spitsbergen, tentatively ascribed a Tatarian age (Ezaki *et al.* 1994). Rugose corals are also encountered in the basal Wordie Creek Formation of north-east Greenland. If they are not reworked (see below), they are probably of late Changxingian age.

Bryozoa

For that other major group of colonial organisms, the Bryozoa, the Late Permian was also a time of crisis only slightly less severe than for the corals. The four major stenolaemate orders of the Palaeozoic – the Cystoporata, Trepostomata, Fenestrata, and Cryptostomata – all suffered major generic-level extinctions although only the fenestrates became extinct (Taylor and Larwood 1988). The timing of the extinction is not well known. In Taylor and Larwood's (1988) view, most genera disappeared in the Maokouan and Wujiapingian, leaving depauperate Changxingian faunas, but, as we have already seen, latest Changxingian reefs contain a diverse and abundant bryozoan fauna. In a more recent analysis, Ross (1995) concluded that a major extinction event occurred in the late Maokouan followed by 'a burst of diversity in the Tethyan faunal realm' (Ross 1995: p. 208). However, this radiation was confined to the Wujiapingian, and Ross, like Taylor and Larwood, considers the Changxingian to be an interval of decline. None the less, several bryozoan genera are also known from the early Griesbachian (the age-equivalent of the uppermost Changxingian) of Greenland – although Teichert and Kummel (1976) considered that they were reworked – while Nakrem (1994) recorded 30 genera of trepostomes, fenestrates, and cryptostomes from the poorly dated highest Permian strata of Spitsbergen.

The Triassic radiation of bryozoans was slow indeed; the Scythian marks a prolonged 'outage' for most groups (only two trepostome genera are known from the Olenekian), 4 genera appeared in the Anisian, 7 in the Ladinian, and 34 in the Carnian (Sakagami 1985). The dominant Palaeozoic orders never really recovered from their late Palaeozoic crisis and the final few cryptostomes and trepostomes disappeared at the end of the Triassic, while the cystoporates trickled on into the Early Jurassic before fading away. As with many groups that suffered severely during mass extinction events, the subsequent radiation was by previously insignificant components of the pre-extinction fauna, in this case the Cyclostomata and Ctenostomata orders (Taylor and Larwood 1988).

Echinoderms

For echinoderms the Late Permian was a time of almost but not quite complete annihilation, although the timing of the mass extinctions is poorly known. The crinoids lost two subclasses, and probably only one genus of cladids survived to found the post-Palaeozoic Articulata subclass (Paul 1988; Simms and Sevastopulo 1993). Similarly, only one echinoid genus (*Miocidaris*) is known to have survived the extinction to ensure that post-Palaeozoic echinoids were quite distinct from their Palaeozoic forebears (Kier 1984; Erwin 1994). Blastoids were not so lucky, and their last representatives are known from the Wujiapingian (Simms *et al.* 1993).

Echinoids are common in the late Changxingian of the Dolomites (Italy) but only *Miocidaris* has been recorded (Broglio-Loriga *et al.* 1988). If this monogeneric assemblage is a true reflection of the diversity, then it suggests that the main echinoid crisis was a pre-Changxingian event (Fig. 5.3) – a conclusion supported by the compilation of Jin *et al.* (1994c) – but the details of the timing is seriously hampered by the poor dating of many Permian echinoderm assemblages. Jin *et al.* (1994c) suggested that the late Maokouan crisis was essentially a low-latitude affair, a conclusion that appears supported by the echinoderm data. Thus the last blastoids and camerate crinoids disappeared from equatorial Tethys at this time, but they are known from younger Permian strata in the high southern palaeolatitudes of Australia (Simms *et al.* 1993).

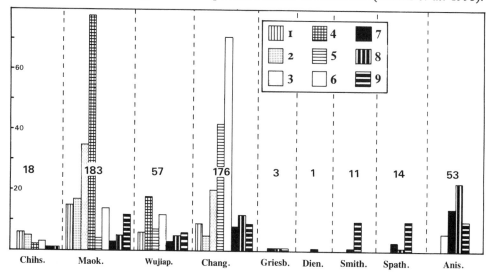

Fig 5.3 Bar charts showing the numbers of families going extinct in each stage between the Maokouan and Anisian: 1, sponges; 2, corals; 3, bryozoans; 4, echinoderms; 5, foraminifers; 6, brachiopods; 7, gastropods, 8, bivalves; 9, ammonoids. After Jin *et al.* (1994b,c).

Echinoderms are common in the Early Triassic, and we have found abundant crinoid and ophiuroid remains throughout the Scythian of North America, the Dolomites and South China. Nothing is known of the phylogenetic significance of this fauna, with the sole exception of Schubert *et al.*'s (1992) study of Spathian crinoids from Nevada. This showed them to consist of a single Palaeozoic holdover taxon, probably the stem group for all later crinoids. The implication here is that Scythian echinoderms, while being common, consisted of low-diversity, pre-radiation Palaeozoic survivors.

Foraminifera

Foraminifera provide one of the best and certainly the most informative fossil records during the P-Tr crisis. This is because they are common throughout this interval and, unlike the echinoderms, they have received adequate attention; but, most importantly, the extinction was highly selective, allowing us to study which ecological–environmental attributes were best suited to survival. The Late Permian was none the less the worst crisis in foraminifer history – numerous families and an entire suborder, the Fusulinina, disappeared at this time (Tappan and Loeblich 1988).

Foraminifera appear to have suffered two mass extinctions in the Late Permian, one in the late Maokouan, which primarily affected fusulinids, and a more severe one in the latest Changxingian, which affected all groups (Jin *et al.* 1994c; Stanley and Yang 1994). The late Maokouan event was marked by the gradual loss of around 45 genera of the larger, more complex fusulinids (Ross and Ross 1995a). In particular, all the fusulinid taxa with a distinctive honeycomb wall structure became extinct (Stanley and Yang 1994). The radiation of numerous foraminifera lineages to produce larger, more complex descendants in the succeeding Lopingian supports the existence of a late Maokouan crisis followed by a recovery and radiation to fill the niches left by the departed fusulinids. Thus the nodosarioids and miliolids produced complex, partitioned genera such as *Pachyphloia* and *Shanita*, respectively (Brasier 1988). The fusulinids failed to fully recover from their late Maokouan crisis and, although the 14 surviving genera were joined by 5 new ones in the Lopingian, they never again produced complex, large forms (Stanley and Yang 1994).

Wujiapingian foraminifera assemblages are diverse, and several distinct provinces can be identified within Tethys (Okimura *et al.* 1985). The data from the Changxingian are less good, but it appears that there was a decline in endemicity as equatorial assemblages from eastern and western Tethys share many genera and even species in common (Noé 1988). Many foraminiferal lineages continued to diversify in the Changxingian and the nodosarioid *Colaniella*, the biseriamminid *Paradagmarita* and the hemigordiopsid miliolids all become common at this time, indeed the last group were producing new species in the *latidentatus* Zone (Pasini 1985; Noé 1988; Tong 1993; Fig. 5.4). The foraminifers were clearly unaware of the impending crisis and in many sections they maintained their diversity to within a metre of the P–Tr boundary.

The mass extinction, when it struck in the latest Changxingian, primarily affected the tropical, architecturally complex forms (the fusulinaceans, partitioned nodosarioids, miliolids, and uncoiled palaeotextulariids) and included most of the groups which had

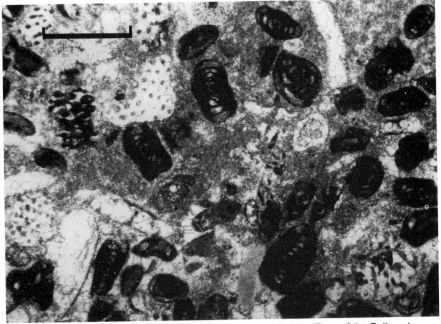

Fig 5.4 Photomicrograph of *Hemigordius* species from the *latidentatus* Zone of the Bellerophon Formation. Tesero section, southern Dolomites. This foraminifera lineage was diversifying until almost the end of the Permian. Scale bar is 0.5 mm wide.

been radiating during the Lopingian (Brasier 1988). The extinction was highly selective with regard to shell structure as calcareous, microgranular tests almost entirely disappear from the fossil record at this time. In South China, of the 66 species with this shell structure in the late Changxingian all but one disappeared (Tong 1993). Curiously, several foraminifera families with microgranular tests survived the extinction by converting to hyaline tests. As shell structure is used in the higher-level taxonomy of foraminifers, this change also produces a change of name. Thus, the Palaeozoic nodosarioids became the Mesozoic Nodosariina, while the archaediscids appear to have produced the Spirillina and Involutinina (Brasier 1988). The Fusulinina found no escape route and the last of these mostly large, tropical, and specialised foraminifera are found in the topmost decimetres of many Tethyan sections. The sole survivor of the suborder, *Earlandia*, was a simple, long-ranging eurytopic form which is locally very abundant in many Induan sections. It gave rise to a modest species-level radiation in the Middle Triassic before finally going extinct in the Late Triassic.

By the standards of the Fusulinina, the Miliolina suborder survived the P-Tr crisis relatively well: 'only' 50% of genera became extinct, mostly the larger, more complex forms. The agglutinating Textulariina survived better still, losing only 33% of its genera in the Late Permian (Tappan and Loeblich 1988). For example, in south China, 34 genera of textulariids crossed the boundary and only 6 large, complex forms disappeared (Tong 1993). Even so, this modest extinction of agglutinating forms represents the only crisis in their long history (Tappan and Loeblich 1988). The record for foraminifera 'survivability' goes to the Lagenina, which suffered no major extinction in their entire history. They seemed unaware of the P-Tr crisis and, during this interval, the nodosarioids (the most common group of lageninids) underwent a change in name (see above) and a modest radiation (Pande and Kalia 1994).

So what can the relative survival of the textulariids and the absolute success of the nodosariids during the P-Tr crisis tell us of the extinction mechanism? Tappan and Loeblich (1988) have noted that the flattened, lenticular, and low trochoid tests of the lageninids are characteristic of infaunal detrital feeders. They therefore speculated that these would not be susceptible to crashes of primary productivity in the same way as other groups more directly dependent on the food chain. We would also note that foraminifera assemblages dominated by nodosariids and textulariids are typical of many Mesozoic dysaerobic biofacies (see, for example, Wignall 1990), with the implication that dysoxia may have been a factor in the extinction (see below).

The Scythian radiation of foraminifera was modest indeed, and was essentially restricted to the dysaerobic groups noted above (Tosk and Anderson 1988). It was not until the Spathian and Anisian that a major radiation began with the appearance of the Robertinina suborder (Tappan and Loeblich 1988).

Ostracods

Ostracods are common in many Late Permian and Scythian sections, and it is therefore unfortunate that they have received little study. Shallow-water taxa appear to have undergone a substantial turnover at this time (Kozur 1985; Teichert 1990), while the deep-water, long-ranging, cosmopolitan ostracods of the palaeopsychrospheric fauna were unaffected (Kozur 1991). This distinctive fauna consisted of very thin-shelled, spinose ostracods belonging to several enigmatic families that may have been adapted to

life within the oxygen-minimum zone (Lethiers and Whatley 1994). Thus, like foraminifers, ability to withstand dysoxia may have conferred an ability to survive the
P-Tr crisis.

Brachiopods

Brachiopods were abundant and diverse in nearly all late Palaeozoic marine environments, but the collector of Mesozoic fossils will generally find relatively few brachiopods
(but plenty of bivalves). This wholesale changeover in the dominant benthic taxa is the
fundamental legacy of the Late Permian mass extinction.

The Late Permian history of brachiopods begins with a significant and protracted
extinction of tropical taxa in the late Maokouan (Jin *et al.* 1994c; Fig. 5.3), followed by a
Wujiapingian radiation of the specialised Oldhamminidae, Strophalosiacea, and
Richthofenacea superfamilies (Waterhouse and Bonham-Carter 1976). The significance
of the Maokouan event is well seen in South China where a diverse, indigenous fauna of
82 genera and 169 species was reduced to 29 genera and 57 species in the Wujiapingian
(Shen and Shi 1996). Many of the Wujapingian brachiopods belong to new taxa,
indicating that the extinction was even more severe than these figures suggest. The
event appears abrupt in Chinese sections because late Wujiapingian strata are every-
where unconformably overlain by Wujiapingian strata, although of course herein may
lie a reason (regression) for the cause of the crisis (Shen and Shi 1996).

At one time the brachiopods of the succeeding Changxingian were thought to show a
significant diversity decline, but new Chinese data have done much to enhance recorded
diversity values at this time (Xu and Grant 1992, 1994; Shen and Shi 1996). Late

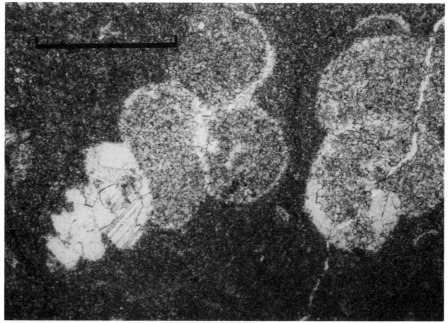

Fig 5.5 Microgastropod seen in thin section from a limestone at the top of the Mazzin Member (mid-
Griesbachian) from l'Uomo, near Passo San Pellegrino, Dolomites. Such dwarfed gastropods are
abundant in low palaeolatitude. Early Triassic sections. Scale bar is 1.5 mm wide.

Changxingian brachiopod diversity could be further increased by including the genera recorded from the basal Wordie Creek Formation of north-east Greenland. Widely regarded as reworked Middle Permian forms (see, for example, Waterhouse 1972; Teichert and Kummel 1976), new conodont data suggest that they are of latest Permian age (p. 117).

The mass extinction struck brachiopods late in the Changxingian and was the worst in their history – 90% of families and 95% of genera became extinct (Carlson 1991; Erwin 1994). It caused the highly provincial late Changxingian faunas to be replaced by the widespread *Lingula–Crurithyris* assemblage in the basal Triassic (Xu and Grant 1994). The extinction was also severe at the superfamily level, but, as Carlson (1991) has shown, many of these are probably artificial extinctions of paraphyletic taxa; of Carlson's 26 major clades only 10 disappear in the Late Permian, a considerable lessening of the impact of the extinction. In Italy, Pakistan, and Greenland the last brachiopods are found high in the Changxingian but not right up to the P-Tr boundary. In South China 81% of brachiopod genera (51 of 63) were lost in the late Changxingian while eight genera lingered on into the basal Triassic *parvus* and *isarcica* Zones, notably *Crurithyris* spp., where they were joined by abundant *Lingula* (Xu and Grant 1992, 1994; Dagys 1993). It is noteworthy that both *Crurithyris* and *Lingula* are long-ranging, dysaerobic genera found in many Palaeozoic black shales (Kammer *et al.* 1986). Ability to survive low oxygen levels may have conferred an advantage during the P-Tr crisis although, in the case of *Crurithyris*, this was not for long as it is unknown after the *isarcica* Zone. *Lingula* is a classic disaster taxon able to survive in a wide range of conditions, including brackish waters (Pasini 1985). Its abundance in Early Triassic sections throughout the world points to the widespread and prolonged occurrence of stressful conditions.

The radiation of Triassic brachiopods shows the familiar pattern of no recovery in the Scythian (only 14 genera known) followed by the abrupt appearance of 50 genera in the Anisian (Dagys 1993). Terebratulids and rhynchonellids, two minor groups in the Palaeozoic but the main post-Palaezoic success story of the brachiopods, did not become common until the Spathian (Perry and Chatterton 1979). However, their radiation was really a Jurassic affair (Dagys 1993) following the demise of a further 11 of Carlson's clades at the end of the Triassic.

Gastropods

As we move our coverage to the first of the major mollusc classes, we at last begin to deal with groups which managed to weather the Late Permian storm with a modicum of success (foraminifers excepted). In an influential paper, Batten (1973: p. 603) considered the gastropod fossil record from the Middle Permian to the Late Triassic, where he found that 'Dzhulfian and Scythian faunas lack genera which occur in the Guadalupian, *and* reappear in the Ladinian!' The failure to mention the Dorashamian/Changxingian stage is a reflection of the confusion prevailing in Permian stratigraphy at the time (it is hardly better today) but the paper was most significant as the first explicit mention of the importance of Lazarus taxa (although this term came later). More recent collecting, especially in south China, has shown that many endemic gastropods became extinct sometime in the Yanghsingian (Erwin and Pan 1996) but that the rather more cosmopolitan survivors radiated until almost the end of the Permian (Pan and Erwin 1994).

The final end-Changxingian mass extinction was severe at the generic level (90% extinction) but only 3 of 16 families disappeared. Previous clade history had no influence on survivorship and relatively young clades such as the pleurotomariids were as badly affected as the older clades such as the bellerophontids (Erwin 1990). The only patterns to emerge from the extinction are the unsurprising ones that broad geographic range and environmental distribution conferred some degree of immunity (Erwin 1990: p. 187).

The P-Tr boundary marks a curious, temporary change in the nature of the gastropod fossil record. They were a diverse but generally rather rare component of Late Permian assemblages, while in the Early Triassic they became less diverse, but they were prolific and tiny. Microgastropod grainstones composed of millimetre-sized species are found throughout all equatorial Tethyan sections (see, for example, Fig. 5.5) and extend into the Perigondwanan sections of Pakistan. *Coelostyolina* is commonly recorded, but this is rather a dustbin taxon and the systematics of these curious assemblages is virtually unknown. Only Batten and Stokes (1987) have undertaken a taxonomic study and that has shown that the Scythian gastropods are tiny, adult species rather than juveniles. Gastropods returned to 'normal' in the Anisian as they once again became rather rare and attained sizes more typical for the group. The Anisian is the true start of post-Palaeozoic radiation as nearly 40 new genera appeared at this time, although the majority belong to Palaeozoic Lazarus families (Pan and Erwin 1994).

Bivalves

The P-Tr crisis exacted its modest toll of bivalves but on the whole it marked a favourable turning point in their fortunes as they subsequently became the dominant shelly benthos of the Mesozoic. Like gastropods, Late Permian bivalves are a generally rare but reasonably diverse group. They have received most attention in South China (see, for example, Yin 1982) where they are mostly found in cherty shale facies alongside a low-diversity dysaerobic brachiopod fauna dominated by chonetids and *Crurithyris*. This distribution is typical of late Palaeozoic bivalves as a whole: they are only common in offshore, deep-water and/or low-oxygen settings (Nakazawa and Runnegar 1973; Jablonski *et al.* 1983).

Pectinaceans are the most diverse Late Permian bivalves, and their diversity changes are well constrained. The late Maokouan was a significant crisis, with 13 of 24 genera disappearing – an extinction that, in terms of absolute numbers, was greater than the late Changxingian event when 8 of 12 pectinacean genera went extinct (Yin 1982, 1985b; Newell and Boyd 1995). Many other bivalve genera disappeared in the latest Permian, although only three families – the Chaenicardiinae, Euchondriinae, and Deltopectinidae – went truly extinct at this time (Yin 1985b). For many genera and families the event is a pseudo-extinction as there is a tendency for many bivalves simply to change their name at the Palaeozoic–Mesozoic divide (Nakazawa and Runnegar 1973). For many other groups the base Triassic marks the start of a prolonged absence from the fossil record. This includes many familiar Mesozoic genera (such as *Thracia*, *Pinna*, and *Chlamys*) which first make their rare appearance in the Late Permian but are not seen again until the Middle Triassic.

Despite these Lazarus genera, bivalves are abundant Scythian fossils, and they dominate nearly all assemblages of this age (see, for example, Kummel 1957; Schubert and Bottjer 1995). Their composition is uniquely Scythian, consisting of neither

Palaeozoic survivors nor groups that subsequently radiated in the Mesozoic. Four genera are ubiquitous – *Claraia, Eumorphotis, Unionites,* and *Promyalina* – while *Leptochondria* is almost as common (Fig. 5.6). Nearly a hundred species are known, although the majority belong to *Claraia* (a broadly defined genus that probably represents several genera as it includes a broad range of shell ornament types) and *Eumorphotis* (Yin 1985b). The oldest *Claraia* dates from the Changxingian (Nakazawa *et al.* 1980; Yin 1985b) but the lineage began an explosive radiation in the later Griesbachian (*isarcica–carinata* Zones) and continued to produce new species through the Dienerian and Smithian. No convincing Permian *Eumorphotis* records are known, and the genus appeared and began a rapid radiation at the same time as *Claraia*. *Promyalina* also appeared at the same time – although Nakazawa and Runnegar (1973) considered it congeneric with Palaeozoic *Myalina* – but did not produce more than two or three species. The fourth member of the unique Scythian bivalve assemblage, *Unionites,* again appeared late in the *latidentatus* Zone and began to proliferate, although many species may be variants of a single, morphologically flexible species. The ecological and phylogenetic significance of *Unionites* is unclear. It is considered to be the first of the Pachycardiidae, but this makes it the only Scythian representative of a brackish-water family that is otherwise not known before the Ladinian.

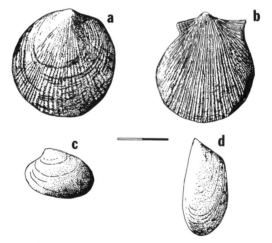

Fig 5.6 Conquerors of the Early Triassic: four species that are prolific in the Induan:(a) *Claraia stachei*, (b) *Eumorphotis multiformis*, (c) *Unionites fassaensis*, (d) *Promyalina* sp. This assemblage is found at all palaeolatitudes and could claim to be the most cosmopolitan assemblage of all time. Scale bar is 2 cm long.

These four Scythian bivalves are encountered in marine sections from throughout the world and constitute one of the most cosmopolitan faunas of all time (Fig. 5.6). The only evidence for provinciality is provided by the slightly higher species diversity of tropical assemblages; concentrically ornamented *Claraia* is restricted to Tethys, for example (Yin 1985b). This cosmopolitanism is all the more remarkable when compared to Late Permian bivalve assemblages which are readily divided into tropical, temperate, and high-latitude realms.

The hegemony of the Scythian bivalves appears to have gradually come to an end in the late Smithian or early Spathian and, by the Anisian *Claraia, Eumorphotis,* and *Promylina* have gone extinct; only *Unionites* has a few questionable post-Scythian records.

Ammonoids and nautiloids

Ammonoids are the perpetual weather-vanes of the fossil record, and it is therefore unsurprising to find that they suffered grievously in the Late Permian. However, most recent studies have tended to downplay the significance of the event and, as Wiedmann (1973) showed, there are many pseudo-extinctions of paraphyletic taxa at this time. The end-Permian ammonoid extinction was certainly not as severe as their end-Triassic crisis nor their end-Cretaceous finale. None the less, Wiedmann's (1973) conclusion that as few as five genera became extinct at the end of Permian is probably a serious under-estimate in the light of recent Chinese discoveries. Yang (1993) recorded the loss of 20 out of 21 genera and 102 out of 103 species in the latest Changxingian of South China. Even allowing for paraphyly and over-splitting of species, this is a considerable extinction event. The ammonoids radiated rapidly in the Scythian only to decimated by a Spathian extinction of comparable magnitude to the end-Permian event (Benton 1986; Yang 1993).

This characteristic boom-and-bust history contrasts markedly with the more serene progress of the nautiloids at this time: Teichert (1990) could detect no nautiloid extinction event during the P-Tr interval, although they become very rare fossils in Scythian strata. This remarkable response to the extinction crisis may be related to the nautiloid's position within the food chain. If the carnivory of modern *Nautilus* is typical of the class as a whole then the ancient representatives presumably occupied a fairly elevated position within the trophic hierarchy and, as we shall see in the following section, the marine predators of the Late Permian ecosystems were curiously immune to this extinction crisis.

Marine chordates

Conodonts are common but not particularly diverse in the Changxingian: Clark (1987) estimated that probably no more than eight species were extant. Several species (but no genera) disappeared around the P-Tr boundary and the event hardly merits being called an 'extinction crisis'. Of more significance for conodont evolution is their subsequent radiation in the Scythian, the last in their history, when 35 species appeared (Clark 1987).

That the extinction did not affect marine predators is shown more unequivocally by the diverse fish groups (for example, sharks, coelacanths, lungfishes, and palaeonisci-formes) which suffered little or no extinction (Schaeffer 1973; Patterson and Smith 1987). Indeed, in north-east Greenland, which is the only area to yield a well-preserved and common fish fauna from both latest Permian and earliest Triassic strata, fish diversity increased across the boundary, with the elasmobranchs in particular radiating considerably (Bendix-Almgreen 1976). The most 'severe' extinctions were suffered by the basal actinopterygians, the most diverse of the Late Permian fish groups, which lost two of eight families in the Late Permian, but nine new families appeared in the Scythian (Gardiner 1993). Schaeffer (1973) suggested that the apparent Scythian radiation may be a product of the excellent fossil record during this interval relative to the poorer Late

Permian record. However, the diversity changes are essentially identical to those seen in the conodont record, suggesting that the apparent changes of fish diversity are an accurate reflection of the true state of affairs.

The Scythian fish radiation is also paralleled by a radiation of the first of the terrestrial vertebrate lineages to return to the sea. Trematosaurid amphibians, a fully aquatic, marine group, are rare Griesbachian predators (Milner 1990), while the more familiar ichthyosaurs appeared in the late Smithian and had produced eight genera by the Spathian. Several further marine reptiles (such as nothosaurs) appeared in the Anisian (Lucas 1994), and by this stage of the Triassic the uppermost elements of the food web were rapidly attaining their typical, complex Mesozoic appearance, with reptiles joining sharks as the top predators.

Plankton

If we turn to examine the base of the marine food web, the perspective on the P-Tr crisis becomes wholly different. Interpretation of the carbon isotope record suggests a rapid decline of primary productivity in the late *changxingensis* Zone (p. 128), but radiolarians provide our only direct measure of the health of the late Palaeozoic plankton populations. Radiolarian cherts are common in deep marine sections of South China, Japan, and western Canada, but they disappear abruptly at the end of the Permian and only reappear in the Middle Triassic after a 'chert gap' of 7–8 m.y. This disappearance is associated with the almost total extinction of all genera and species of radiolarians (Isozaki 1994). It represents the only significant crisis in a long history that stretches back to the base of the Cambrian (Casey *et al.* 1983).

Palynological samples from earliest Triassic marine sections frequently contain unprecedented numbers of acritarchs (see, for example, Balme 1970) and also commonly tintinnids (Eshet 1992). Generally, acritarchs are encountered in shallow-marine, eutrophic conditions typically associated with poorly oxygenated waters, and their presence at this time points to the prolongation of stressed conditions after the main extinction event.

Terrestrial tetrapods

As we move our coverage on to the land we encounter a story that has changed dramatically in recent years. Much of the literature of the past few decades has tended to downplay the existence of an end-Permian tetrapod extinction. Initial appraisals, derived from data in Romer (1966), recognised only a minor faunal crisis (Colbert 1973) or no crisis at all (Pitrat 1973). The unimportance of the event continues to be reiterated (see, for example, Maxwell 1992) despite growing evidence to the contrary. The recognition of a cosmopolitan, earliest Triassic fauna overwhelmingly dominated by *Lystrosaurus* was an early indication that something had happened to the terrestrial tetrapods in the preceding latest Permian (Anderson and Cruickshank 1978). Olson (1982) returned to Romer's data set and this time found what Colbert and Pitrat had failed to find, a major Late Permian diversity decline. Benton's (1988, 1989) examination of more up-to-date, cladistically validated data confirmed the presence of a major end-Permian mass extinction involving the loss of 21 terrestrial tetrapod families or 63% of the total, a figure that is actually greater than the percentage loss of marine families (Benton 1995).

The earlier failures to detect the huge end-Permian mass extinction are a salutary warning of over-reliance on the literature rather than going into the field and collecting the data first hand (cf. Lucas 1994). This is well demonstrated by Colbert's (1973: p. 481) assertion that 'there is little sedimentary evidence to distinguish between the top of the Permian *Daptocephalus* Zone and the base of the Triassic *Lystrosaurus* Zone' in the Karoo Basin of South Africa. This interval in fact marks a major, long-term change of sedimentary environment according to the detailed fieldwork of Smith (1995).

As currently understood, the last few million years of the Permian witnessed a prolonged crisis in terrestrial tetrapod diversity (King 1991) terminated by a rapid extinction event (Olson 1989) which wiped out a broad spectrum of life-styles. Thus small omnivores (such as the Millerettidae), all large herbivores (including the dominant pareiasaurs), gliding reptiles (the Weigeltisauridae), and six out of nine amphibian families all became extinct (Benton 1988, 1989). The only ecological grouping to survive the extinction with moderate success were the small, carnivorous Therocephalia – five genera cross the P-Tr boundary in the Karoo (Smith 1995). Tetrapods remain common fossils in the succeeding Scythian but, as noted, over 90% of the assemblages consist of the single genus, *Lystrosaurus*, a medium-sized dicynodont.

The remainder of the fauna consists mostly of semi-aquatic stereospondyl amphibians, which underwent 'a vast adaptive radiation of families' (Milner 1990: p. 338). Numerous parallels can be drawn between the *Lystrosaurus* and *Claraia*-dominated assemblages. Both are extraordinarily cosmopolitan and both are post-extinction, disaster-taxa assemblages that were evolutionary dead ends. Thus, although the dicynodonts subsequently radiated, this was from Lazarus lineages and not directly from *Lystrosaurus* (Smith 1995). In contrast to the Scythian-long duration of the *Claraia* assemblage, the reign of the *Lystrosaurus* assemblage spans only the early Induan and, by the later Induan the therapsids (particularly the cynodonts, dicynodonts, and therocephalians) were rapidly diversifying and 'reinventing' many of the lost bodyforms of the Late Permian.

A further extinction crisis struck the tetrapods at the end of the Scythian and primarily removed amphibian families, although seven reptile families also disappeared around this time (Benton 1988, 1989). The event may relate to a loss of the estuarine and freshwater habitats preferred by the amphibians, but the necessary facies analysis and detailed timing of the extinctions remains to be done.

Terrestrial flora

Our understanding of the floral changes during the P-Tr crisis has undergone a similar volte-face to the tetrapod story as new investigations have overthrown existing dogma. The Permian to Triassic interval was considered to show the diachronous replacement of palaeophytic with mesophytic floras (Knoll 1984) and an 'increasing unity of floras in the Lower Triassic' (Meyen 1973: p. 622). The apparently prolonged nature of this changeover, spread over 30 m.y., precluded any comparison with the more rapid changes of the animal kingdom (Traverse 1990). However, the palynostratigraphic record has long hinted at more dramatic changes as diverse Late Permian spore and pollen assemblages are rapidly replaced in the Early Triassic by assemblages dominated by spores of lycopods (Balme 1970; Balme and Helby 1973; Visscher and Brugman 1988; Eshet 1992). The failure to recognise a corresponding mass extinction of plant megafossils across the same interval derived from poor dating of terrestrial sections and the use of the arbitrarily defined Palaeophytic and Mesophytic terms. Thus, the Angaran floral province of Siberia witnessed the replacement of a high-diversity *Cordaites* flora by a low-diversity, fern-dominated Korvunchana flora at the P-Tr boundary (Dobruskina 1987). The ferns were previously insignificant components of the Palaeophytic flora, but this appellation masks a major generic-level extinction event.

A combination of detailed sampling and better radiometric dating of P-Tr boundary sections in the Sydney Basin, Australia, has allowed Retallack (1995a) to demonstrate a major and sudden floral mass extinction at the boundary in high southerly palaeolatitudes. The peat-forming glossopterids, a long-established Gondwanan flora, were suddenly replaced by a low-diversity conifer–lycopod assemblage. Elsewhere in the world, peat-forming trees, such as *Cordaites* in Angara and *Gigantopteris* in North China, were also lost (Dobruskina 1987), producing a 'coal gap' in the geological record that lasted until well into the Middle Triassic (Veevers *et al.* 1994). This is despite the presence of suitable conditions (high water table/humid conditions) in the Scythian of both Australia and Antarctica (Retallack *et al.* 1996).

The floral mass extinction selectively removed the dominant large plants of the Late Permian and left small, weedy survivors. One such weed, the quillwort *Isoetes*, a small pioneering plant of oligotrophic lakes and ponds, formed the stem group for a moderately successful Triassic radiation (Retallack 1995b; Fig. 5.7). The seed ferns, typified by *Dicroidium*, similarly proliferated. *Isoetes* and *Dicroidium* can clearly be added to the lists of the *Claraia* and *Lystrosaurus* assemblages as Scythian disaster taxa but the most striking comparison with the animal fossil record is the similarly cosmopolitan nature of Scythian floral assemblages (Meyen 1973; Schopf 1973; Dobruskina 1987). 'The Lower Triassic differs from the Palaeozoic as well as from the Mesozoic by the vast expansion of lycopods' (Dobruskina 1987: p. 75), particularly *Pleuromeia* which was encountered in virtually all coastal habitats (see, for example, Wang 1996). Inland, in drier conditions, the conifer *Voltzia* and *Dicroidium* are similarly encountered at all palaeolatitudes (Dobruskina 1987; Veevers *et al.* 1994; Retallack 1995a), a pandemic distribution suggestive of an extraordinarily low climatic gradient (Ziegler *et al.* 1994).

The final, fascinating component of the floral story has been derived from palynological studies of P-Tr boundary sections. Fungal spores are generally very rare components of palynological samples but, in the Changxingian, they become increasingly common and reach unprecedented abundances towards the top of the stage. In the best-dated sections (for example, northern Italy and South China) the fungal spike occurs late in the *changxingensis* Zone (Visscher and Brugman 1988; Ouyang and Utting 1990; Eshet *et al.* 1996), although in other sections (notably those without conodont control) the spike is commonly and mistakenly assigned a base Triassic age (Balme 1979; Eshet 1992). The fungi are extraordinarily widespread and occur in all sections 'irrespective of depositional environment (marine, lacustrine, fluviatile), floral provinciality, and climatic zonation' (Visscher *et al.* 1996: p. 2155).

The significance of a world covered in fungi has yet to be addressed, although presumably a lot of rotting vegetation was available in the late Changxingian. Insects are the main destroyers of plant matter at present and were presumably equally important in the past. Could the end-Permian insect extinction, noted at the start of this section, have

TRIASSIC ADAPTIVE RADIATION OF QUILLWORTS

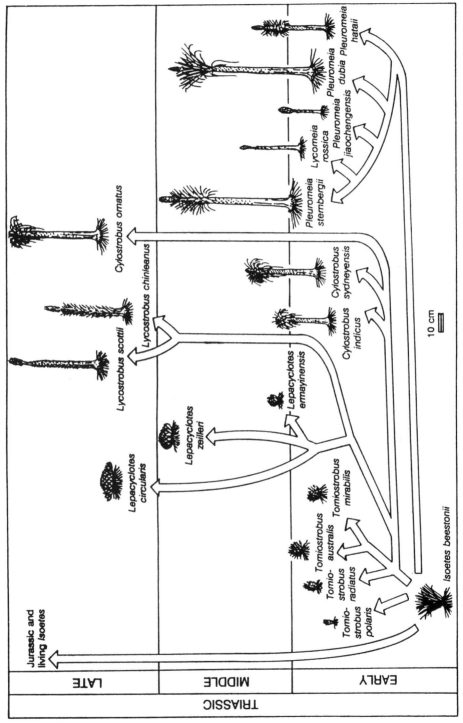

Fig 5.7 *Isoetes* radiation in the immediate aftermath of the end-Permian mass extinction (after Retallack 1995b).

allowed fungi to exploit the increased availability of dead vegetation? In a thought experiment on what would happen if present-day insects were to undergo a mass extinction, the eminent sociobiologist Ed Wilson (1992: p. 125) noted:

The land surface would literally rot. As dead vegetation piled up and dried out, closing the channels of the nutrient cycles, other complex forms of vegetation would die off . . . The free-living fungi, after enjoying a population explosion of stupendous proportions, would decline precipitously, and most species [of plants and animals] would perish.

These chilling predictions are remarkably similar to those we see in the latest Permian; perhaps this near-future *Götterdämmerung* scenario happened just over 250 m.y. ago.

Summary

This review of the fauna and flora of the P-Tr interval is by no means exhaustive, but sufficient is known of the crisis for the most salient issues to be identified (with the inevitable proviso that more work needs to be done). Five distinct stages can be recognised.

Late Maokouan crisis

It is clear that the end-Permian crisis was in fact two distinct events separated by a period of recovery in the Lopingian. The lesser and, until a few years ago, little-known earlier crisis is vaguely dated as late Maokouan for many groups (for example, corals, bryozoans, honeycombe fusulinids, brachiopods, pectinaceans, and ammonoids). For the echinoderms the crisis was probably prolonged into the Wujiapingian. The extinction is clear at the family and generic level, but the blastoids are the only higher taxonomic group to have gone extinct at this time. Mostly Tethyan faunas were lost, while Boreal faunas appear not to have noticed the event (Jin *et al.* 1994c).

Lopingian radiation

With the exception of Chinese faunas, Lopingian taxa are generally poorly known and little studied (Erwin 1993). None the less, it is evident that several major groups were radiating during the interval (sphinctozoans, foraminifera, bryozoans, gastropods, ammonoids, and, to a lesser extent, brachiopods and pectinacean bivalves), providing compelling evidence that the entire Late Permian was not an era of prolonged crisis. However, for the tabulate corals and echinoderms the late Maokouan crisis appears to have taken the wind out of their sails as they showed little signs of recovery in the Lopingian. There is some evidence for a decline of provinciality in the Changxingian among Tethyan foraminifers, but this was a minor change compared to the forthcoming Scythian pandemics.

The terrestrial tetrapod record appears out of step with that of the marine invertebrates in showing a protracted Lopingian diversity decline.

Late Changxingian disaster

The late Changxingian mass extinction evidently came as a surprise to most of the world's biota as groups as diverse as large, chambered foraminifers, calcisponges, and Gondwanan glossopterids were diversifying to almost the end of the era (Fig. 5.8). The best-dated sections from equatorial Tethys reveal the mass extinction to be entirely confined to the *changxingensis* Zone, while in the Perigondwanan sections of Pakistan the crisis was a little later but no less sudden (p. 122). Our interpretation of a rapid but not instantaneous extinction (it is usually spread over a few decimetres to metres of stratum) differs considerably from that of many (such as Holser and Magaritz 1992: p. 3297). Teichert (1990: p. 231) memorably remarked:

The way in which many Paleozoic life forms disappeared towards the end of the Permian Period brings to mind Joseph Haydn's Farewell Symphony where, during the last movement, one musician after the other takes his instrument and leaves the stage until, at the end, none is left.

We prefer to envisage all the musicians bolting for the door after playing Tchaikovsky's 1812 Overture. Erwin (1994: p. 231) covered all eventualities:

The pattern of disappearance across the Permo-Triassic boundary is complex, with some clades disappearing well below the boundary [in the late Maokouan?], others diversifying right up to the boundary, and still others seemingly oblivious to extinction.

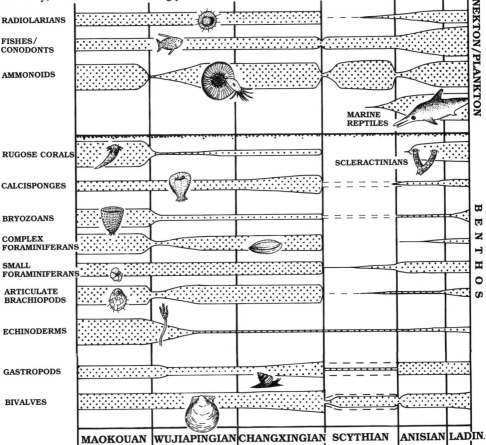

Fig 5.8 Summary of the relative fortunes of the marine biota during the Permo-Triassic crisis. Width of columns for each group denotes the relative changes of diversity within each group and is not intended to convey diversity changes between groups. Note that pelagic groups fared extremely well compared to the bulk of the benthos.

By scrutinising the members of the 'oblivious' category we can begin to see some selectivity in the extinction. The only benthic groups to weather the crisis were the nodosarioids, agglutinating foraminifera, palaeopsychrospheric ostracods, and several 'paper pecten' bivalve lineages; notably, all are Palaeozoic dysaerobic taxa. Indeed the agglutinating foraminifera and paper pectens were (uniquely) diversifying during the extinction crisis. Both groups would come to dominate Mesozoic black shale assemblages. The widespread Triassic holdover brachiopod *Crurithyris* is also notable for its tolerance to lowered oxygen levels in the Palaeozoic. However, the other survivors (gastropods and other bivalve lineages) were not members of pre-Triassic dysaerobic assemblages.

One of the greater mysteries of the P-Tr crisis is the survival, unscathed, of all marine nektonic forms (fish, conodonts, and nautiloids) – nektobenthic (?) ammonoids excluded. While swimming around in the upper water column can confer an immunity to bottom-water dysoxia, it is difficult to see how such diverse predators and scavengers could have survived the collapse of the invertebrate food chain on which they relied.

Scythian nirvana: back from the brink

To a Buddhist, the rather intangible concept of nirvana signifies an end of struggle and the attainment of a refuge from pain in 'hidden and incomprehensible peace' (Conze 1959: p. 40). These are all concepts that appear rather appropriate to a Scythian world that was characterised by several features:

1. Low-diversity assemblages composed of distinct Scythian lineages which included four bivalve genera and some rather poorly known microgastropods and echinoderms. On land *Lystrosaurus* and three plants (*Pleuromeia*, *Voltzia*, and *Dicroidium*) constituted most of the faunal and floral assemblages in the earliest Scythian.

2. A remarkable cosmopolitanism of the aforementioned fauna and flora.

3. An absence from the fossil record of numerous Lazarus taxa of gastropods, bivalves, brachiopods, bryozoans, calcisponges, and calcareous algae – all benthic groups.

4. An absence of reefs, coals, and biogenic cherts from the geological record.

5. A prolonged failure of the marine benthos to radiate in the Scythian – an interval of perhaps 7–8 m.y. (Hallam 1991). In contrast, marine nekton and the terrestrial biota recovered rapidly during the Scythian.

Post-Scythian radiation

There is tentative evidence, discussed further in the next chapter, that the late Scythian may have witnessed a further biotic crisis, probably at the Smithian–Spathian boundary. Both marine taxa and amphibians display elevated extinction rates at this time (Sepkoski 1986; Benton 1986, 1995). The Anisian finally marks the return of normal benthic conditions with the reappearance of most Lazarus lineages. Diverse groups, including marine reptiles, foraminifera, and scleractinians, radiated rapidly while algal–sponge reefs became re-established. The bivalve, brachiopod, and bryozoan radiation was delayed until the Ladinian or even later and was derived from previously insignificant Palaeozoic lineages: a fundamental legacy of the Late Permian crisis.

Boundary sections

Greenland

Throughout the Late Permian and Early Triassic a narrow seaway extended down the north-east coast of Greenland from the Boreal Ocean and thereby provided a valuable record of conditions from these northerly palaeolatitudes (Fig. 5.1). The regional stratigraphy consists of the Foldvik Creek Formation, a heterolithic association of limestone, shale, and gypsum, unconformably overlain by the clastic sediment of the Wordie Creek Formation (Grasmück and Trümpy 1969; Teichert and Kummel 1976). The age of both formations is controversial. On the basis of its brachiopod and coral fauna the Foldvik Creek Formation is considered to be no younger than the Maokouan (Teichert 1990). The Wordie Creek Formation, on the other hand, contains several ammonoid zones in its basal part, indicating a Griesbachian age and thus a substantial hiatus at the formation boundary (Trümpy 1969). All would be well were it not for the presence of abundant and diverse Permian benthic fossils in the *Glyptophiceras triviale* Zone up to 100 m above the base of the Wordie Creek. These include productid and spiriferid brachiopods, rugose corals, large foraminifers (*Colaniella*), crinoids, echinoids, and fenestellid and trepostome bryozoans (Grasmück and Trümpy 1969). The fauna is found in a range of lithologies and varying states of preservation. Grasmück and Trümpy (1969: p. 22) noted that the 'delicate bryozoan colonies are quite fresh and do not appear to be derived' and are preserved in life position alongside productids with their spines still attached. Grasmück and Trümpy came to the apparently unavoidable conclusion that a diverse Permian benthic community survived in the Boreal seas well into the Triassic.

Teichert and Kummel (1973, 1976) were having none of this and drew attention to the fact that many of the taxa were closely comparable at generic and even specific level to forms from the underlying Foldvik Creek Formation. They therefore suggested that they were reworked, but this raised an alternative dilemma: how to transport delicate bryozoans and spiny brachiopods without causing damage? The problem was 'solved' by attaching them to armoured mudballs and rolling them out to sea. Unfortunately for Teichert and Kummel, no examples of these rare sedimentological phenomena have been recovered and they therefore proposed that the 'argillaceous boulders, once coming to rest, dissolved, leaving well-preserved fossils that were rapidly buried [in life position!] in the coarse sediment' (Teichert and Kummel 1973: p. 269). This frankly ridiculous scenario, which includes mysteriously dissolving clay minerals, is probably not necessary to explain the presence of diverse Permian holdovers as recent work on conodont stratigraphy may have provided some answers. As discussed above (p. 98), the lower Griesbachian ammonoid zones of the Boreal realm are likely to be equivalent to the upper Changxingian Stage of lower latitudes. The diverse Permian benthos of north-east Greenland is therefore latest Permian in age, and there is no need for any special pleading to explain their occurrence in Triassic strata.

The Greenland fauna is of fundamental importance to understanding the P-Tr extinction and is in urgent need of taphonomic analysis (a branch of geology that was not invented the last time it was examined in the late 1960s). The available evidence suggests that diverse benthic communities survived until late in the Changxingian before succumbing at around the same time as the Tethyan faunas. Curiously though, the

Boreal fauna appears to have been remarkably archaic as it is congeneric (and even conspecific) with the pre-Maokouan faunas of the region. Perhaps this supports the conclusion of Jin *et al.* (1994c) that the Tethyan faunas underwent an end-Maokouan extinction, and subsequent radiation of new genera, while the Boreal faunas escaped this extinction and therefore did not produce many new genera in the later Permian.

The Dolomites, Italy

The Dolomites of northern Italy provide not only some of the most spectacular mountain scenery in the world but also an excellent record of P-Tr boundary conditions from the western cul-de-sac of the Tethyan Ocean (Fig. 5.1, location 2). The crucial interval is recorded in the upper part of the Bellerophon Formation and the lower part of the Werfen Formation although, until recently, the precise position of the boundary was unknown due to the absence of ammonoids. Recent conodont collecting has revealed that the *H. latidentatus–parvus* transition is located close to the base of the Mazzin Member, the basal member of the Werfen (Perri 1991; Wignall *et al.* 1996). This level, and the underlying 10–20 m of strata, records a fascinating sequence of lithofacies developments (Fig. 5.9).

The best-studied and certainly the biostratigraphically best-constrained sections occurs at Tesero, an easily accessible roadside outcrop located towards the south-west margin of the depositional basin (Neri *et al.* 1986; Noé 1987; Wignall and Hallam 1992), and at Gartnerkofel – the site of a borehole in southernmost Austria – a more basinal location (Holser *et al.* 1991). At Tesero most of the exposed Bellerophon Formation consists of silty dolomicrites, a supratidal flat facies, but, in the topmost 50 cm of the formation, they are sharply (disconformably) overlain by wackestones with a diverse early Changxingian marine fauna (Fig. 5.9). These pass gradationally up into the first of several beds of oolitic grainstone. These constitute the Tesero Oolite Horizon (TOH), a markedly retrogradational sediment package. Thus, 9 m of massive oolite on the basin margin passes offshore into the interbedded oolites and micrites seen at Tesero and ultimately is lost altogether in the basin centre where the Mazzin Member rests directly (and conformably) on the Bellerophon Formation (Brandner 1988, Fig. 2). Holser *et al.* (1991: p. 226) considered that 'the occurrence of oolites suggests the sea had withdrawn'. We disagree; while oolites undoubtedly indicate shallow water, the fact that such facies were rapidly retreating towards the basin margin testifies to an advancing not a withdrawing sea.

The deepening recorded in the TOH continues uninterrupted into the overlying Mazzin Member as oolite beds thin and disappear, to be replaced by micritic limestones. The lowest micrite, in the TOH, is uniquely characterised by abundant fungal spores and a clay with aviculopectinids and several brachiopod genera (Neri *et al.* 1986; Wignall *et al.* 1996). This is also the only micrite to display pervasive bioturbation; all higher examples are finely laminated with only a few small burrows developed (Wignall and Hallam 1992: Fig. 11; Wignall and Twitchett 1996; Fig. 5.9). The transition to laminated strata marks the disappearance of the last of the high-diversity fauna. The presence of pyrite framboids, well-preserved kerogen, high sulphur/carbon ratios, and elevated authigenic uranium concentrations in the micrites all testify to anoxic depositional conditions (Holser *et al.* 1991; Wignall and Hallam 1992; Wolbach *et al.* 1994; Wignall and Twitchett 1996). Higher in the TOH and in the Mazzin Member, three

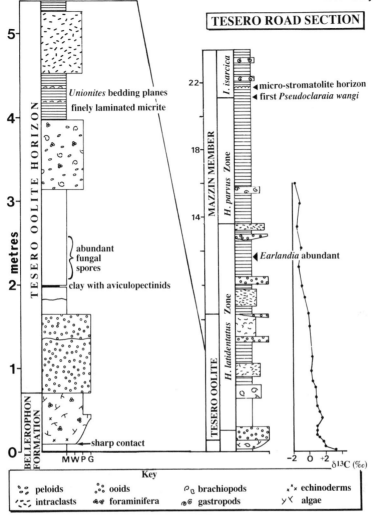

Fig 5.9 Graphic sedimentary log of the P-Tr boundary section at the Tesero road outcrop in the Dolomites of northern Italy. The boundary occurs low in the Mazzin Member at the base of the *H. parvus* Zone, while the fungal spike occurs below this level, within the *H. latidentatus* Zone. The last diverse Permian benthos occur immediately below the first bed of finely laminated micrites within the Tesero Oolite Horizon. After Wignall *et al.* (1996); the carbon isotope curve is from Holser *et al.* (1989).

additional facies types are developed: intraclastic, flat-pebble grainstones (probably the storm-eroded product of laminated micrites), microgastropod grainstones (mostly composed of *Coelostylina werfensis*, again probably concentrated by storm activity), and stromatolitic beds. The facies range indicates a low-energy, dysoxic/anoxic environment developed in relatively shallow water (that is, within the photic zone) and subject to infrequent storm events.

Sosio Valley, Sicily

The structurally complex geology of the Sosio Valley in western Sicily has recently yielded evidence of deep-water (bathyal?) facies changes in western Tethys. Conodonts are fortunately prolific in the condensed sections, allowing the detailed timing of facies changes to be constrained with precision (Gullo and Kozur 1993). Late Changxingian strata consist of red clays containing a rich fauna of foraminifers, radiolaria, palaeo-pyschrospheric ostracods, and the deep-water conodont *Clarkina sosioensis* (Kozur 1991). The diversity abruptly drops to a single species, the ubiquitous *H. parvus*, at the base of a finely laminated, pyritic, brown-weathering shale. Limestones appear in the *isarcica* Zone but diversity shows little recovery and the persistence of fine lamination and pyrite in the beds suggests a prolongation of the anoxic conditions begun in the *parvus* Zone (Gullo and Kozur 1993).

South China

During the P-Tr interval South China constituted an equatorially situated microcontinent in the eastern regions of Tethys (Fig. 5.1). Due to the relatively low continental hypsometry, typical of small continents, much of the area was flooded throughout the Late Permian and Early Triassic, thereby providing an extensive and thick sedimentary record.

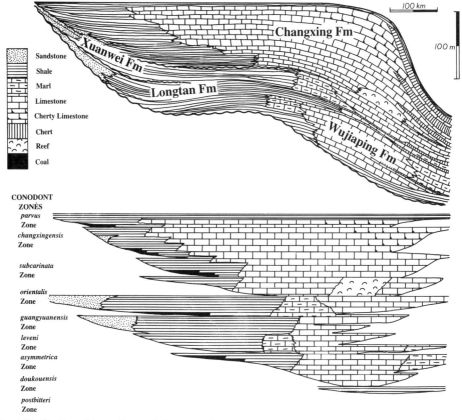

Fig 5.10 Stratigraphic section and chronostratigraphic chart for the Permo-Triassic basins of central Sichuan, south-west China. Note that the principal sequence boundary and regressive interval occurs at the end of the Wujiapingian (*orientalis* Zone) and base of the Changxingian (*subcarinata* Zone). Somewhat simplified from Jin *et al.* (1994b).

Environmental heterogeneity was high in the Late Permian of South China and a diverse range of basinal, shelf, ramp, and platform environments are recorded in the carbonate-dominated successions (Yang and Li 1992). Sequence stratigraphic analysis of individual basins has revealed a consistent picture of onlap punctuated by several regressions, the most significant of which occurs at the Wujiapingian–Changxingian boundary (Jin *et al.* 1994b). Thus basal Changxingian *subcarinata* Zone sediments are of extremely limited extent but the ensuing phase of rapid onlap ensured that, by the late Changxingian, marine deposition was extensive (Fig. 5.10). Subsequent regression and erosion have been reported from a few P-Tr boundary sections in Sichuan Province (Reinhardt 1989) but there is little supporting field evidence and the continued rapid onlap at this time indicates that the boundary occurs within a transgressive interval (Wignall and Hallam 1996). The extremely uniform and fine-grained nature of Griesbachian and Dienerian strata is probably testimony to the widespread establishment of deep-water conditions at this time.

Good P-Tr boundary sections are encountered throughout southern China (Yang and Li 1992). The most intensively studied and likely choice for the international boundary stratotype section occurs at Meishan in northernmost Zhejiang Province in a series of hillside quarries (Yin *et al.* 1994; Fig. 5.11). A diverse late Changxingian fauna of conodonts, brachiopods, ammonoids, and foraminifera occurs to within 20 cm of the

Fig 5.11 P-Tr boundary beds at Meishan, the most likely locality for the eventual choice of the international stratotype. The most commonly utilised bed numbers are shown.

boundary before abruptly disappearing at the base of a smectitic white clay 5 cm thick (Bed 25 of Chinese workers). This ash, the product of acidic volcanism, is bounded by pyrite laminae with a trace-metal enrichment signature characteristic of anoxic conditions (He 1989). The overlying 'black clay' of Bed 26 (Yin *et al.* 1994) is a dark grey shale with abundant burrows, Permian brachiopods and foraminifera, Changxingian conodonts, and the first *Claraia*. The distinct black-on-white double clay bed has been recognised at other Chinese localities and provides an excellent marker bed independent of biostratigraphic criteria (Wignall *et al.* 1995). The overlying Bed 27, a wackestone, contains the first primitive *H. parvus* 8 cm above the base, while the first *I. isarcica* appear 8 cm above that, making for a very thin *parvus* Zone. Several Permian brachiopods persist into Bed 27 where they are joined by *Unionites* and further *Claraia* species: *Ophiceras* appears at the base of the *isarcica* Zone. Bed 27 is the highest level to show burrow mottling. All higher beds are finely laminated and consist of centimetre-scale interbeds of micrites, marls, and dark grey shales (Wignall and Hallam 1993: Fig. 19). The only evidence for benthic life above Bed 27 consists of occasional levels with modest *Planolites* burrow mottling, rare *Claraia*, and even rarer *Neowellerella* – the last Permian holdover brachiopod.

The mass extinction at Meishan therefore occurs within approximately 1 m of strata recording the change from pure carbonates to a more marl-dominated succession. The majority of taxa make their last appearance immediately below an ash band.

Salt Range, Pakistan

Marine conditions on the Perigondwanan margin, the southern arm of Pangaea, are recorded in the sediments of one of the larger embayments of the supercontinent which extended from Pakistan and north-west India to Madagascar (Fig. 5.1). The succession in the Salt and Surghar ranges of northern Pakistan is among the best exposed and best known. The crucial boundary interval occurs in the Kathwai Member, a highly condensed unit dominated by echinoderm calcarenites (Kummel and Teichert 1970) which sits erosively on the (Wujiapingian?) Chhidru Formation (Fig. 5.12). Ophiceratids are restricted to the upper third of the Kathwai Member, while the lower two-thirds lack ammonoids – an absence that has led to considerable discussion of the position of the P-Tr boundary (Sweet 1992; Wignall and Hallam 1993). The basal Kathwai contains a Changxingian brachiopod fauna but initial examination of the conodont fauna suggested a basal Triassic age (Sweet 1970), thereby raising the spectre of reworking of the brachiopod fauna once again. Re-examination of the conodonts has revealed that the *latidentatus* to *parvus* transition is in the middle of the Kathwai, allowing the brachiopods to return to the Permian (Wignall *et al.* 1996). Interestingly, however, the abundant echinoderm and common foraminifera fauna of the Kathwai are not lost until higher in the member, at a level marked by a transition to finely laminated micrites and marls (within the *carinata* Zone). This late disappearance of a typical Permian assemblage points to a considerably later timing of the extinction in these southerly palaeolatitudes.

Japan

The foregoing sections provide ample information on the P-Tr transition in shelf seas from a range of palaeolatitudes around the margin of Pangaea. Only the southernmost seas around Australia and New Zealand are poorly known due to a combination of lack

Fig 5.12 Outcrop of the Kathwai Member at Nammal Nala in the Salt Range of northern Pakistan: 1, unconformity at the base of the Kathwai, 2, P-Tr boundary (as defined by conodonts); 3 level of mass extinction coinciding with the transition to finely laminated strata. Tony Hallam for scale.

of outcrop and poor dating, particularly of the Permian. However, until recently the major gap in our knowledge has been of events in the vast Panthalassa Ocean. Oceanic crust of this age has long since been subducted, but fortunately sufficient of its thin veneer of oceanic sediments has been skimmed off and preserved in the accretionary complexes of Japan (Isozaki *et al.* 1990). Intensive study of these sediments since the early 1990s has allowed us a glimpse of the extraordinary palaeoceanographic changes taking place at the time.

Both the Mino-Tanba Terrane of south-west Honshu and the Chichibu Terrane of eastern Shikoku contain tectonic slices 100–200 m thick and several kilometres long incorporated into a Jurassic accretionary *mélange* (Ishiga and Yamakita 1993). These record pelagic sedimentation mostly of red radiolarian cherts spanning the Late Carboniferous to Early Jurassic interval. The monotony of chert deposition is broken at two places in the succession by a thick black shale tens of metres thick straddling the P-Tr boundary and a much thinner bed in the early Toarcian (cf. Chapter 7). The detailed lithofacies and geochemical changes associated with the P-Tr example show a fascinating symmetrical succession (Fig. 5.13). Initially, Late Permian red cherts become dark grey as they lose their haematite content. This is followed by a change to light grey, finely laminated siliceous shales with a high pyrite content known as 'Toishi-type' siliceous shales (Musashino 1993). The Toishi shales in turn grade into a black, finely laminated organic-rich claystone with original total organic carbon (TOC) values of 4–10% (Isozaki 1994; Kakuwa 1996). Biomarkers and carbon/nitrogen ratios indicate a marine planktonic source for the organic matter (Suzuki *et al.* 1993). The higher succession is a mirror image of the lower stratigraphy as the black claystone is succeeded by a Toishi shale, dark grey cherts and finally red cherts once again (Fig. 5.13).

The development of the lower Toishi shale coincides with the demise of all the Permian radiolarian species (Isozaki 1994, 1997). The disappearance coincides with the development of anoxic deposition as indicated by the fine lamination and abundance of pyrite and well-preserved marine organic matter (Isozaki 1994, 1997). Enrichment of rare earth elements and uranium in the Toishi shales and black claystone adds further

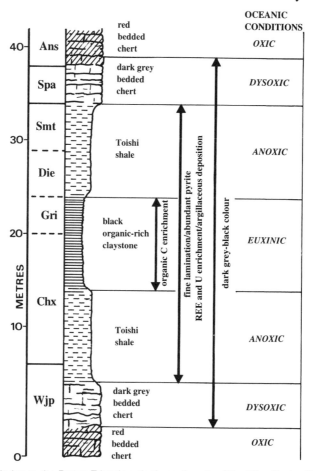

Fig 5.13 Graphic log to the Permo-Triassic pelagic sediments of the Mino-Tanba Belt, south-west Japan, showing the gradual intensification and then amelioration of oxygen deficiency during this interval. Based on Isozaki (1994).

weight to this interpretation (Musashino 1993), although it has been argued otherwise by Kajiwara *et al.* (1994) on the basis of their sulphur isotope analyses. These showed that relatively light values of $\delta^{34}S_{pyr.}$ in the Toishi shales become even lighter (by around 20‰) in the black claystone. While agreeing that the S isotope values of the Toishi shales are diagnostic of anoxic conditions, Kajiwara *et al.* argued that the extreme fractionation of S isotopes in the claystone is diagnostic of repeated reoxidation reactions typical of oxic environments. They thus envisaged an oxic event punctuating a long-term anoxic event across the P-Tr boundary. In fact comparison with modern-day environments and experimental work suggests that the interpretation of Kajiwara *et al.* (1994) is wrong. The greatest fractionation occurs in intensely anoxic events such as the Black Sea where sulphide-disproportionating bacteria repeatedly process elemental sulphur in the water column (Canfield and Thamdrup 1994). Correctly interpreted, the data of Kajiwara *et al.* show that the peak intensity of anoxia occurred during black claystone deposition when, like the present-day Black Sea, much of the lower water column of the Panthalassa Ocean was anoxic and H_2S-bearing (euxinic conditions).

The occurrence of euxinic conditions probably accounts for a total absence of fossils in the black claystone, but this unfortunately makes it difficult to date the peak of the anoxic event. The highest radiolaria in the lower Toishi shale indicate a Changxingian age, while Smithian conodonts are found in the upper Toishi shale. As there is no evidence for a break in deposition the black claystone was probably deposited during the late Changxingian to late Dienerian interval.

Normal, red radiolarian chert deposition did not resume until well into the Anisian. Oxygen-deficient to euxinic deposition therefore characterised the Panthalassa Ocean for around 10 m.y., a considerable duration that led Isozaki (1994, 1997) to talk of a 'superanoxic' event.

Sea-level changes

Sea-level variations during the P-Tr transition have long been a subject of interest because of both the substantial base-level changes in the geological record at this time and their implication for the mass extinction. The Permian witnessed the major Phanerozoic lowstand, with the lowest point typically placed at the very end of the period (Schopf 1974; Forney 1975; Holser and Magaritz 1987; Baud *et al.* 1989), followed by a rapid basal Triassic transgression (Schopf 1974; Holser and Magaritz 1987; Embry 1988; Hallam 1992; Paull and Paull 1994b). So entrenched is the concept of an end-Permian regression that it is routinely incorporated in discussions of the mass extinction. However, while examination of the geological record does indeed reveal a series of major Permian regressions, the youngest example occurs around the Wujiapingian–Changxingian boundary substantially before the Late Permian crisis and several million years before the end of the Permian.

The biostratigraphically best constrained sections of South China reveal a major downward shift of onlap and sequence boundary generation in the *orientalis* Zone (Fig. 5.10). The early *subcarinata* Zone sediments (containing the *Phisonites* ammonoid fauna in Tethys) show the characteristically limited depositional extent of a lowstand systems tract. The succeeding Changxingian sediments rapidly onlap the basin margins, and by the *changxingensis* Zone deep-water facies, characterised by radiolarian-bearing cherty limestones, are encountered in some of the deeper basins of the region (Wu *et al.* 1993). Importantly, the onlap and deepening continued across the P-Tr boundary with no evidence of a downward shift of onlap at this time.

The broad picture of sea-level changes in South China appears to be developed in all the other regions of the world (where biostratigraphic dating allows reasonable correlation) pointing to a eustatic origin. Thus the sequence boundaries at the base of the Wordie Creek Formation in Greenland, close to the top of the Bellerophon Formation in Italy and at the base of the Kathwai Member in Pakistan are probably all of early Changxingian age, while the overlying sediments all show the upward deepening and onlap associated with transgression. The deepening event is also known from Iran where it was responsible for the loss of the shallow-water dwelling Rugosa in the region (Ezaki 1993a,b). The late *changxingensis* Zone sediments show some evidence for shallowing in more marginal locations, such as those seen in Italy, but this is typical of early highstand rather than lowstand conditions as onlap continues unabated, and renewed deepening marks all basal Triassic succession.

The biotic crisis therefore occurred several million years *after* a regression and during a major transgression and, when viewed in adequate detail, the extinction event is seen to be unrelated to any major base-level change in the marine realm.

Geochemical evidence

Carbon isotopes

Thanks to the joint endeavours of the cosmopolitan team of Bill Holser, the late Mordecai Magaritz, and Aymon Baud, the $\delta^{13}C_{carbonate}$ variations of the Late Permian to Early Triassic interval in the Tethyan realm are well known. Their results have consistently shown the same trends – a minor positive peak in the Late Permian followed by a prolonged negative shift of 5–6‰ culminating in a low point at or close to the P-Tr boundary (Baud *et al.* 1989; Holser *et al.* 1991; Fig. 5.14). An identical trend has been observed in the terrestrial organic carbon record of Australia and, remarkably, in the tooth apatite of herbivorous therapsids from the Karoo (Thackeray *et al.* 1990; Morante *et al.* 1994). The carbonate carbon isotopic signature from the Boreal sections of Spitsbergen has been recorded from brachiopods and rugose corals, and it too shows a similar trend, although of greater magnitude (Gruszczynski *et al.* 1989). There is therefore little doubt that the Late Permian record is a primary one that offers a much needed possibility of achieving correlation between marine and terrestrial sections (Morante *et al.* 1994). The positive peak in the Late Permian has been variously dated, on biostratigraphic criteria, as early Changxingian (Holser and Magaritz 1985; Faure *et al.* 1995), early Wujiapingian (Holser *et al.* 1986) or, for the Boreal record, as early as the Maokouan (Nakrem *et al.* 1992). The most reliable dating suggests that the youngest age is correct.

Intense sampling from across the P-Tr boundary at Meishan and in the Gartnerkofel core has shown that there is a dramatic acceleration of the negative swing within a few centimetres of the boundary (Holser *et al.* 1991; Xu and Yan 1993). At Meishan there are two negative spikes of 8‰ magnitude in Beds 26 and 27, separated by an equally large positive spike at the base of Bed 27. The lower spike occurs in the dysaerobic, dark grey boundary clay and is likely a diagenetic signal of carbonates generated in the sulphate-

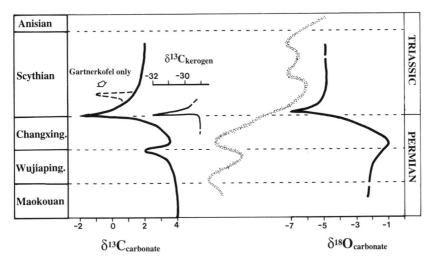

Fig 5.14 Stable isotope variations from the mid-Permian to the Middle Triassic. The δ^{13}C carb curve, from Holser *et al.* (1991), is identical to that seen in many carbonate sections throughout the world, although the Gartnerkofel core is unique in showing a second negative excursion in the early Scythian. The δ^{13}C kerogen trend from the immediate vicinity of the boundary is also shown (after Wang *et al.* 1994), a signature that has proved useful in the correlation of terrestrial with marine sections (Morante 1996). The δ^{18}O carbonate curve is also from the Gartnerkofel core. The eustatic sea-level curve is shown (stippled) for comparison with transgression occurring to the right: Scythian fluctuations are from Paull and Paull (1994b).

reduction zone (Grossman 1994). The upper spike occurs within a micritic limestone precisely at the conodont-defined P-Tr boundary; any diagenetic effects should therefore be rock-buffered and the δ^{13}C$_{carbonate}$ value is thus presumably a primary one. A similar negative spike straddles the P-Tr boundary at Gartnerkofel, although it is of lesser magnitude (2–3‰) and spread over a few metres rather than a few centimetres of section – probably a function of higher sedimentation rates in the Austrian section. Uniquely, a second negative excursion is seen in the Gartnerkofel core 35 m above the P-Tr boundary, in the *isarcica* Zone, but this is probably a product of diagenesis (Scholle 1995).

These complex long- and short-term variations in the δ^{13}C$_{carbonate}$ are clearly telling us something about the contemporary biotic crisis and have figured centrally in many mass extinction mechanisms. The gradual decline of δ^{13}C$_{carbonate}$ through the Changxingian indicates that ^{12}C-enriched CO_2 was being supplied to the ocean–atmosphere system throughout this interval. The obvious and most favoured source is from oxidation of ^{12}C-rich coals exposed during the supposed Late Permian regression (Holser and Magaritz 1987; Baud *et al.* 1989; Holser *et al.* 1991). Unfortunately, the non-synchrony of sea-level changes and the δ^{13}C$_{carbonate}$ curve argues against this link. As already demonstrated, a major Late Permian lowstand occurred around the Wujiapingian–Changxingian boundary at a time when carbon isotopes became heavier not lighter (Fig. 5.14). During the succeeding transgression, as the area of exposed coals became less, the isotopic ratios then became progressively lighter.

Alternatively, Berner (1989) has suggested that a decline of glossopterid coal swamps in the Changxingian (due to high-latitude aridity increase) could be a salient factor.

However, Retallack's (1995a) work has demonstrated that the decline of glossopterids was a good deal more sudden than previously realised and was essentially confined to the very end of the Changxingian. Faure *et al.* (1995) may have supplied the answer to this dilemma: the Permo-Triassic interval witnessed the widespread uplift and erosion of the southern margins of Gondwana, particularly in the area to the south-west of South Africa. This was by no means a major orogenic episode but, importantly, it uplifted and exposed to erosion an area where, previously, thick glossopterid coals had accumulated. Therefore the negative carbon isotope swing may be the product of orogeny not regression.

The long-term $\delta^{13}C_{carbonate}$ shift of the Changxingian has a rate and magnitude characteristic of large-scale changes in the exogenic carbon cycle (cf. Magaritz *et al.* 1992). On the other hand, the rapid acceleration of the negative excursion at the boundary suggests changes in the biosphere on a time-scale of thousands of years. Productivity collapse is an obvious candidate for a negative $\delta^{13}C_{carbonate}$ spike, and this possibility has been investigated in the Gartnerkofel core by comparing differences between $\delta^{13}C_{carbonate}$ and $\delta^{13}C_{organic}$ – the $\Delta^{13}C$ value – but with no conclusive result (Magaritz *et al.* 1992). However $\Delta^{13}C$ can also be affected by changes in the composition of $\delta^{13}C_{organic}$ caused by the changing proportion of terrestrial as opposed to marine organic matter. Such variations, which are independent of productivity changes, were not assessed by Magaritz *et al.* (1992), which somewhat invalidates their results. The requisite analysis was undertaken by Wang *et al.* (1994) on kerogen from a P-Tr boundary section in British Columbia. Nitrogen/carbon ratios characteristic of marine organic matter were seen throughout the section, and the influence of terrestrial organic matter on the isotopic signal was therefore considered negligible. The $\delta^{13}C_{kerogen}$ trend showed a negative swing of 1.5‰ at the boundary before returning to pre-boundary values 3 m higher up – changes that Wang *et al.* (1994) attributed to the catastrophic collapse of marine primary productivity at the base of the Triassic. Unfortunately things are never simple when interpreting carbon isotope ratios, and a change of this magnitude can also be caused by a transition from a mixed, oxic ocean to a stratified ocean with an anoxic lower water column. In the latter situation plankton derive their carbon from the ^{12}C-rich upper water column and thus give a light $\delta^{13}C_{kerogen}$ signature (Lewan 1986). That such a change occurred at the P-Tr boundary is supported by a host of evidence (pp. 138–9), including some of Wang *et al.*'s own data. Their analysis of H/C ratios from the British Columbia section showed an increase across the boundary, indicating an enhancement of organic matter preservation characteristic of anoxic deposition. This observation does not invalidate the interpretation of Wang *et al.* (1994) as productivity decline and stratification could both have happened. In support of their interpretation is the fact that $\delta^{13}C_{kerogen}$ values returned to their original values 3 m above the base of the Triassic while H/C remained high. Further, independent evidence for productivity collapse comes from the cessation of biogenic silica flux and the radiolarian crisis late in the Changxingian (p. 110).

Oxygen isotopes

The $\delta^{18}O_{carbonate}$ signature of the P-Tr interval has been obtained from the dolomites of the Gartnerkofel core and the bioclasts of the Spitsbergen sections (Gruszczynski *et al.* 1989; Holser *et al.* 1991). Although they differ in magnitude, both show a similar trend that is remarkably like the $\delta^{13}C_{carbonate}$ curve (Fig. 5.14): a positive excursion followed by a major negative one that peaks in the earliest Triassic slightly after the the carbon isotope peak. Explaining changes of 7‰ in the oxygen isotope record is not easy: the huge atmosphere–ocean reservoir of oxygen is little affected by changes in the carbon cycle and so the $\delta^{18}O_{carbonate}$ fluctuations may show only a coincidental propinquity with the carbon isotope curve. However, this conclusion is built on the assumption that the $\delta^{18}O_{carbonate}$ record is in isotopic equilibrium with the atmosphere – a precept challenged by Hoffman *et al.* (1991). Notwithstanding these uncertainties, the negative shift could be the product of a major global temperature increase, of around 6°C, or a decreased evaporation of seawater (Holser *et al.* 1991). Independent sedimentological and palaeobotanical evidence tends to favour the former scenario (see, for example, Retallack 1995a, 1996; p. 137).

Sulphur isotopes

Seawater $\delta^{34}S_{sulphate}$ values recorded in anhydrites show a prolonged decline during the Permian to an all-time Phanerozoic low around the Wujiapingian–Changxingian boundary (Holser 1977; Claypool *et al.* 1980). This was followed by a spectularly rapid positive swing from the Changxingian to the Scythian, culminating in a +28‰ high point late in the Scythian (Fig. 5.15). The timing of the switch to increasingly heavy values is not well constrained due to a paucity of values, particularly in the earliest Scythian, but Kramm and Wedepohl's (1991) data from the Zechstein Basin show that it was already well under way in the later Changxingian.

The rapid rise of $\delta^{34}S$ points to the widespread precipitation and burial of ^{32}S-rich pyrite during the P-Tr interval. However, Holser (1977) and Claypool *et al.* (1980) considered that the amount of pyrite precipitation required to achieve a 10‰ swing of seawater sulphate would have been inconceivably large. They therefore postulated that the trend is recording the signature of ^{34}S-enriched brines formed in hypothetical marginal marine basins in the Changxingian. These are envisaged to have begun leaking into the world's oceans (to the detriment of marine life) to produce the heavy

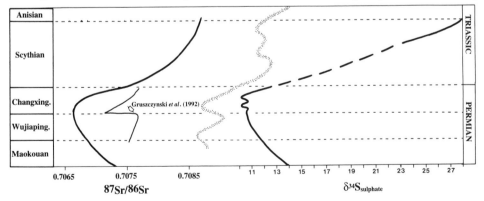

Fig 5.15 Strontium isotope and sulphur isotope trends during the Late Permian and Early Triassic from Martin and Macdougall (1995) and, for the S curve, Claypool *et al.* (1980) and Kramm and Wedepohl (1991). The distinctly different $^{87}Sr/^{86}Sr$ curve obtained by Gruszczynski *et al.* (1992) from brachiopods is probably affected by diagenetic alteration. The sea-level curve (stippled), shown for comparison, illustrates transgression to the right.

signature recorded from numerous anhydrites from around the world. This rather *ad hoc* notion receives no support from the geological record: the Zechstein Basin, one of the few potential sources of heavy brines – if not the only one – became infilled in the latest Permian. Moreover, the subsequent discovery in Japan of pyrite-rich anoxic sediments with extremely light S isotopes (Kajiwara *et al.* 1994) is an obvious sink of ^{32}S-rich sulphides at this time.

The seminal study of Claypool *et al.* (1980) in fact contains a long-overlooked section that provides further evidence for widespread oceanic anoxia in the P-Tr interval. In addition to the sulphur isotopic studies, for which their paper is a justifiable bench-mark in the field, Claypool *et al.* also measured the oxygen isotope ratios of their anhydrite samples. Their results showed that $\delta^{18}O_{sulphate}$ varies little if at all throughout the Phanerozoic. This led them to conclude 'that most deposition of sulfide must have been accompanied (in time although not necessarily in place) by the deposition of sulfate', otherwise the preferential reaction of ^{16}O sulphate during sulphide formation would be recorded in the $\delta^{18}O_{sulphate}$ record. The exception to this Phanerozoic rule occurs from the latest Permian through to the Early Triassic when the $\delta^{18}O_{sulphate}$ curve undergoes a rapid swing of $+5‰$. The clear implication here is that this interval was marked by one of the most intensive periods of sulphide formation in the entire Phanerozoic.

Strontium isotopes

The most recently published $^{87}Sr/^{86}Sr$ curve of Martin and Macdougall (1995) enjoys the twin advantage over previous curves of being measured on conodont apatite (considered relatively immune from diagenetic alteration) taken from stratigraphically well-constrained samples (Fig. 5.15). The curve none the less shows essentially the same trends as those previously recorded in anhydrites (Kramm and Wedepohl 1991) and carbonates (Holser and Magaritz 1987; Denison *et al.* 1994; Morante 1996): only the curve recorded in Spitsbergen brachiopod shells (Gruszczynski *et al.* 1992) appear distinctly out of phase, presumably due to their diagenetic alteration.

The $^{87}Sr/^{86}Sr$ trend during the Permian was one of long-term decline culminating in lowest values of 0.7066 late in the Wujiapingian followed by an increasingly rapid rise through the Changxingian and into the Early Triassic (Fig. 5.15). The rate of increase in the early Scythian is the highest known from the Phanerozoic (Holser and Magaritz 1992; Martin and Macdougall 1995) and for once there is remarkable unanimity as to the ultimate cause of these changes. All confrères are agreed that the rapid rise of $^{87}Sr/^{86}Sr$ records a major increase in the global rate of continental weathering – a conclusion further supported by Nd isotopic data (Martin and Macdougall 1995). There are two alternatives for this increase: global regression and erosion of the increased land area (Holser and Magaritz 1987, 1992) and/or enhanced chemical weathering triggered by increased humidity and atmospheric CO_2 levels (Erwin 1993; Martin and Macdougall 1995). These are not necessarily mutually exclusive alternatives, but comparison of the Sr isotope curve with the sea-level curve suggests that the former mechanism was not operating. The Late Permian sea-level low occurred in the early Changxingian and may have contributed to the gradual rise of $^{87}Sr/^{86}Sr$ values at this time, but the succeeding, accelerated rise occurred during a major transgression. Indeed the peak rate of rise corresponds to the extremely rapid transgression in the basal Scythian.

Trace metals

The search for an Ir anomaly at the P-Tr boundary has been a curious case of 'now you see it now you don't'. Initially an Ir spike of 8 ppb was reported from Bed 25, the volcanic ash in the late *changxingensis* Zone of the Meishan section (Sun *et al.* 1984; Fig. 5.11) and 2 ppb from the equivalent horizon in a section at Shangsi in Sichuan Province (Xu *et al.* 1985). Subsequent analyses of these boundary clays only found a barely detectable 0.002 ppb Ir or less (Clark *et al.* 1986; Zhou and Kyte 1988). Reanalysing the Meishan sediments yet again, Chai *et al.* (1992) found 0.123 ppb Ir and an Ir/Au ratio characteristic of mixing between crustal and cosmic sources. This led them to the (inescapable?) conclusion that volcanism had been triggered by a meteorite impact! In fact the concentration of Ir is well below the reported 1 ppb detection limit of their instrument.

Despite diligent searching (see, for example, Holser and Magaritz 1992) the only other Ir spikes have been detected in the Gartnerkofel core where they coincide with the twin peaks of the $\delta^{13}C_{carbonate}$ curve. The Ir values do not exceed 0.233 ppb and Attrep *et al.* (1991: p. 124) concluded:

Although we can not completely preclude an impact source for the lower Ir anomaly [at the P-Tr boundary], the absence of other impact signatures (chondrite ratios for Ir/Co etc., microspheres and shocked mineral grains) and the accompanying increase in sulfide content suggest an enrichment mechanism associated with reducing conditions in the paleo sea floor.

This conclusion probably also applies to Meishan, where the Ir anomaly is similarly associated with a pyrite-rich level and trace-metal enrichment typical of anoxic conditions (He 1989).

Evidence for anoxia at Meishan also includes rare earth element concentrations which are not depleted in Ce (Chai *et al.* 1992) – an attribute of anoxic oceans where Ce is not preferentially scavenged by iron oxides (Wright *et al.* 1984). Relative oceanic abundances of rare earths are thought to be faithfully recorded in the biogenic apatite of fish and conodonts. Analysis of such material has revealed that early Palaeozoic and Early Triassic samples lack a negative Ce anomaly, while it is present at other times of the Phanerozoic (Holser *et al.* 1986; Wright *et al.* 1987). The implication here is that the Early Triassic witnessed the re-establishment of anoxic oceans not seen since the early Palaeozoic.

In the Early Triassic pelagic shales of Japan the negative Ce anomaly not only disappears but becomes enriched relative to other rare earth elements, producing a positive Ce anomaly. This exceptional phenomenon was explained by Musashino (1990: p. 175) as indicative of 'reductive regeneration of Ce from continental shelves'. Anoxia was therefore considered widespread both in the shelf seas and the oceans of the Early Triassic.

Extinction mechanisms

In summarising the state of knowledge 30 years ago, Rhodes (1967: p. 72) concluded that the end-Permian mass extinction was probably caused by 'the multiple interactions of a wide variety of physical and biological factors' – more a statement of ignorance than an extinction mechanism. More recently, Erwin (1994: p. 231) has similarly suggested that the 'mass extinction appears to involve a tangled web rather than a single mechanism'.

There have been persistent attempts to untangle this web over the last 40 years or so, and at least nine distinct extinction scenarios have been proposed. Many of these are mutually contradictory, which leads to the inevitable conclusion that the majority must be wrong, as we will see below.

Cosmic radiation

Schindewolf (1954) started the current debate on end-Permian extinction mechanisms with his observations in the Salt Range of Pakistan. He was struck by the abrupt nature of the extinction and how it was unrelated to any major break in sedimentation. He therefore postulated that Earth-bound causes, such as Newell's (1952) regression hypothesis, could not be held responsible and suggested that the extinction was caused by the sudden bombardment of the Earth by cosmic radiation from a nearby supernova.

This dramatic suggestion has not gained much favour, and in an early exegesis Newell (1962: p. 606) noted that Schindewolf's 'failure to discover physical evidence that an unconformity exists is taken . . . as proof that one does not exist'. It seems a little harsh to complain that an unconformity was not found when there is no evidence for one! In fact Schindewolf has been vindicated (and Newell proved wrong) by subsequent sequence stratigraphic (Haq *et al.* 1988; Wignall and Hallam 1993) and biostratigraphic analysis (Wignall *et al.* 1996) in Pakistan, which has revealed no stratigraphic break at the level of the mass extinction (Fig. 5.12).

The cosmic radiation hypothesis has perhaps received little favour because it has no independent supporting data and it fails to explain a great deal of the geochemical data, such as the δ^{34}S and ^{87}Sr/^{86}Sr trends. However, there is little doubt that it would provide a lethal extinction mechanism, principally by destroying photosynthesisers on land and in the sea (Ellis and Schramm 1995).

Brackish oceans

The possibility that the extinction was caused by the development of low-salinity oceans in the Late Permian has a long track record that dates back to Beurlen (1956) who noted that stenohaline taxa (brachiopods, bryozoans, corals) were more heavily affected than euryhaline groups (fish, gastropods). The extensive formation of salt deposits in the Permian was cited as the means of reducing oceanic salinity. However, both Fischer (1964) and Benson (1984) have calculated that the known volumes of Permian salt are utterly inadequate to produce this effect. As an alternative, Fischer postulated that surface-water salinities were reduced by the formation of deep-ocean brines, thereby obviating the need for inconceivably large volumes of salt to be formed. Fischer's model has been used to explain the δ^{34}S trends (Holser 1977).

The brackish-ocean hypothesis continues to receive sporadic support (see, for example, Stevens 1977; Posenato 1991) and it is equally regularly debunked (Boucot and Gray 1978; Benson 1984). There are several problems with the model; Boucot and Gray (1978) noted that the extinction was not nearly as stenohaline-selective as Beurlen originally proposed and that stenohaline groups (notably ammonoids) thrived immediately after the extinction. The source of deep-ocean brines is also problematical (cf. p. 130).

Regression

Until recently the single most favoured cause of the end-Permian extinction has been regression, with the withdrawal of shallow seas from the margins of Pangaea producing an all-time Phanerozoic sea-level low (see, for example, Newell 1952, 1967; Valentine and Moores 1973; Schopf 1974; Forney 1975; Nakazawa 1985; Sweet *et al.* 1992). The actual kill mechanism for marine taxa is tied up in the loss of habitat area and is thus somehow related to the species–area effect.

The idea of an end-Permian regression has become so established in the literature that it is a matter of dogmatic assertion; evidence is rarely presented. Thus characteristic features of major sequence boundaries such as incised valleys and karstic surfaces have rarely been found and, where reported (see, for example, Reinhardt 1988), have not been confirmed (Wignall and Hallam 1996). In fact the underlying rationale behind the regression lies in the absence of *Otoceras*-bearing boundary sediments in low-palaeo-latitude Tethyan sections (Baud *et al.* 1989), but this absence is related to the cold-water preference of this genus. Curiously, it has long been overlooked that in those sections with *Otoceras*, such as the Arctic Canadian type sections, it first appears within a transgressive sequence some distance above an unconformity that is poorly dated but probably of base Changxingian age.

Regression is a much better candidate for the late Maokouan extinctions. A major regression terminated reef formation in the celebrated Guadalupian sections of Texas and the loss of vast areas of shallow-marine, tropical carbonate habitats both in this area and elsewhere may have had a profound effect on the indigenous faunas (Ross and Ross 1995b). As already related, the Maokouan–Wujiapingian boundary is a major sequence boundary in South China.

Bolide impact

Geochemical evidence for meteorite impact at the P-Tr boundary is extremely tenuous (p. 131). Vanishingly small Ir anomalies provide little confirmation, shocked quartz has not been detected, while common ferruginous spherules from Chinese boundary clays are ascribed a volcanic origin (Yin *et al.* 1992). Of course, this lack of evidence does not categorically rule out bolide impact, because a non-chondritic meteorite would not produce a tell-tale Ir anomaly. Rampino *et al.* (1994) have argued that the abrupt extinction of the terrestrial flora and the cessation of phytoplankton productivity are both characteristic of impact events. However, as shown below, there are other ways to achieve the rapid near-destruction of the biosphere.

Global refrigeration

For a decade or so S.M. Stanley (1984, 1988) has championed global cooling as a mechanism for most, if not all, mass extinctions. The diagnostic evidence for the end-Permian (and other) mass extinctions is the preferential loss of tropical taxa, the gradual nature of the extinction (an attribute of Plio-Pleistocene extinctions and therefore supposedly an attribute of other glacially induced extinctions) and the lack of limestone formation in the post-extinction interval. Stanley (1988) cited the presence of glacio-marine tillites in Siberia as further supporting evidence although, as he admits, these contain ammonoids no younger than the Maokouan.

The evidence for Late Permian cooling is rather contradictory. The Gondwanan glaciation was over by the beginning of the Late Permian; the last dropstones of early

Wujiapingian age occur in Australia (Veevers *et al.* 1994). Evidence that cold but not glacial conditions persisted in high latitudes is, however, found in Australia (Conaghan *et al.* 1993) and Arctic Canada (Beauchamp 1994). However, as both these regions were at high palaeolatitudes in the Late Permian this is not so surprising.

Stanley's hypothesis starts to unravel when his three principal lines of evidence are examined in detail. The extinction was not at its most severe in the tropics – quite the opposite in fact, as seen in the abrupt demise of the glossopterid flora in high southern latitudes (Retallack 1995a). Neither was the extinction a prolonged and gradual affair. Most surprising of all is the purported lack of Early Triassic carbonates when in fact thick successions of limestones are widespread in low-latitude sites of this age.

Despite this exegesis, Stanley's global refrigeration model need not necessarily be discarded altogether as the earlier late Maokouan extinction event possesses many of the attributes to be expected from death by cooling. Thus, only tropical taxa were affected while temperate and polar faunas escaped unharmed (Jin *et al.* 1994a), carbonates were rare in the immediate post-extinction interval, and evidence for glaciation in both Siberia and Gondwana is well known at this time (Caputo and Crowell 1985). Furthermore, the end-Maokouan was marked by a widespread (glacially induced?) regression that saw the termination of the celebrated Texan Guadalupian reefs and their constituent taxa.

Hypercania and global cooling

Despite the lack of evidence for cooling in the later part of the Late Permian, it has recently figured in a new end-Permian extinction mechanism involving death by CO_2 poisoning (hypercania). Knoll *et al.* (1996) suggested that Late Permian oceans were anoxic and therefore sites of organic matter burial. The twofold consequence of this situation would be to reduce atmospheric CO_2 levels and to enrich the deeper waters of the oceans in the by-products of organic matter mineralisation, namely CO_2 and H_2S. Ultimately it is proposed that the world switched to icehouse conditions, as the atmospheric concentration of greenhouse gases declined, causing the flushing of the oceans by cold, oxygenated waters and the spilling of CO_2-rich deep waters into shelf areas, to the detriment of the resident fauna. The sequence of events is identical to that proposed earlier by Malkowski *et al.* (1989), although they suggested that a drastic productivity decline would be the main kill mechanism, and is closely comparable to Wilde and Berry's (1986) model for the end-Ordovician crisis. The hypercania kill mechanism is the new component of the Knoll *et al.* (1996) scenario, and interestingly it is one of few models to predict directly the survival of fish during the P-Tr crisis because such metabolically active organisms are the best adapted to withstand an over-abundance of CO_2.

Unfortunately for the model, the available evidence indicates a warming trend and a switch from oxic to anoxic oceans in the Late Permian, precisely the opposite of the changes proposed by Knoll *et al.* (1996).

Volcanic winter

Stanley's extinction by prolonged refrigeration has not gained much favour as a mechanism for the end-Permian event (although he may have got it right for the Maokouan extinction) but it lives on as part of a more complex scenario involving the eruption of the tholeiitic flood basalts of western Siberia. The original volume of this vast trap province is estimated to have been 1.5–2.5 × 10^6 km^3 (Renne and Basu 1991;

Renne *et al.* 1995), making it by far the largest continental flood basalt province of the Phanerozoic.

For a long time radiometric dating indicated that the basalts were erupted over a considerable period of time spanning the Early and Middle Triassic, thereby making them too young to be implicated in the end-Permian extinction. However, recently new dates have shown that the flood basalts are considerably older and were erupted over a much shorter time interval. Renne and Basu (1991) suggested that all the Siberian Traps were erupted over 900 000 ($\pm$ 800 000) years at 248 Ma – again just slightly too young when compared to the 251.2 ($\pm$ 3.4) Ma age obtained from zircons from Bed 25 in the Meishan boundary section (Claoué-Long *et al.* 1991). However, by recalibrating their standard Renne *et al.* (1995) obtained a 250.0 ($\pm$ 1.6) Ma age for the onset of volcanism. Renne *et al.* (1995) also redated Bed 25 using argon–argon dating of feldspars and produced an age of 249.98 ($\pm$ 0.20) Ma. It is now clear that the eruption of the Siberian Traps exactly coincided with the end-Permian mass extinction. With such a massive smoking gun now available it has not taken long for a volcanically induced global refrigeration model to gain favour as one of the most popular extinction mechanisms.

Campbell *et al.* (1992) noted five reasons why the Siberian Traps could have been an effective cause of global darkness:

(1) they are the largest subaerial volcanic event of the Phanerozoic;

(2) They are associated with voluminous Cu-Ni-sulphide ore bodies and are therefore likely to have had SO_2-rich eruptions;

(3) SO_2 levels are likely to have been further enhanced as the ascending magma passed through anhydrite-rich rocks;

(4) they contain an unusually high proportion of tuffs compared to 'normal' flood basalt provinces;

(5) the tuffs contain an (even more) unusually high proportion of lithic fragments.

Using reasons (4) and (5) Campbell *et al.* (1992) argued that the eruptions were exceptionally explosive by flood basalt standards and that they were therefore capable of injecting vast amounts of dust and sulphate aerosols into the stratosphere, hence global darkness! The kill mechanism is caused by the shutdown of photosynthetic activity (as suggested by the rapid negative swing of $\delta^{13}C$) and the consequent demise of the food chain it supports (Fig 5.14). The inevitable temperature drop is also postulated to have triggered a major glaciation and thus a major regression, leading to exposure of all the continental shelves, to the further detriment of marine invertebrates (Campbell *et al.* 1992; Conaghan *et al.* 1993; Renne *et al.* 1995). Evidence for this cooling includes 'stone rolls' – a product of freezing – in the seatearths beneath the latest Permian coals of Australia (Conaghan *et al.* 1994). The absence of tillites is argued away on the grounds that the exposed continental shelves produced by the (supposed) major regression were unfavourable for their preservation (Campbell *et al.* 1992).

The darkness is likely to have been short-lived – no longer than the duration of the eruptions – as the dust and sulphate aerosols would have been rapidly rained out. The high acidity of the rain is likely to have been a further suffering inflicted on the world's biota (Liu and Schmitt 1992). After their intense starved refrigeration the pitiable survivors may then have been cooked during a global warming episode due to the input of large amounts of volcanic CO_2 into the atmosphere (Conaghan *et al.* 1993).

The volcanic winter hypothesis clearly has no shortage of kill mechanisms, and the proposed chain of events provides convincing explanation for many of the isotopic trends (Fig. 5.16). For example, Conaghan *et al.* (1993: p. 793) noted that the 'rapid leaching by acid rain of the erupting Siberian Traps [may have caused] the long term rise of $^{87}Sr/^{86}Sr$ from its minimum'. Despite this, we feel that the scenario is an unlikely one for several reasons. A potential problem lies in the timing of the isotopic trends. Carbon isotope ratios began falling around the Wujiapingian – Changxingian boundary (others would have the trend starting even earlier), while $^{87}Sr/^{86}Sr$ began rising about the same time – both begin long before the onset of eruptions in Siberia.

More fundamental flaws in the hypothesis include the complete lack of evidence for a cooling event (see below). The stone rolls beneath Australian coals are not impressive evidence; these formed at 70°S, and it would be surprising *not* to find evidence of freezing at such latitudes. More significant is the disappearance of stone rolls and the cold-adapted glossopterid forests in this region at the end of the Permian, suggesting a major warming trend associated with the mass extinction (Retallack 1995a). The argument that the cooling event was too short-lived to leave any evidence, including tillites (Campbell *et al.* 1992), is hardly a scientific one.

Yet more problems occur when the role of the Siberian Traps is considered. The supposedly less than 1 m.y. eruption interval does not accord with the plant megafossil stratigraphy of the inter-trap sediments, which have long been known to record a complete Scythian succession (see, for example, Sukhov *et al.* 1966). Clearly, considerably more radiometric ages from throughout the province are needed to demonstrate the precise contemporaneity of the eruptions. Indeed, it is already known that the Maimecha-Kotui area, in the north-east of the province, is characterised by alkali ultrabasic volcanics that are considerably older (253.3 $\pm$ 2.6 m.y.) than the main tholeiitic flows (Basu *et al.* 1995).

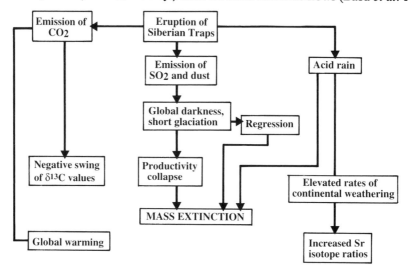

Fig. 5.16 Chain of events and the supporting evidence for the volcanic winter model for the P-Tr mass extinction Based on Campbell *et al.* (1992), Conaghan *et al.* (1993), and Renne *et al.* (1995).

A further problem lies in the unlikelihood that the fissure eruptions responsible for the basalt flows were explosive enough to inject dust and aerosols into the stratosphere. Campbell *et al.* (1992: p. 1761) argued that the high proportion of tuffs and 'the widespread occurrence of lithic fragments . . . testifies to the violent nature of the eruptions'. In fact, when examined in the field, the 'tuffs' are commonly cross-bedded, sometimes on a metre scale, and appear to be fluvially reworked (C. Mitchell pers. comm.). It therefore appears more than probable that the 'tuffs' record periods of erosion of the basalt lava flows by rivers that were also eroding sedimentary rocks and depositing a heterolithic clast assemblage that Campbell *et al.* found so unusual. The significance of the so-called tuffs lies not so much in their evidence of explosive activity as in their record of significance intervals of erosion between eruption events.

Global warning

Rather than cooling, the preponderance of evidence indicates warming at the end of the Permian. The negative oxygen isotope excursion in the Gartnerkofel data suggests at 6°C warming in this tropical western Tethys location, while the temperature rise in higher latitudes was probably even greater. Thus the latest Permian peats in Antarctica and Australia at 70°S are comparable to those formed at similarly high latitudes today where intense winter freezing occurs, but they are rapidly replaced by warm, temperate palaeosols in the Triassic (Retallack 1996). In the slightly lower palaeolatitudes of the Karoo Basin sedimentary evidence indicates a transition from a humid, temperate to a hot, semi-arid climate (Smith 1995).

The floral evidence for temperature rise is even more overwhelming. The cold-adapted glossopterids went extinct to be replaced by warm, temperate floras (Dobruskina 1987; Veevers *et al.* 1994; Retallack 1995a; Retallack *et al.* 1996); cold temperate and polar floras are completely unknown at this time (Ziegler *et al.* 1994). The Scythian is also the only interval of the Phanerozoic for which there is no evidence of ice; even that rather ambiguous line of evidence, dropstones, are unknown at this time (Worsley *et al.* 1994).

There is no shortage of the potential sources of CO_2 required to elevate atmospheric concentrations and produce global warming (Erwin 1993). Brandner (1988) was the first to highlight the significance of the Siberian Traps as a major source of CO_2 and thus the global warming scenario. However, the negative carbon isotope shift begins significantly earlier than the onset of volcanism and, as Erwin (1993) has pointed out, volcanic CO_2 is insufficiently light to cause the major negative swing no matter how much is emitted. This paradox is explained (or at least side-stepped) by having an alternative, additional source of isotopically light CO_2. This source was probably the glossopterid coals of southern Gondwana which underwent uplift and oxidation during the orogeny of the Gondwana margin in the Late Permian (Faure *et al.* 1995).

Thus during the Changxingian Stage, atmospheric CO_2 levels were rising at a pace that the normal carbon sinks were unable to counterbalance. Continental shelves, the main repository for marine organic C (Berner 1989), were of limited extent (although increasing in area) at this time. The methane of gas hydrates (cf. Erwin 1993, 1994; Morante 1996) may also have exsolved as temperatures rose, thus further exacerbating the greenhouse effect (methane rapidly oxidises to CO_2 in the atmosphere). The rapid rise of continental weathering rates at this time, indicated by Sr and Nd isotopes (Martin and Macdougall 1995), is probably a further indication of the increasing atmospheric CO_2 levels.

Whether the amount of CO_2 added to the atmosphere by coal oxidation (and gas hydrate melting) would have been sufficient to cause the end-Changxingian crisis is a moot point, but the sudden eruption of the Siberian Traps and the voluminous emission of yet more CO_2 could have turned a crisis into a catastrophe. On land, the high-latitude flora was totally eradicated and a remarkably uniform, warm to hot global climate became established (Veevers *et al.* 1994; Ziegler *et al.* 1994). The crisis among terrestrial tetrapods which produced a single, pandemic, low-diversity assemblage is presumably a response to the loss of environmental heterogeneity. Interestingly, Smith (1995) has suggested that most of the constituent taxa of this assemblage, including *Lystrosaurus*, were able to burrow, thereby raising the possibility that they aestivated during long, hot, dry seasons.

While providing a kill mechanism on land, global warming does not directly explain the extinction of the marine fauna except perhaps in equatorial latitudes where the water temperature may have exceeded the lethal threshold of 32°C (cf. Thompson and Newton 1989). At higher latitudes, where temperatures are unlikely to have reached this level, the warming appears to have produced warm temperate conditions that should have favoured an increase of marine diversity (all other things being equal). To explain the marine extinction we must therefore turn to the final and, for us, probably the single most important piece in the extinction jigsaw: global marine anoxia.

Marine anoxia

Evidence for anoxia in the seas and oceans of the latest Permian and Early Triassic comes from numerous, disparate geological disciplines. The preferential survival of dysaerobic benthic groups and nektic taxa strongly implicate low bottom-water oxygen levels as a cause of the extinction. Where it has been undertaken, facies analysis of boundary sections consistently shows that the disappearance of diverse Permian assemblages precisely coincides with the change from burrowed aerobic strata to finely laminated anaerobic strata containing large amounts of framboidal pyrite. This transition also coincides with the development of organic-rich sediments in abyssal and bathyal sections, although such black shales are only sporadically developed in shelf settings where organic-poor micritic limestones are more common. None the less, although the actual amount of organic C preserved in post-extinction sediments is variable, the quality of the organic C is high, pointing to anoxic depositional conditions (Magaritz *et al.* 1992; Suzuki *et al.* 1993; Wang *et al.* 1994; Wolbach *et al.* 1994).

The most consistently developed feature of all boundary sections is pyrite enrichment. Sulphur isotope analyses of pyrite from the abyssal Japanese sections reveal exceptionally light ratios (Kajiwara *et al.* 1994) the like of which is only seen in the present-day Black Sea, implying that the world's oceans became euxinic at this time. The trace-metal enrichment of the pyrite is a further pointer to intensely anoxic conditions (He 1989; Holser *et al.* 1991), while the huge magnitude of this pyrite-forming event comes from the interpretation of the major changes in the S and O isotope ratios of evaporites. That the world's oceans became euxinic (that is, with free H_2S in the lower water column) in the latest Permian is further shown by the loss of a Ce anomaly in the real earth element abundance patterns of biogenic apatite and in whole-rock analyses of pelagic strata

(Wright *et al.* 1987; Musashino 1993). Authigenic uranium enrichment provides yet further evidence (Wignall and Twitchett 1996).

Clearly there is considerable evidence of a major end-Permian to Scythian anoxic event, although not all are convinced. Erwin (1994), in particular, has argued against a death-by-anoxia mechanism. However, he is wide of the mark when stating that 'The geochemical data advanced to support the [anoxia] hypothesis actually provides little support; indeed the shift in $\delta^{13}C$ permits only a moderate shift in atmospheric oxygen levels, limiting the extent of marine anoxia' (Erwin 1994: p. 234). In fact, modelling suggests that the intense oxidation of coals may have lowered atmospheric oxygen levels to as low as 15% in the latest Permian (Berner 1989; Berner and Canfield 1989; Graham *et al.* 1995). However, this is not the crucial point as the extent of marine anoxia has little to do with atmospheric oxygen levels; global circulation, biological oxygen demand and temperature are all much more salient issues in oceanic redox chemistry.

It is the apparent failure of marine anoxia to produce the observed extinction patterns that Erwin (1994: p. 234) considered to be the most telling argument against the mechanism when he stated that

global anoxia is a diversity-independent mode of extinction, thus survival should be enhanced by broad oxygen tolerance and large population size, yet many of the surviving taxa had small population sizes [before the extinction?] . . . and closer look reveals no association between survival and [low] oxygen tolerance.

Neither of these two lines of reasoning stands up to scrutiny. As we have shown, the link between dysoxia tolerance and survival is a good one. The logic of Erwin's argument concerning population sizes is unclear as there no reason why species with large populations should preferentially survive a global crisis. It is true that many of the successful Scythian survivors had low population sizes prior to the extinction (for example, latest Permian *Claraia* are very rare) and subsequently became abundant in the post-extinction interval. We see this as a flourishing of a few taxa able to survive the harsh conditions of the extinction interval.

Changes of oceanic oxygen levels have also been invoked in a model proposed by a Polish group based on their C, O, and Sr isotopic data from Spitsbergen bioclasts (Hoffman *et al.* 1991; Gruszczynski *et al.* 1992). In this model they suggested that Permian oceans were euxinic and therefore the lower water column was isolated and isotopically distinct from the upper water column. Mixing and oxygenation of the world's oceans at the end of the Permian are held responsible for the isotopic shifts of this interval. Kajiwara *et al.* (1994) suggested a similar but shorter-term mixing event at the P-Tr boundary. The mixing event is suggested to have reduced oceanic productivity (due to the greater efficiency of phosphorus removal in oxic oceans) and therefore caused the marine mass extinction. Elegant though the Polish group's modelling is, it has failed to predict the abundant sedimentary and geochemical evidence that has been discovered subsequently, particularly from Japan where the pelagic sediments record a change from oxic to anoxic conditions – not the other way around – in the latest Permian.

Catastrophic productivity decline constitutes the crux of both the volcanic winter and the Polish group's extinction mechanisms, but in fact the available evidence indicates a considerably more complex series of productivity changes during the P-Tr transition. Carbon isotope evidence from marine sections in China and Canada indicates a rapid decline in productivity precisely at the erathem boundary, followed by a rapid recovery in the basal Triassic, perhaps only a few tens of thousands of years later. The recovery is also marked by a proliferation of acritarchs in Griesbachian sections from throughout the world. Abundant acritarchs generally characterise stressed, eutrophic environments and their appearance shortly after the productivity crash indicates that it was followed by a productivity boom.

So, how can this complex series of events be reconciled with the marine anoxia hypothesis? Recent modelling of the influence of oceanic circulation and redox conditions on nutrient fluxes may provide the answer (cf. van Cappellen and Ingall 1994; Fig. 5.17). In fully oxygenated oceans, such as those pertaining today, circulation is vigorous due to a high equator-to-pole temperature gradient but productivity is low due to intense phosphate limitation of primary productivity – the Polish group spoke of 'hungry' oxic oceans (Malkowski *et al.* 1989). In contrast, in euxinic oceans, where a substantial portion of the lower water column is anoxic, P is much more abundant because of the 'enhanced benthic regeneration of phosphorus from organic matter and ferric oxyhydroxides' (van Cappellen and Ingall 1994: p. 677). Thus, despite the much more sluggish circulation of euxinic oceans, this is more than compensated for by the greatly increased availability of reactive phosphorus: euxinic oceans are therefore 'overfed'. The crucial factor in van Cappellen and Ingall's modelling occurs at the transition from hungry to overfed oceans. The reduction in circulation necessary to produce a stratified, euxinic ocean produces a catastrophic crash in primary productivity due to the decrease in nutrient circulation (Wignall and Twitchett 1996; Fig. 5.17). The counteracting increase in reactive P takes a finite amount of time (a few tens of thousands of years) to compensate for this effect. Therefore the lag time taken to establish a steady-state, high-productivity euxinic ocean would clearly mark a crisis for the marine biota – this is probably what is observed at the P-Tr boundary.

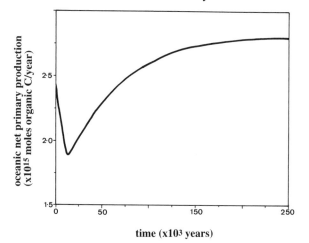

Fig 5.17 Time evolution of net primary productivity in the oceans after an instantaneous decrease in circulation. In this case vertical overturn is modelled to have decreased from 3 to 2 m/yr – a 33% decrease. More drastic declines in circulation would produce even more catastrophic productivity crashes. After van Cappellen and Ingall (1994).

The full scenario for the combined global warming – marine anoxia extinction mechanism is thus a complex chain of events triggered by the increased flux of CO_2 into the atmosphere, initially from coal oxidation and later from the Siberian Traps (Fig. 5.18). The resultant global warming would have led to a decline in oxygen concentrations in marine waters because of two factors: the declining solubility of oxygen in water as temperature increases – this becomes particularly critical above 30°C (Truesdale *et al.* 1955) – and the decreased circulation of the oceans as the equator-to-pole gradient declined. Eventually a threshold point was reached whereby the temperature-stratified ocean became stagnant, leading to a collapse of primary productivity as nutrient circulation declined drastically (Fig. 5.18).

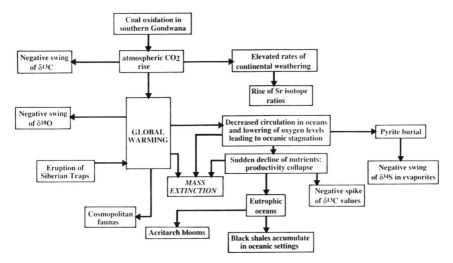

Fig 5.18 Flow chart for the global warming–marine anoxia model for the P-Tr mass extinction.

The productivity collapse occurred at precisely the P-Tr boundary, but in many sections dysaerobic and anaerobic strata are seen slightly before this. The rapid positive swing of the $\delta^{13}S$ curve was also well under way in the late Changxingian, indicating that large amounts of pyrite were already accumulating by this time. Many Permian taxa appear to have gone extinct in this late Changxingian interval rather than precisely at the P-Tr boundary. Therefore for many organisms it was the development of oxygen-poor conditions that directly caused their extinction, while the productivity collapse marked the final *coup de grâce* for already depauperate Permian ecosystems. Productivity rapidly recovered in the earliest Triassic, but it was too late by this time to save the Permian invertebrates.

The cessation of the Early Triassic 'superanoxic' event has received little study, but the available data (principally the Japanese sections and the $\delta^{34}S$ curve) indicate that it was not until the Middle Triassic that oceanic oxygen levels began to improve. This was probably caused by the reversal of the global warming trend as atmospheric CO_2 levels were lowered by the combined effects of high rates of continental weathering (cf. the Sr isotope curve), the formation of thick carbonate successions in Tethys, and the burial of large amounts of organic C in pelagic, abyssal sections.

6 Extinctions within and at the close of the Triassic

One of the 'big five' Phanerozoic marine extinctions (Raup and Sepkoski 1982) occurred at or close to the end of the Triassic, and mass extinction on the continents has also been claimed for this time. Other notable marine and continental extinctions have been reported for earlier in the Late Triassic, in the Carnian, and so a little must be said initially about correlation between Upper Triassic marine and continental strata.

The standard marine Triassic zones are based on ammonite successions in North America and Alpine Europe (Tozer 1974). There has been a long-standing dispute among biostratigraphers about whether the Rhaetian should be considered to be the youngest Triassic stage or, as Tozer (1974, 1979) prefers, a mere substage of the upper Norian, because he attributes to it only one zone, the *crickmayi*, whereas the Norian has a total of eight, including the *crickmayi*, and the Carnian five (Fig. 6.1). Furthermore, fully marine ammonite-bearing Rhaetian is known from only a few sections in the world. More recently, however, Hillebrandt (1990, 1994) has established four Rhaetian zones in South America, established from successions in Chile and Peru, three of them, including *crickmayi*, based on species of the distinctive heteromorph genus *Choristoceras*. In recent years there has been a movement of opinion in favour of distinguishing the Rhaetian as a stage rather than a substage, as a consequence of further biostratigraphic work by Europeans on the ammonite, conodont, and brachiopod faunas. Dagys and Dagys (1994), in reviewing this work, make a good case for distinguishing three successive ammonite zones in the Rhaetian, the *reticulatus*, *sturzenbaumi*, and *marshi*. The conodont genus *Misikella* enters the succession at the base of the *reticulatus* and disappears at the end of the *marshi* zone, and hence provides a good marker for the stage thus defined.

The biostratigraphy of continental sections is based on pollen and spores, and tetrapods. As Fig. 6.1 indicates, stratigraphic subdivision is less refined than for the ammonite succession, and precise correlation between marine and continental sections remains difficult in the absence of a good magnetostratigraphic record. The best correlation involves the use of pollen and spores that have been carried into nearshore marine facies (Visscher and Brugman 1981; Litwin *et al.* 1991).

Less work has been done on the Late Triassic extinctions than on other major extinction events, which largely reflects the paucity of accessible, complete or near-complete richly fossiliferous and easily correlatable sections, either in marine or

		Ammonoid zones	Palynological zones	Tetrapod zones
NORIAN	(Rhaetian)	Crickmayi	(Rhaetian)	NOR L2
	U.	Amoenum	Sevatian	
		Cordilleranus		NOR L1
	M.	Columbianus	Upper Norian (Alaunian)	NOR M2
		Rutherfordi		NOR M1
		Magnus	Lower Norian (Lacian)	
	L.	Dawsoni		NOR E
		Kerri		
CARNIAN	U.	Macrolobatus	Tuvalian	CRN L2
		Welleri		CRN L1
		Dilleri		
	L.	Nanseni	Julian	CRN M
		Obesum	Cordevolian	CRN E
LADINIAN	U.	Sutherlandi	Langobardian	LAD
		Macleami		
		Meginae		

Fig 6.1 Zonation of the Upper Triassic, by means of ammonites, palynomorphs, and tetrapods. Simplified from Benton (1994).

continental facies. Before they are described and discussed, however, attention must be drawn to a minor event shortly before the end of the Early Triassic, which has hitherto passed generally unnoticed.

The Smithian–Spathian boundary event

In Tozer's (1967) scheme the traditional single Lower Triassic stage, the Scythian, was subdivided on the basis of North American successions into four stages, the Griesbachian, Dienerian, Smithian, and Spathian; an alternative scheme unites the Dienerian and Smithian as the Nammalian.

Ammonites and conodonts are the two best-studied groups from a stratigraphic point of view, and both record a notable extinction event at the stage boundary. According to Tozer (in Hallam 1995a) there was a nearly total extinction of ammonites, with the disappearance of the Xenodiscaceae and near-disappearance of the Noritaceae and Dinoritaceae, followed by a major radiation in the Middle Triassic (Tozer 1981). The biggest Triassic conodont diversity was in the Smithian, with a big drop to almost nothing at the Smithian–Spathian boundary. There was indeed a species diversity reduction at each of the four stage boundaries of the Lower Triassic. After this subperiod conodont diversity declined, with final extinction at the end of the Triassic

(Clark 1987). Relevant information is available for one other group, the Bivalvia. *Claraia* and *Eumorphotis* are probably the two most important Lower Triassic bivalve genera. *Claraia* is abundant in the Griesbachian and attained its acme in the Dienerian. Many species vanished at the top of the Dienerian or in the lower Smithian. The genus is missing from the Spathian, except for the doubtful occurrence of one species in small quantities. *Eumorphotis* reached its acme in the Dienerian and lower Smithian; nearly half the species disappear at the Smithian–Spathian boundary (Yin 1990).

With regard to non-marine biota, Benton (1986) records a major extinction peak for tetrapods at the Lower–Middle Triassic boundary. The lack of stratigraphic precision for non-marine strata is such that this hitherto unremarked event could well correspond to the slightly older Smithian–Spathian boundary event in the marine realm. According to Balme (1970), the available palynological evidence in Europe suggests that the end of the Early Triassic marks a more critical floral change than that at the beginning of the Mesozoic.

More detailed information about the organic turnover and facies changes up stratigraphic sections in various parts of the world is required before any serious attempt at interpretation of possible causal factors can be undertaken. Meanwhile, the postulated extinction event presents a major challenge for future research.

Extinction within the Carnian

Marine realm

The lower–upper Carnian boundary marks an important extinction event among the ammonites, with a fairly large turnover. The earliest Late Carnian faunas are not very well known, but they are considerably different from those of the latest Middle Triassic, or Ladinian (Tozer, in Hallam 1995a). Though the stratigraphic data are no more precise than mid-Carnian there is evidence of an important phase of extinction among several other marine groups; further research may enable this to be pinned down more precisely to the lower–upper Carnian boundary.

Thus, among Triassic crinoids the Encrinidae are the most conspicuous elements of Anisian to lower Carnian faunas. They reached a peak of diversity in the early Carnian but disappeared completely by the mid-Carnian (Johnson and Simms 1989; Simms 1991). Similar patterns of an abrupt decline in diversity following an early Carnian peak have also been documented among echinoids (Smith 1990) and scallops (Johnson and Simms 1989). However, while it is true that the bivalve species of the lower Carnian Cassian Formation of the Alpine Dolomites are generally very different from those of the Upper Norian Kössen Formation of the Northern Calcareous Alps of Austria and Bavaria, Newton *et al.* (1987) have found that a large number of bivalve species in the lower Norian of Oregon are also present in the Cassian Formation, suggesting that their extinction was of regional, not global, significance.

Other groups are more poorly documented. Bryozoans appear to show a similar pattern to that of crinoids and the other groups cited in that, from a peak in the early Carnian, the group declined from 22 to 13 species in the late Carnian (Schäfer and Fois 1987). However, gastropods appear to have experienced a steady increase in extinction rate from Ladinian to Norian times without any significant peak in the Carnian (Erwin 1990). There is clearly a need for more precise data and for more groups, at the level of substage rather than stage. With regard to reef taxa, there is some evidence for a lack of synchrony in change up the stratigraphic succession. One of the dominant reef builders, *Tubiphytes*, underwent a major decline from the early to the late Carnian, roughly contemporaneous with calcisponges, bryozoans, and rhodophyte and chlorophyte algae, and roughly synchronous with the diversification of foraminifera, tabulozoans, spongimorphs and corals (G.D. Stanley 1988, Fig. 3). Stanley considers that this major turnover within the Carnian in reef biota was global in extent. Turning to the vertebrate record, Benton (1988) points out that four out of six families of marine tetrapods became extinct at the end of the Carnian.

Continental realm

In detailed work on the Newark Supergroup of eastern North America, Cornet and Olsen (1990) identified an episode of elevated palynofloral turnover near or at the end of the mid-Carnian. Terrestrial plants also show a peak in extinctions at the Carnian–Norian boundary (Boulter *et al.* 1988). Olsen and Sues (1986) also record a late Carnian extinction peak for pollen and spores in the Newark Supergroup.

With regard to terrestrial vertebrates, nine diverse families of tetrapods (13, including families with only one species) died out during the late Carnian, compared to six in the latest Triassic, according to Benton (1991); see also Benton (1994). These included such important reptile groups as the rhynchosaurs, traversodontids, and kanneymeyeriids (Fig. 6.2). Benton considers that at species level the late Carnian event far exceeded the end-Triassic event in magnitude. Many of the tetrapod families disappearing were relatively diverse, averaging four species each, while the six families disappearing at the end of the Triassic had a mean terminal diversity of only 1.3 species each. The post-Carnian diversity reduction is thought by Benton to be almost certainly the result of depauperate faunas, not poor preservation or inadequate collecting.

Benton's views have been challenged by some terrestrial vertebrate researchers, with no extinction event being recognised at the end of the Carnian (Hunt and Lucas 1992; Weems 1992; Lucas 1994). Hunt and Lucas point out that the claim of an extinction event at this time is based on areas where they maintain there is a major facies change at the Carnian–Norian boundary, as in Germany, or a significant unconformity that spans the early Norian, as in Argentina. The Chinle Group of the North American Western Interior exhibits, however, a relatively continuous depositional record across the boundary, with no major facies change; no significant extinction event is recorded. The most significant extinction precedes the end of the Carnian; three families that supposedly went extinct are monospecific, while others survived into the Norian. Within Benton's level of resolution, all the last occurrences of rhynchosaurs are merely late Carnian (Lucas 1994). Although the data are somewhat poorer, the pattern of change in deposits in India is believed to be similar to that exhibited within the Chinle Group.

An examination of the further data provided by Benton (1994) indicates, however, that Hunt and Lucas's arguments fail to address the global pattern of a profound shift from non-dinosaurian to dinosaurian communities, and the disappearance of the key herbivore groups mentioned above. Benton also disputes their claim of a major facies change at the Carnian–Norian boundary in Germany.

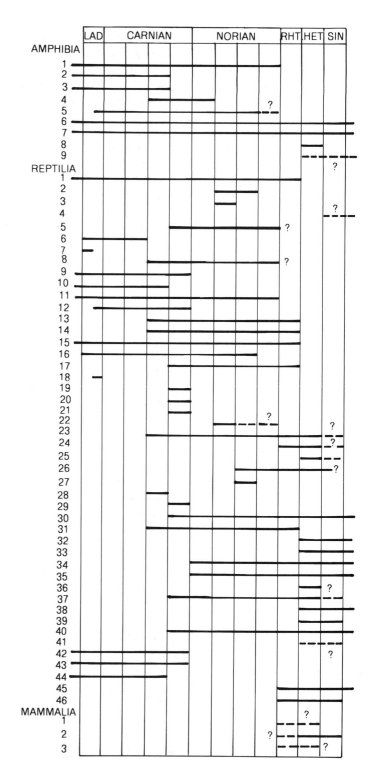

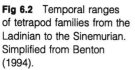

Fig 6.2 Temporal ranges
of tetrapod families from the
Ladinian to the Sinemurian.
Simplified from Benton
(1994).

To summarise, the evidence for Carnian extinctions in either the marine or continental realms is currently somewhat equivocal. The magnitude of any event within the Carnian is uncertain, but there is no reason as yet to consider it a veritable mass extinction. Furthermore, a case for some kind of event of possible global magnitude has only been made for reef organisms, and there is a decided lack of stratigraphic precision. At present all that is suggested is an episode of accelerated biotic turnover in the later part of the stage. The presumed event affecting bivalves and crinoids is evidently only of regional extent, having not been established for anywhere other than the Alps, and once more there is no stratigraphic precision better than between lower Carnian and upper Norian. Since the stratigraphically intermediate deposits include a non-marine unit, the Raibl Formation, there is a strong suggestion that the species extinctions were caused by reduction in habitat area consequent upon marine regression. If one is to accept a late Carnian extinction event among terrestrial vertebrates and plants within Laurasia, it could possibly be bound up with environmental stresses produced by an episode of climatic change involving a wet monsoonal phase in much of Europe and North America (Simms and Ruffell 1990). Such a 'pluvial event' has, however, been disputed by palynologists (Visscher *et al.* 1993).

The end-Triassic event
Marine realm

The marine mass extinction at the end of the Triassic is generally recognised as being one of the five biggest in the Phanerozoic (Hallam 1990a). In their analysis of extinction rate (families/m.y.), Raup and Sepkoski (1982) place it fourth in importance behind the end-Permian, end-Ordovician, and end-Cretaceous mass extinctions. Using a different extinction metric, percentage extinction of genera per million years, the end-Triassic extinction peak (0.06) is higher than that of the end-Cretaceous (0.04) (Raup and Sepkoski 1988) and second only to that of the end-Permian. However, as Benton (1995) points out, the end-Triassic extinction rate appears high because of the short estimated time-span of the youngest Triassic unit, the Rhaetian stage or substage. Judging by the number of ammonite zones, its duration might have been as short as 1–2 m.y. according to Tozer's (1979) scheme, but somewhat longer, perhaps 3–4 m.y., according to the scheme of Dagys and Dagys (1994). According to Sepkoski (1986) there was a 48% extinction of invertebrate genera, largely among cephalopods, bivalves, gastropods, and brachiopods. Evidence is fullest for the two most abundant and diverse macroinvertebrate groups, the ammonites and bivalves.

Cephalopods

The extinction event is expressed most dramatically on a global scale by the ammonites, with no fewer than six superfamilies becoming extinct either at or shortly before the Triassic–Jurassic boundary (Fig. 6.3). After diversity had reached a high level in the Norian, possibly no more than one genus survived the boundary (Tozer 1981). From this sole survivor there followed another dramatic radiation; Jurassic and Cretaceous biostratigraphers therefore owe it an enormous debt. According to Dagys and Dagys (1994), there were important extinctions at the end of the Norian, with the disappearance of several families including the Metasibiritidae, Tibetidae, and Arpatidae.

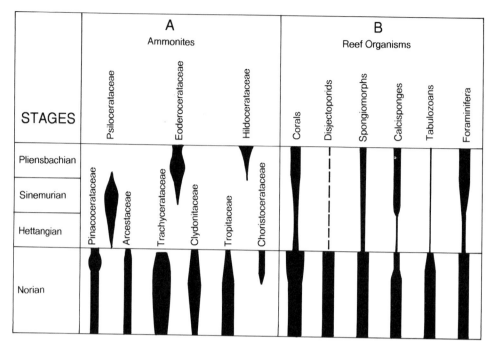

Fig 6.3 End-Triassic (Norian–Hettangian boundary) extinction in important marine groups: A, ammonites; B, reef organisms. After Hallam (1990a).

More limited information is available from another cephalopod group, the Aulaco-cerida, which are belemnite-like coleoids which first appeared in the Devonian and attained their greatest diversity in the Late Triassic, with a geographic distribution largely confined to the Tethys and the American western cordillera. These Late Triassic forms belong to two families, the Aulacoceratidae and Xiphoteuthidae, and nine genera. By the beginning of the Jurassic virtually the only surviving genus, and certainly the only common one, was the xiphoteuthid *Atractites* (Doyle 1990). Unfortunately, there is no precise stratigraphic information available on when the other genera went extinct.

Bivalves

The bivalves do not show any significant change at family level, but nearly half the genera disappeared (Fig. 6.4). The trigonioid, unionoid, and hippuritoid orders were especially strongly affected but most of the pholadomyoid genera survived. Three major ecological categories can be distinguished: a shallow neritic group, to which most taxa belong; a euryhaline group that could tolerate marginal marine and lagoonal conditions; and a deep neritic group, notably *Halobia* and *Monotis*. It should be noted, however, that these two important genera disappeared at the end of the Norian rather the end of the Rhaetian. All groups were strongly affected by the extinction but species of the deep neritic group were apparently the most susceptible. Endemic genera prove to have been much more subject to extinction than cosmopolitans (Hallam 1981a).

The only continent where adequate species-level data are available is Europe, and Hallam (1981a) made a tentative attempt, based partly on the literature and partly on a rather cursory examination of museum specimens, to establish an extinction percentage

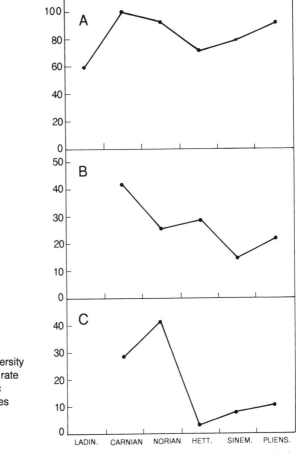

Fig 6.4 Graphical plots of generic diversity (A), origination rate (B), and extinction rate (C) for Late Triassic and Early Jurassic bivalves. Origination and extinction rates plotted as percentages. After Hallam (1981a).

for species that failed to survive the end of the Rhaetian in Great Britain and the Northern Calcareous Alps, where some of the best exposed marine sections occur. Hallam made a provisional estimate of 92% (and 42% genera), suggesting an almost complete turnover. A very similar figure of 88% was estimated for bivalve species in the Southern Alps by Allasinaz (1992). However, he suggested that much of the taxonomic loss can be attributed either to pseudo-extinction or an apparent loss based on chronospecies boundaries rather than the termination of lineages. For the same region McRoberts *et al.* (1995) recognised a significant decline in species richness in the lower Rhaetian, when 51% disappeared, and a more severe taxonomic loss in the middle part of the Rhaetian, with 71% disappearance, including all infaunal and 50% of epifaunal species. The loss of species in the upper Rhaetian appears to relate to a facies change associated with a relative sea-level fall, with the onset of peritidal micrites, and shifting ooid shoals (McRoberts 1994).

A more thorough evaluation than attempted in Hallam (1981a) of generic and species loss at or near the end of the Rhaetian has now been undertaken, both for the north-west European (England, Wales, Germany) marginal marine facies and the Alpine (Austria, Bavaria, Italy) fully marine facies. For the former region 4 out of 27 genera went extinct at or close to the Triassic–Jurassic boundary: *Lyriomyophoria*, *Palaeocardita*, *Palaeopharus*, and *Rhaetavicula*. Of a total of 28 well-demarcated species, 8 definitely went extinct and 16 definitely persisted into the Jurassic. For the Alpine region, of 29 Norian (including Rhaetian) genera, 9 went extinct: *Rhaetavicula, Myophoria, Neoschizodus, Palaeocardita, Schafhaeutlia, Myophoriopsis, Dicerocardium, Neomegalodon* and *Conchodon*. Distinctive species that went extinct (among genera that survived) include *Gervillia inflata* Schaf., *Pinna miliari* Stopp., *Lopha haidingeriana* (Emm.) and *Homomya lariana* (Stopp.). Other species are more dubious, but there are a number of possibilities. Evidently, therefore, the European species turnover at or near the Triassic–Jurassic boundary was less marked than originally inferred by Hallam (1981a).

Reef organisms

In the Norian (including Rhaetian) of the Northern Calcareous Alps there was a substantial diversity increase of scleractinian corals, following the establishment of this group as major reef components in the Carnian, a phenomenon quite possibly bound up with the establishment of a symbiotic relationship with zooxanthellae (Stanley and Swart 1995). At the end of the Triassic, however, there was a dramatic disappearance of scleractinians and associated reef organisms, most notably sponges, which appears to reflect mass extinction on a global scale, though nowhere else in the world is there known a rich Upper Triassic reef fauna (G.D. Stanley 1988; Fig. 6.3). Of 50 scleractinian genera recognised globally in the Late Triassic, only 11 survived into the Liassic (Beauvais 1984).

Other macroinvertebrates

Among the brachiopods and gastropods, taxa dominant in the Palaeozoic that survived the end-Palaeozoic mass extinction finally succumbed at the end of the Triassic. Brachiopods underwent a severe reduction of athyridaceans, spiriferoids and dielasmatids. Thus of 12 Norian–Rhaetian dielasmatid genera, only one survived into the Jurassic. There was some turnover of brachiopod taxa at the Norian–Rhaetian boundary (Dagys and Dagys 1994). Among groups that flourished in the Jurassic, the rhynchonellids and terebratulids, *Austrirhynchia, Rhaetina,* and *Zugmayeria* are important elements of the youngest Triassic faunas of the Alps that became extinct at the end of the period. The murchisoniacean gastropods also disappeared; according to Batten (1973), the gastropod turnover was much more significant than that at the end of the Permian. On the other hand, the main extinction event among echinoids and crinoids had taken place in the Carnian and there is little notable change in the low-diversity faunas across the Triassic–Jurassic boundary (Smith 1990; Simms 1991).

Microorganisms

Conodonts are another important Palaeozoic group that survived the end-Permian but not the end-Triassic crisis. Although diversity had declined since the Early Triassic (Clark 1987) there were, according to Mostler *et al.* (1978), no fewer than ten genera occurring in the upper Norian Kössen Formation of the Northern Calcareous Alps and five genera survived until the *Choristoceras*-bearing *crickmayi* zone. The ostracods also underwent a major mass extinction, with the virtual disappearance of the Palaeocopida and severe reductions in the Cytheracea and Bairdacea (Whatley 1988). No such striking change seems to have taken place among the foraminifera, with Late Triassic species in the Boreal regions showing close affinities with Hettangian assemblages (Jenkins and

Murray 1981) but only in Europe has detailed work been undertaken on these two groups for Triassic–Jurassic boundary beds. El-Shaarawy (1981) made a study of foraminifera and ostracods in both the Northern Calcareous Alps of Austria and in England and Wales. He found that many of the species in the upper Norian (including Rhaetian) Kössen and Zlambach Formations of Austria occur also in the oldest Jurassic (Hettangian) of England and Wales, though not in the latest Triassic marginal marine Penarth Group, having migrated with the marine transgression in that region. Taking into account data from both regions, he estimated an end-Triassic species extinction of 21% for the foraminifera and 23% for the ostracods. Non-septate, involute, aragonitic-walled Involutinina reached a climax in the Norian (including Rhaetian) with the partitioned form of *Triasina*. These reefal forms did not survive the end of the Triassic, though simpler, non-septate *Involutina* and *Trocholina* did (Brasier 1988).

In marked contrast to the end of the Cretaceous, there was no calcareous plankton other than a few species of coccoliths. For the Austrian Alps, the only region for which data are available, the group experienced an almost complete turnover at the Triassic–Jurassic boundary, with only one species surviving (Bown and Lord 1990). Dinoflagellates are another planktonic group, although the fossilisable cysts belong to a non-motile benthic resting stage; they also experienced a marked increase in extinctions at the system boundary (Riding, in Hallam 1995a). On the other hand, there is no evidence of an end-Triassic extinction event among the radiolarians (E. Pessagno, pers. comm.).

Geochemical evidence

In striking contrast to other mass extinction horizons, most notably at the Cretaceous–Tertiary boundary, there is a paucity of geochemical data for Triassic–Jurassic boundary strata, which largely reflects the lack of abundant sections suitable for analysis. Intensive investigations have indeed been confined to two in Europe, St Audries in Somerset, England, and Kendelbach in the Austrian Salzkammergut. Geochemical results in the former section have been strongly influenced by diagenesis but Kendelbach has allowed original environmental signals to be detected both for carbon and oxygen and for strontium isotopes (Hallam and Goodfellow 1990; Hallam 1994a; Morante and Hallam 1996).

The Triassic–Jurassic boundary in the Kendelbach section is taken at the boundary between the Kössen Formation and Grenzmergel (the German term for boundary marl), which is marked by the disappearance of conodonts, appearance of Jurassic bivalves and major turnover in coccoliths and palynomorphs (Hallam 1990b; Fig. 6.5). Figure 6.5 shows that the only perceptible change up the succession is at the level of the Grenzmergel, with a pronounced negative shift in $\delta^{18}O$ and $\delta^{13}C_{carb.}$ and a positive shift in $\delta^{13}C_{org.}$ which is a mirror image of the negative shift in $\delta^{13}C_{carb.}$. These changes are all attributable to diagenesis and hence have no bearing on the original depositional environment (Hallam and Goodfellow 1990; Morante and Hallam 1996). The same is true for the strontium isotope results, which show no significant change across the system boundary apart from that attributable to diagenesis within the Grenzmergel (Hallam 1994a). The only other boundary section that has been subjected to carbon and oxygen isotope analysis is the classic one in New York Canyon, western Nevada. Unpublished analyses by the late Mordecai Magaritz likewise failed to reveal any change up the succession not attributable to diagenesis (W.T. Holser, pers. comm.).

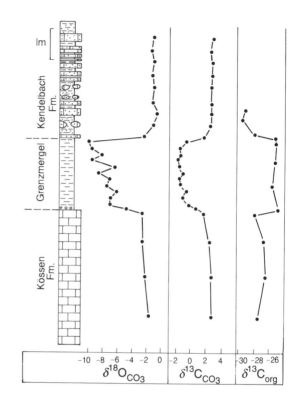

Fig 6.5 Lithostratigraphy of the section across the Triassic–Jurassic boundary at Kendelbach, Austria, with oxygen and carbon isotope results. The T-J boundary is taken at the base of the Grenzmergel. After Morante and Hallam (1996).

Iridium analyses have been undertaken independently for the Kendelbach and St Audries sections by Carl Orth and Frank Asaro. No primary iridium anomaly was discovered by either researcher (Orth *et al.* 1990; Hallam 1990a).

Terrestrial realm

Plants

It has been stated that the terrestrial plant record is less affected by mass extinction events than the marine invertebrate record (Knoll 1984), specifically that of the Triassic–Jurassic boundary by Ash (1986) and Traverse (1988); neither macrofossils nor palynomorphs show much change across the boundary. Close examination of certain classic sections suggests, however, that there might have been a significant extinction event. Many of the changes involved seed ferns, with the families Glossopteridaceae, Peltaspermaceae and Corystospermaceae becoming extinct at or near the boundary. In east Greenland it has been well established for many years that the Rhaetian *Lepidopteris* and Hettangian *Thaumatopteris* macrofloral zones are quite distinct from each other. According to Pedersen and Lund (1980), there is a very low percentage of seven species common to the two zones, whereas the spores and pollen have many species in common. They suggested that some indistinguishable palynomorphs have been produced by different macrofossil plants, which implies that palynologists are likely to underestimate the degree of floral turnover at a given boundary. It is therefore noteworthy that, although the change in pollen and spores through the Late Triassic to Early Jurassic interval in Europe is not especially marked, the biggest 'species' turnover takes place exactly at the Triassic–Jurassic boundary (Visscher and Brugman

1981; Traverse 1988). At Kendelbach this turnover indeed corresponds to the base of the Grenzmergel (Morbey 1975).

An even more pronounced palynofloral turnover at the Triassic–Jurassic boundary has been recognised in North America, as first recorded in the Newark Supergroup of the north-eastern United States (Olsen and Sues 1986; Olsen *et al.* 1990). The careful study of Fowell and Olsen (1993) has provided more details of this change. A diverse assemblage of monosulcates and bisaccates is replaced by *Corollina*-dominated palynofloras. The system boundary falls at about 201 Ma, as determined by argon–argon

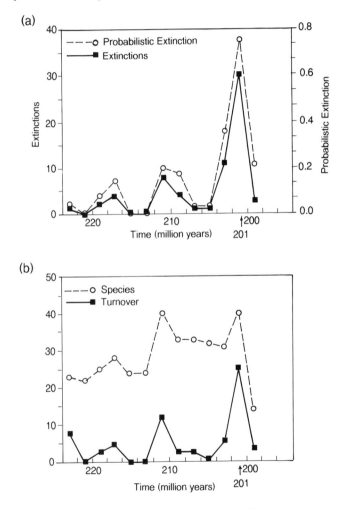

Fig 6.6 Extinction, diversity and turnover of palynomorphs in the Newark Supergroup during successive 2 m.y. intervals. (a) Extinction is measured by two different metrics: the total number of extinctions per interval and the probabilistic extinction rate, defined as the number of extinctions divided by the number of taxa present in the interval. (b) Diversity is defined as the total number of species present in the interval (species richness) and turnover is defined as the total number of originations and extinctions defined by the length of the interval. Arrow points to the T-J boundary. After Fowell and Olsen (1993).

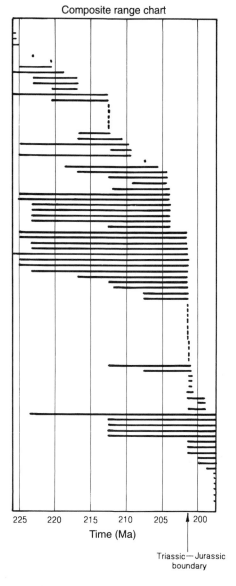

Composite range chart

225   220   215   210   205   200

Time (Ma)

Triassic—Jurassic
boundary

Fig 6.7 Palynospecies ranges in the Newark
Supergroup. Simplified from Fowell and Olsen
(1993).

dating of interbedded basalt; the last appearances of 60% of Late Triassic species are
localised at this horizon (Figs 6.6 and 6.7). The boundary horizon of this turnover, which
is essentially synchronous throughout the depositional area of the Newark Supergroup,
is dominated by fern spores, suggesting a comparison with the well known end-
Cretaceous 'fern spike'. The boundary horizon is evidently unique in the basin, with
no other rocks having such a high spore/pollen ratio.

A pronounced palynofloral turnover is also recorded from the Sverdrup Basin of
Arctic Canada, with the abrupt disappearance of many common Norian–Rhaetian
palynomorphs including *Riccisporites, Limbosporites,* and *Rhaetipollis,* an influx of
Jurassic bisaccates, and the first appearance of typical Lower Jurassic forms such as
Cerebropollenites thiergartii (Embry and Suneby 1994). In other parts of the world no
palynofloral extinction event has been recognised at the system boundary. In Australia
there appears to have been an extinction event within the Hettangian but there is
evidently no change from the Triassic to the Jurassic (Helby *et al.* 1987). Likewise no

change has been recognised in northern Siberia, where there is as good a palynomorph record as in Australia (Sarjeant *et al*. 1992: Table 13.5).

Vertebrates

On the basis of data from both skeletons and footprints, it has been claimed that pronounced turnover took place in the Newark Supergroup at the system boundary, from a Late Triassic fauna dominated by archosauromorph reptiles, labyrinthodont amphibians, and mammal-like reptiles to an Early Jurassic fauna with dinosaurs, crocodilomorphs, mammals, and essentially modern amphibians and small reptiles; at least ten families are claimed to have gone extinct during the most striking event within the whole Carnian–Sinemurian succession of the supergroup (Olsen *et al*. 1987, 1990). This interpretation has been strongly challenged by Benton (1991), Weems (1992), and Lucas (1994), who all deny a mass extinction among terrestrial vertebrates at the end of the Triassic. Both Benton (1994) and Lucas (1994) point out that Norian–Rhaetian tetrapod skeletal remains are almost absent from the Newark deposits. Lucas also remarks that there is no place where Rhaetian and Hettangian tetrapod faunas are clearly superposed. Indeed, the only demonstrably Hettangian faunas in a straightforward stratigraphic succession are in the Fundy Basin of Nova Scotia, but there are virtually no late Norian or Rhaetian tetrapods there.

With regard to ichnotaxa, six disappeared just below the Triassic–Jurassic boundary in Pennsylvania, four continue across and three appear at or just above. This may be the most precise stratigraphic information available. However, the relationship between tetrapod ichnotaxa and body fossils is neither clear nor precise. Most Liassic faunas are Sinemurian or Pliensbachian in age; these are clearly very different from Norian faunas (Lucas 1994). It must be concluded at the least that an end-Triassic mass extinction event among terrestrial vertebrates has not been conclusively demonstrated.

How catastrophic was the end-Triassic event?

The best evidence for catastrophic change comes from the study of microfossils, but unfortunately there has so far been insufficient study of the marine record to allow any categoric general statements to be made. The situation is much better for the terrestrial record because of the meticulously detailed and careful work on the Newark Supergroup by Paul Olsen and his colleagues. If correlation by Milankovitch cyclicity is allowed, the palynofloral extinction event, unanticipated by earlier change, could have taken place in less than 40 000 years (Olsen *et al*. 1990). A more conservative estimate is within 500 000 years (Olsen and Sues 1986).

The best available marine record is in the Northern Calcareous Alps of Austria. Whereas some diminution of molluscan diversity up the late Norian succession is probably attributable to facies change, there is no hint of any temporal diversity reduction of the spectacular reef assemblages until their sudden disappearance at the end of the Triassic (E. Flügel, pers. comm.). Indeed, the loss of mountain-building reef limestones at the end of the Triassic is as dramatic as the much-discussed disappearance of reef ecosystems at the end of the Frasnian, and might have been as sudden. The palynological research of Morbey (1975) on the Kendelbach succession similarly indicates almost complete floral stability through the Rhaetian (Kössen Formation), preceding an episode of marked turnover in the oldest Jurassic strata. The coccoliths tell the same story (Bown and Lord 1990). If the base of the Grenzmergel is taken as the mass extinction horizon, which seems reasonable on the available evidence, the duration of the event need have been no longer than the temporally shorter estimate of Olsen *et al.* (1990) for the Newark Basin.

Turning to other regions, McRoberts *et al.* (1995) record a progressive decline in bivalve species diversity through the Rhaetian succession of the Southern Alps, which is at least partly attributable to facies change. Similarly, Newton (in Hallam 1995a) notes a decline in bivalve diversity up through the Norian succession in a number of displaced terranes of western North America. In a global survey of bivalve genera, Hallam (1981a) concluded that there was some diversity decline through the Norian succession but that there was nevertheless a mass extinction at the end of the Triassic.

To summarise, good evidence of catastrophic change is found only in eastern North America and the Austrian Alps and there is a great need for more intensive investigations at this horizon in other parts of the world.

A discussion of possible causes

As with other mass extinction horizons there are only four serious contenders: climate, volcanism, bolide impact, and sea-level changes with associated anoxia. These will be reviewed in turn.

Climate

Disappearance of the Alpine reef ecosystem at the end of the Triassic has been attributed by Fabricius *et al.* (1970) to a sharp fall of water temperature, but this interpretation receives no support from oxygen isotope analysis of the Kendelbach section (Hallam and Goodfellow 1990). As indicated in Fig. 6.5 and mentioned earlier, there is no evidence of any change apart from that attributable to diagenesis in the Grenzmergel. There is nothing in facies changes across the system boundary in other parts of the world that appears to demand any climatic change, and in the survey by Frakes *et al.* (1992) of climatic modes in the Phanerozoic both the Late Triassic and Early Jurassic are portrayed as belonging to a warm mode, with no change at their boundary.

Volcanism

Since both the Permian–Triassic and Cretaceous–Tertiary boundary mass extinctions correlate closely in time with large-scale eruptions of flood basalt, respectively the Siberian and Deccan Traps, one needs to explore a possible correlation with the most substantial flood basalt eruptions that appear to be roughly associated in time with the Triassic–Jurassic boundary event, the Karoo volcanics of southern Africa. Numerous radiometric dates indicate, however, a Pliensbachian–Toarcian peak (Fitch and Miller 1984), and palynological dating indicates an onset of volcancity no earlier than late Sinemurian (Aldiss *et al.* 1984). Therefore the eruptions took place several million years after the extinctions, clearly ruling them out as a possible causal factor.

However, the initiation of a significant episode of basaltic volcanicity in eastern North America, and probably elsewhere on the North Atlantic margins, can now be precisely dated both palynologically and radiometrically as coinciding almost exactly with the beginning of the Jurassic, but slightly post-dating the Triassic–Jurassic boundary (Olsen *et al.* 1987). Using a correlation based on Milankovitch cyclicity, Olsen *et al.* (1990) suggested that the volcanicity began about 60 000 years after the boundary. It would therefore appear that the North Atlantic volcanic province is also too young to be implicated directly in the extinctions, but there may be an indirect connection through tensional tectonic activity, as discussed in the subsection below on sea-level change.

Bolide impact

Because the large Manicouagan crater in Quebec has been dated as Late Triassic, with an error embracing the Triassic–Jurassic boundary, the claim was made by Olsen *et al.* (1987) that an impact event might have been responsible for the end-Triassic extinctions. The most obvious place to look for supporting evidence such as iridium anomalies and shocked quartz is in the Newark Supergroup of eastern North America, the most proximal well-studied succession, where the Triassic–Jurassic boundary is quite well constrained by palynological data. No such evidence has yet been found, despite prolonged search for shocked quartz by Mark Anders, and iridium analyses by Frank Asaro (pers. comm.). New U-Pb zircon dating of the Manicouagan impact event, apparently more reliable than any previous effort, gives a date of 214 ± 1 Ma (Hodych and Dunning 1992). This is appreciably older than a recent date for the Triassic–Jurassic boundary, using the same method, of 202 ± 1 Ma (Dunning and Hodych 1990). The earlier date falls within the early Norian (Gradstein *et al.* 1994), a time for which no mass extinctions have been claimed.

Other claims for bolide impact have been made on the basis of examination of stratigraphic sections in Europe. Bice *et al.* (1992) recognised what they interpreted as shocked quartz at the Triassic–Jurassic boundary at a section in northern Italy. In fact the purported shocked quartz has been found at no fewer than three horizons, none of which can be firmly established on biostratigraphic evidence as being precisely at the system boundary. The grains in question do not have more than four sets of planar deformation features, and most in fact have only single sets, and the angular distribution of these is rather diffuse. As Bice *et al.* (1992: p. 445) concede, 'these differences (from classic K–T shocked quartz) make it impossible to demonstrate unambiguously that the grains at the T–J boundary have a shock-metamorphic origin'. An alternative hypothesis would be that these grains contain highly unusual Böhm lamellae, presumably produced by normal Earth-bound tectonism (cf. Benton 1994)

It is pertinent to point out that a similar claim for impact-induced shocked quartz in the Kendelbach Grenzmergel was made a few years ago by Badjukov *et al.* (1988), but this claim was refuted by the impact expert Richard Grieve (in Hallam 1990a). Nor has an iridium anomaly been found in the Kendelbach boundary section, as mentioned earlier.

A claim has been made recently, on the basis of a palaeoecological study of latest Triassic bivalves, for a reduction in primary productivity that could have been caused by a bolide impact (McRoberts and Newton 1995). However, neither the carbonate- nor organic-carbon isotope results from Kendelbach give any indication of such a productivity fall (Morante and Hallam 1996; Fig. 6.5). Furthermore, the palaeoecological arguments put forward by McRoberts and Newton are unconvincing. Thus their claim that most low-oxygen bivalve faunas are dominated by infaunal detritus and filter feeders does not correspond with the fact that the vast majority of dysaerobic assemblages consist of epifaunal bivalves (Wignall 1994). Furthermore, they have not demonstrated that 'body size, larval dispersion, and geographic distribution do not account for the selective extinction of infaunal elements, because these characters are similarly distributed between both infaunal and epifaunal life-habit groups' (McRoberts and Newton 1995: p. 103). It is more likely that the taxa which survived the extinction were strongly biased towards those with a widespread distribution, which in turn may tie up with an opportunistic life stategy involving widespread larval dispersal.

Sea-level changes and anoxia

In his analysis of bivalve extinction Hallam (1981a) tentatively concluded that the end-Triassic mass extinction in the marine realm was probably a result of reduction in shelf-sea habitat area produced as a consequence of a eustatic sea-level fall giving rise to widespread regression, rapidly followed by transgression with extensive anoxic or dysoxic bottom waters. This interpretation can now be re-examined in the light of further evidence.

There is indeed good evidence for a geographically extensive and significant regression at the end of the Triassic, with the best evidence coming from Europe and the North American Arctic (Embry 1988; Embry and Suneby 1994). In the northern and southern European stratigraphic successions, where the relevant record is best known, the regression at the end of the Triassic appears to have been relatively abrupt, and was immediately followed by a rapid transgression and marine deepening event at the beginning of the Jurassic. Such a pattern emerges clearly from facies analysis of the Alpine sequence, where the abrupt sea-level fall and subsequent rise is perceived as being one of the most marked changes in the whole Upper Permian to Lower Jurassic succession (Brandner 1984). In the Northern Calcareous Alps the end-Triassic regression led to widespread emergence and karstification of reef complexes (Hallam and Goodfellow 1990; Satterley *et al.* 1994). In the Southern Alps there was an episode of shallowing in the carbonate platform environment at about the system boundary (McRoberts 1994).

Across a large area of northern Europe, from Great Britain to Germany, there are also clear indications of a relatively short-lived, major end-Triassic regression rapidly followed by an earliest Jurassic transgression. The basal Hettangian marine limestones and shales contain benthos-free laminated, organic-rich deposits alternating with beds containing high-density, low-diversity faunas suggestive of opportunistic colonisation in dysaerobic conditions (Hallam 1995b). Such deposits indicate a spread of anoxic or dysoxic bottom waters associated with marine transgression.

Because in the latest Triassic and earliest Jurassic the extent of epicontinental seas was very restricted as a consequence of the relatively low sea-level stand at that time (Fig. 6.8) the number of marine sections crossing the system boundary is very restricted, and the only well-studied examples outside Europe are on the western margins of North and South America. At the top of the Triassic succession in the classic New York Canyon in western Nevada, the Muller Canyon Member is a silty shale that was deposited in anoxic or dysoxic bottom waters, effectively a highly weathered 'black shale' (Hallam 1990a). A closely comparable succession has recently been reported in Sonora, Mexico (González-Leon *et al.* 1996). Similarly, in both northern Chile and Peru, shallow-water Norian limestones rich in molluscs and corals pass up abruptly into black ammonite-bearing laminated shales of Upper Rhaetian age, virtually barren of benthos, a facies that continues in the basal Jurassic (Hillebrandt 1990, 1994). Thus the evidence from the western Americas could be held to signify that there is no evidence of an end-Triassic regression preceding an earliest-Jurassic transgression, rather that the transgression or sea-deepening event began already in the late Rhaetian. In the celebrated Utcubamba Valley sections in Peru, however, Hillebrandt (1994) reports a facies change from north to south, with silty shales indicative of deeper water passing southwards into shallow-water siltstones and limestones. Whereas there is no facies change across the Triassic–Jurassic boundary in the Aramachay and Suta Formations in the North, there is a hiatus in the South, in the area of the Chambara and Chilingote Formations, with the oldest Hettangian resting unconformably on the Rhaetian, and containing reworked fragments of the Rhaetian ammonite *Choristoceras* together with Hettangian *Psiloceras*. Thus a marine deepening event starting in the late Rhaetian has not obscured evidence of a preceding shallowing/emergence event in a shallower-water environmental setting.

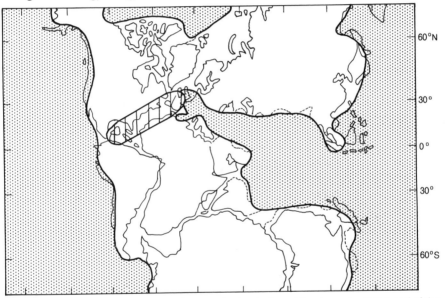

Fig 6.8 The limited extent of epicontinental sea at the beginning of the Jurassic. At the end of the Triassic the extent of fully marine conditions (stippled) was even more restricted, north-west Europe being occupied prior to the end-Triassic regression by a sea of less than normal marine salinity. Note the continental graben zone (vertical lines) in the pre-Atlantic region of Pangaea. After Hallam (1990a).

In a number of regions there is an association between the Triassic–Jurassic boundary regression–transgression couplet and the inception of tensional tectonics. This is most apparent on the North Atlantic margins of eastern North America, north-west Africa and south-west Europe, where it is associated with substantial basaltic volcanicity. It appears to mark a major stage in the rifting that preceded the oceanic opening, which began in the late Middle Jurassic (Manspeizer 1988). In the Austrian Alps, tensional activity at this time is recorded by numerous neptunean dykes of red Jurassic limestone in white Triassic limestone (Bernoulli and Jenkyns 1974). A glance at Fig. 6.8 shows that all these regions are located centrally in Pangaea, and it may not be coincidental that the regression–transgression couplet is most marked here; it could possibly relate to the major tensional event causing rifting and volcanicity in the southern North Atlantic region (Cathles and Hallam 1991).

While a model of sea-level change and associated anoxia therefore seems to provide the best-supported and most plausible explanation of the marine mass extinction, though the effect of a spread of anoxic bottom waters is difficult to disentangle from the immediately preceding regression, such an oceanographic model is of little direct help in explaining the cause of what was evidently an important regional mass extinction among terrestrial plants, manifested over a large area of North America and, to a lesser extent, Europe. Abrupt change of climate offers a possible solution to this problem, but appropriate evidence is so far lacking; there is no indication of such a change from the marine record.

Recovery from the end-Triassic extinction

Only for north-west Europe is there comprehensive information for detailed species-level analysis. As discussed in the next chapter, there is a general pattern for several marine invertebrate groups of a rapid increase through the Hettangian from a very low level after the end-Triassic mass extinction, followed by a slower rate of increase until the Late Pliensbachian (Fig. 7.1). A more general study takes into account data from southern Europe and extra-European localities. The brachiopod pattern is similar to that described from a more limited geographic region, with genera already occurring in the Hettangian and Sinemurian of southern Europe spreading into north-west Europe in the Pliensbachian. Reef ecosystems, which were drastically affected by the end-Triassic mass extinction, did not re-establish themselves until the Pliensbachian, but a small Sinemurian coral reef has been discovered in a displaced terrane in British Columbia. It contains an Upper Triassic holdover species, first described in the Italian Dolomites, that survived the end-Triassic mass extinction (Stanley and Beauvais 1994). This raises the interesting question of how many other displaced terranes provided refugia from the drastic environmental changes affecting Pangaea.

There is a good correlation between the Hettangian to Pliensbachian diversity rise in Europe and rise of sea level, with the very low diversity values of the Early Hettangian being associated with widespread dysoxic and anoxic conditions (Hallam 1996).

7 Minor extinctions of the Jurassic

In their analysis of extinction periodicity, based on family-level data of marine extinctions, Raup and Sepkoski (1984, 1986; Sepkoski and Raup 1986) claimed to have recognised two events in the Jurassic, at about the end of the Pliensbachian and Tithonian stages; Sepkoski's (1992a) update confirms these events. While of lesser significance than the 'big five' Phanerozoic extinction events, they provide key data for the periodicity hypothesis and accordingly must be considered in some detail.

Numerous extinction events can be discerned from the record of Jurassic ammonites, because they have a high rate of turnover compared with all other fossil groups, and thereby provide the basis for biostratigraphy. Ammonite species and genera define zones and stages often in a very clear-cut way, which involves complete extinction of the older taxa. Because of provinciality, however, many if not most such extinctions were only of regional extent, and ammonite extinctions were for the most part not reflected in extinctions of other groups. Therefore, if the ammonite extinctions were the result of environmental perturbations, such perturbations could not have been of great significance for the marine biota as a whole. To qualify as a mass extinction, other groups besides ammonites should have disappeared from the stratigraphic record.

The Pliensbachian–Toarcian event

What Raup and Sepkoski interpreted as a Pliensbachian event was reassessed as probably taking place early in the next stage, the Toarcian, based on the recognition of a clear-cut extinction even in western Europe (Hallam 1986, 1987a). This event, first recognised by Hallam (1961), led to an almost complete species turnover among the benthos, but was thought to have been only regional in extent, being confined essentially to Europe. More recent work suggests, however, that it was of global significance.

The early Toarcian event in western Europe

An analysis of species diversity throughout the Early Jurassic for five important groups – the bivalves, rhynchonellid brachiopods, crinoids, foraminifera, and ostracods – and generic diversity for the taxonomically more finely split ammonites, reveals how a trend of gradually rising diversity through the Hettangian, Sinemurian, and Pliensbachian was sharply reversed in the early Toarcian (Fig. 7.1). This change is clearest from the most species-rich group, the bivalves, with an estimated 84% of species going extinct (Hallam 1986); breakdown of the bivalve species into ecological groups fails to reveal any selectivity in the extinctions. The rhynchonellids show an equally striking change. More generally, among the brachiopods the last of the spiriferids disappeared. Although it is not clear from the diversity analysis, the ostracods underwent a profound extinction event, one of the most important in the whole post-Palaeozoic history of the group, with the loss of the suborder Metacopina (Whatley 1988). Among other groups not shown in Fig. 7.1, both belemnites (Doyle, 1990–92) and dinoflagellates (Riding, in Hallam 1995a) underwent marked extinctions. The ammonites underwent taxonomic turnover, but evidently not more marked than at many other Jurassic horizons. All the groups mentioned so far are benthic or nektobenthic. This is true also for dinoflagellate resting cysts, although the group is during active life a planktonic one (Wille 1982). Other planktonic groups, namely calcareous nanofossils and certain crinoids, experienced no extinction at this time (Hamilton 1982; Simms 1986).

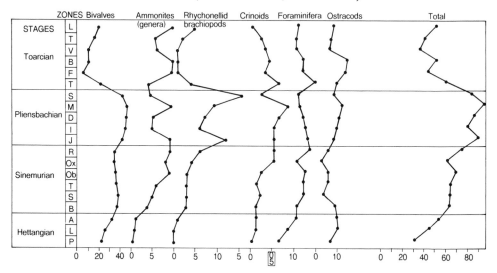

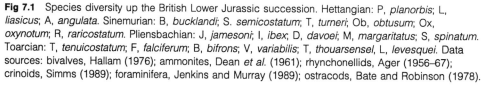

Fig 7.1 Species diversity up the British Lower Jurassic succession. Hettangian: P, *planorbis*; L, *liasicus*; A, *angulata*. Sinemurian: B, *bucklandi*; S. *semicostatum*; T, *turneri*; Ob, *obtusum*; Ox, *oxynotum*; R, *raricostatum*. Pliensbachian: J, *jamesoni*; I, *ibex*; D, *davoei*; M, *margaritatus*; S, *spinatum*. Toarcian: T, *tenuicostatum*; F, *falciferum*; B, *bifrons*; V, *variabilis*; T, *thouarsensel*, L, *levesquei*. Data sources: bivalves, Hallam (1976); ammonites, Dean *et al.* (1961); rhynchonellids, Ager (1956–67); crinoids, Simms (1989); foraminifera, Jenkins and Murray (1989); ostracods, Bate and Robinson (1978).

The extinction event corresponds closely to the widespread occurrence of a unit of laminated organic-rich shales, variously known as Jet Rock (Yorkshire, England), Schistes Cartons (northern France) and Posidonienschiefer (Germany), corresponding to a time of rapid sea-level rise, which is located stratigraphically within the *falciferum* Zone. Although the topmost Pliensbachian *spinatum* Zone has a more diverse fauna than the earliest Toarcian *tenuicostatum* Zone, this appears only to reflect shallower-water facies, and the extinction event can be located precisely at the *tenuicostatum–falciferum* Zone boundary (Hallam 1987a).

By far the best stratigraphic section for studying taxonomic turnover in detail, because of a combination of good exposure, stratigraphic completeness, and fossil-rich facies, is on the coast of northern Yorkshire. Little has undertaken a very thorough study of this section (Little and Benton 1995; Little 1996). He confirms the extinction level indicated above, and records only three epifaunal bivalve species as having survived (Fig. 7.2). Relatively high extinction rates in the *falciferum* Zone and early *bifrons* Zone are attributed to a pattern of pseudoextinctions in rapidly evolving immigrant Tethyan ammonite and belemnite families, rather than true species extinction. The event cannot be recognised at genus level. Only two (amaltheid ammonites and eotomarid gastropods) out of 40 families (5%) had their terminal taxa in the Yorkshire Basin and neither became extinct at the end of the *tenuicostatum* Zone.

The global turnover

There are few other stratigraphic sections outside Europe which are good enough for the kind of detailed analysis undertaken by Little. One such is the Andean Basin of Chile and Argentina. A preliminary survey by Hallam (1986) failed to recognise either a clear-cut extinction event or horizon of laminated organic-rich shales in the Early Toarcian. More recently, however, Aberhan's comprehensive study suggests that such an event is indeed recognisable (Aberhan and Fürsich 1996). Evidently marine bivalves show a marked decrease in diversity across the Pliensbachian–Toarcian boundary, which is

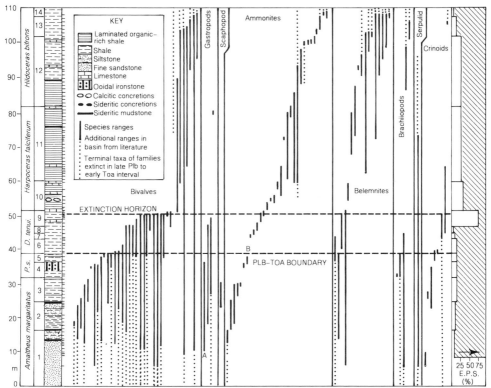

Fig 7.2 Invertebrate macrofaunal species ranges in northern Yorkshire, England, for the upper Pliensbachian to middle Toarcian. Values in extinctions per subzone (EPS) column represent number of species disappearing as percentage of total number of species within each subzone. Zonal abbreviations: P.s. = *Pleuroceras spinatum*; D. tenui = *Dactylioceras tenuicostatum*. Numbers 1 to 14 refer to ammonite subzones. After Benton and Little (1995).

largely due to the extinction of endemics and of some cosmopolitan species, less commonly to pseudo-extinctions and local disappearances of taxa. As in western Europe, the mass extinction correlates with sea-level highstand and widespread oxygen-poor deep-shelf environment, suggesting a causative relationship.

Based on data compiled in the second edition of *The Fossil Record* (Benton 1993), Little and Benton (1995) have produced stage- and zonal-level plots for family extinctions in the Early Jurassic. The stage data are 4 (Hettangian), 10 (Sinemurian), 19 (Pliensbachian), and 22 (Toarcian), suggesting an extended period of extinctions through the latter two stages. The corresponding figures of Sepkoski and Raup (1986) are 3, 11, 17, and 7, and those of Sepkoski (1992a) 3, 11, 22, and 11. The latter two sets of data are obviously the grounds for recognition of the Pliensbachian extinction event in the periodicity analysis, but Little and Benton's inference of an extended Pliensbachian–Toarcian event is confirmed by their zonal-level analysis, which indicates that 33 of the 49 Early Jurassic families died out through a mere five ammonite zones, embracing the Pliensbachian–Toarcian boundary, namely the *margaritatus, spinatum, tenuicostatum, falciferum,* and *bifrons.* There must have been a significant global contribution to this, because only 18 of the 33 families had terminal taxa restricted to western Europe. An event on a global scale is confirmed from research on pyrite-rich beds in deep-sea bedded chert successions in Japan, with a planktonic group, the radiolarians, showing a drastic turnover in the Pliensbachian and early Toarcian (Hori 1993). An early Toarcian extinction event affecting the radiolarians is also recognisable in western North America. Whereas the proportion of species going extinct in the Sinemurian is 23% (4 out of 17) and in the upper Pliensbachian 30% (31 out of 104) in the lower Toarcian it rises sharply to 72% (83 out of 116) (Smith *et al.* 1994). Furthermore, the first larger lituolacean foraminifera of the Tethyan region, *Orbitopsella* and *Cyclorbitopsella,* appeared in the Pliensbachian and disappeared directly thereafter (Brasier 1988).

In conjunction with events in epicontinental seas elsewhere in the world, in Europe and South America, all this suggests a major oceanic event affecting both shallow and deep water. Jenkyns (1988) indeed recognises an oceanic anoxic event in the Early Toarcian, based on the widespread occurrence of organic-rich shales, locally with manganiferous carbonates, in a number of countries surrounding the Mediterranean and elsewhere in the world, and a pronounced positive carbon isotope excursion (Fig. 7.3); see also Jiménez *et al.* (1996). Jenkyns's work confirms the inference that has been made by other cited researchers that the causative agent for the mass extinction was a drastic reduction in habitat area produced by the rapid spread of anoxic bottom waters associated with a sea-level rise. Little and Benton's data show, however, that at family level the extinction event could not be described as catastrophic because it was extended over parts of two successive stages. At this stage the cause of this more temporally extended extinction phase, as opposed to the *falciferum* Zone event, remains obscure.

Recovery from the early Toarcian extinction event in Europe

Shortly before the early Toarcian extinction event, the important Boreal ammonite family Amaltheidae went extinct at the end of the Pliensbachian. Immediately following this north-west Europe was invaded by Tethyan immigrants belonging to the Dactylioceratidae and Hildoceratidae (Hallam 1990c). Correspondingly, the ammonite turnover across the *tenuicostatum–falciferum* Zone boundary was less striking than that

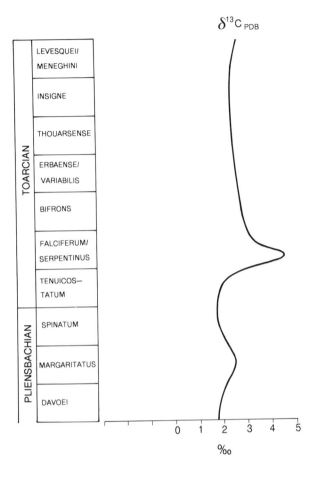

Fig 7.3 Carbon isotope stratigraphy across the Pliensbachian–Toarcian boundary, based on four sections in Italy, Hungary, and Switzerland. Simplified from Jenkyns (1988).

between the *spinatum* and *tenuicostatum* Zone. The bivalve species extinction in the early Toarcian led to a marked fall in diversity, and the Pliensbachian level in Europe was not restored until the beginning of the Middle Jurassic, the Aalenian (Hallam 1976). This renewed rise in diversity was largely a consequence of immigration from elsewhere in the world, presumably filling vacated ecological niches, and evidently Andean South America, or perhaps Pacific borderlands in general, were major sources. Thus a number of important bivalve genera which did not appear in Europe until the late Toarcian or Middle Jurassic, occurred already as early as the Pliensbachian in South America; they include *Cucullaea*, *Lycettia*, *Falcimytilus*, *Trichites*, *Pteroperma*, *Gervillaria*, *Lopha*, *Myophorella*, and *Mesomiltha*. The migration was probably through the so-called Hispanic Corridor across the central part of what is now the Atlantic (Hallam 1996). It is apparent from this fact that the late Pliensbachian–early Toarcian extinction event recorded in South America by Aberhan and Fürsich (1996) was less marked than the Early Toarcian event in Europe.

With regard to the microfauna, agglutinated and nodosariid foraminifera reappeared directly after the anoxic event. There was a marked change in the character of younger nodosariid assemblages, with the place of uniserial *Nodosaria*, *Frondicularia*, and *Lingulina* being largely taken up by coiled *Lenticulina* (Brasier 1988).

The Tithonian event

As throughout the Jurassic, the most abundant and diverse macroinvertebrates in the marine fauna are ammonites and bivalves, and these provide the best available data to study in more detail the extinction event claimed by Raup and Sepkoski. Indeed, if this claim is to be sustained, it should be clearly expressed among the Mollusca. Figure 7.4, based on Sepkoski's (1992a) compilation of Phanerozoic families, shows that this is indeed the case, with 19 families going extinct in the Tithonian, by far the highest number in any of the eight stages portrayed. More generally, Sepkoski and Raup (1986) record a 6.5% fall in marine family diversity, but Benton's (1995) revised figure, based on data in *The Fossil Record*, is slightly less, 5.1–6.1%.

The Raup and Sepkoski cyclicity analysis assumes that family extinctions were concentrated at the end of each stage. Only for the ammonites is there sufficiently precise stratigraphic information to test this assumption. Sepkoski (1992a) records that 7 out of a total of 11 ammonite families became extinct in the Tithonian. However, the

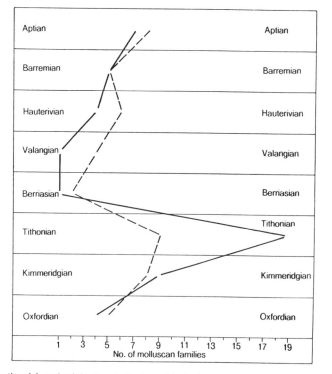

Fig 7.4 Extinction (–) and origination (—) rates of Late Jurassic and Early Cretaceous molluscan families in the world, expressed as times of last and first appearance, respectively. After Hallam (1986).

literature recording more precise stratigraphic information reveals that only three of the seven, the Aspidoceratidae, Ataxioceratidae, and Dorsoplanitidae, persisted to the end of the stage, and that among these only a minority of genera persisted. The duration of the Tithonian is poorly known; the most up-to-date and probably the best time-scale indicates 6 m.y. (Gradstein *et al.* 1994). Thus what may appear superficially to be an important extinction event (7 families out of 11) turns out on closer analysis to be of no great significance compared with other extinction events in ammonite history, and certainly not catastrophic. Indeed, the turnover of taxa at the Tithonian–Berriasian boundary is less marked than that at the Berriasian–Valanginian boundary in both the best-studied regions, Europe and South America (Hallam 1986).

In a review of European bivalve genera, Hallam (1976) established that out of a total of 82 in existence during the Tithonian 22 did not persist into the Cretaceous. Following Kelly's thesis and monograph (1977, 1984–92), this number must be reduced to 18, because he records *Falcimytilus*, *Stegoconcha*, *Liostrea*, and *Nanogyra* as occurring in the Lower Cretaceous. Thus only 22% of the bivalve genera went extinct in the Tithonian. At the species level the extinction was more striking, at least locally. With regard to the fomer Soviet Union, Zakharov and Yanine (1975) point out that the bivalve diversity turnover was greater in shallower-water than in deeper-water facies. In northern Siberia the deepest-water facies is characterised by *Buchia* and thin-shelled *Aequipecten*.

Generally, the Jurassic–Cretaceous turnover is only notable at the species level. In the Crimea and Caucasus there is a succession of shallow-water carbonates rich in bivalves, especially trigoniids and rudists, and a deeper-water facies with lower bivalve diversity, dominated by inoceramids, nuculids, and small, thin-shelled pectinids. A strong turnover is recognised at the Tithonian–Berriasian boundary, with 102 species going extinct and only 5 continuing into the Cretaceous. In contrast, only 2 genera among 61 went extinct at this time. In northern Siberia there is no marked extinction of species at the Volgian–Ryazanian boundary, which is slightly younger than the Tithonian–Berriasian boundary, and the turnover is lower than at the Ryazanian–Valanginian boundary.

An equally detailed study of bivalve species is Jurassic–Cretaceous boundary beds is that of Kelly (1977, 1984–92) who researched on deposits in eastern England. Using Zakharov and Yanine's renewal index (the total number of species becoming extinct plus the total number of new species appearing at a particular horizon) to establish the rate of turnover, Kelly determined that the greatest change was at the junction between the *giganteus* Zone and *oppressus* Zone within the middle Volgian, which is equivalent to the upper Tithonian. For the Boreal realm in general, of 63 species 13 did not persist beyond the middle Volgian, 28 crossed from the middle to the upper Volgian (approximately equivalent to crossing the Tithonian–Berriasian boundary) and 22 entered the succession in the upper Volgian–Ryazanian.

The extinction event, which is only striking in the Crimea and Caucasus, correlates with evidence of a major regression of shallow epicontinental seas which can be traced all the way across Europe. Thus in southern England the shallow-marine Portland beds pass up into the marginal to non-marine Purbeck beds.

In the Andes of central Chile and Argentina, palaeogeographic relationships were very different from those of Europe, with a thick unit of shallow-marine Tithonian to Hauterivian sediments, sandwiched between non-marine volcaniclastic units, testifying to a widespread Tithonian transgression and Barremian–Aptian regression; there is no notable facies change or hiatus across the Jurassic–Cretaceous boundary. Of 38 species of bivalve distinguished in the marine sequence of the Chilean high cordillera, nearly all occur in the Tithonian and none disappear at or near the Tithonian–Berriasian boundary. In fact the bivalve fauna shows no change up the succession that cannot be ascribed to facies differences. Species of such characteristic 'Cretaceous' genera as *Exogyra, Pterotrigonia, Steinmanella, Eriphyla, Ptychomya,* and *Panopaea* are already present in the Tithonian, a situation quite unlike that in Europe. Thus, the bivalve extinction event cannot be recognised in South America, which also lacks evidence of a corresponding regression at the appropriate time (Hallam 1986). Similarly, no significant marine bio-event is recognisable in Japan; many bivalve species survived from the Tithonian into the earliest Cretaceous (Hayami 1989).

Evidence from other invertebrate groups, such as foraminifera and coccoliths, fails to indicate a significant Tithonian extinction event (Hallam 1995a) although Sandy (1988) recognises such an event among brachiopod species in Europe, and Whatley (1988) records an ostracod extinction peak in the 'Purbeckian' – that is, latest Jurassic to earliest Cretaceous – in the same continent, while Ascoli *et al.* (1984) document a 30–35% ostracod species extinction at the Jurassic–Cretaceous boundary on the Scotia Shelf of North America, but the stratigraphic precision is in fact no greater than in Europe, which was an adjacent region at the time and subject to the same regional regression.

To summarise, there is no evidence of a global mass extinction among the marine invertebrate fauna at or near to the Tithonian–Berriasian boundary, but there is evidence, in Europe, of a regional event significant only at species level, among bivalves and a few other groups. As regards marine vertebrates, Bakker (1993) has claimed a near-total extinction of long-necked plesiosaurs (muraenosaurs and cryptocleidids) at the Jurassic–Cretaceous boundary, but the quality of the data is such that this conclusion can be considered no more than tentative, at least as regards stratigraphic precision and global distribution.

In his recent analysis of Phanerozoic diversification and extinction, Benton (1995) recognised a possible Tithonian extinction peak among continental faunas, leading to a fall of familial diversity of between 5.8% and 17.6%, an appreciably greater maximum figure than for marine families (6.1%). On the other hand, there is no evidence of a dinosaur extinction event within the Morrison Formation in the North American Western Interior, which embraces both Upper Jurassic and Lower Cretaceous strata (Hallam 1995a). Furthermore, no end-Jurassic extinction event is recognised in the non-marine megafloras of western Canada (Upchurch and Wolfe 1993). As Benton acknowledges, the Tithonian peak is probably an artefact of the fossil record, produced by exceptional fossil deposits (Lagerstätten), in this case the Kimmeridgian Solnhofen Limestone of Bavaria, being succeeded by a much poorer part of the record.

Other possible extinction events

According to the Raup and Sepkoski periodicity scheme, there should be an extinction event mid-way between those of the Pliensbachian and Tithonian, that is in the Callovian. There is no evidence from the marine invertebrate record of any such event, or anywhere else in the Middle Jurassic (Hallam 1995a). It is worth recording here, however, that the occurrence of either an iridium anomaly or increased concentrations of cosmic dust at certain horizons in the European Jurassic has led to the speculation that bolide impacts could have been the cause. Thus a significant iridium enrichment was found in a ferruginous crust at or near the Lower–Middle Jurassic boundary in the Southern Alps (Rocchia *et al.* 1986) and a concentration of iron–nickel spherules of cosmic origin at the Middle–Upper Jurassic boundary in southern Poland (Brochwicz-Lewinski *et al.* 1984). In neither case do these mark mass extinction horizons, and the same is true for a phosphatic limestone layer at the Jurassic–Cretaceous boundary in northern Siberia, where an anomalously high concentration of iridium has also been found (Zakharov *et al.* 1992). The sedimentary facies in all cases suggests condensation and the alternative explanation considered by the first two sets of authors, and strongly promoted by Zakharov and his colleagues, is likely to be the correct one. Strongly reduced sedimentation rate, as in the deep ocean today, leads to a concentration of cosmic background fallout, including platinum-group metal-bearing minerals. In particular, the topmost Middle Jurassic (Upper Callovian) is a horizon of widespread condensation in Europe, the consequence of a rapid rise of sea level (Norris and Hallam 1995).

8 Minor mass extinctions of the marine Cretaceous

Understanding ancient oceanic conditions becomes fundamentally easier in the Cretaceous due to the plentiful oceanic crust of this age. Coincidentally, the proliferation and radiation of planktonic foraminifera in the Early Cretaceous also allow the direct study of fluctuations in oceanic surface waters for the first time. With this burgeoning of the data base, all post-Jurassic extinction mechanisms have to be in accord with both open-ocean and shelf-sea evidence.

Thanks to the multinational efforts of the Deep Sea Drilling Project and, more lately, the Ocean Drilling Program, much is now known about Cretaceous abyssal and bathyal sediments. One of the earliest and most fascinating discoveries of the Deep Sea Drilling Project were several levels of the Cretaceous characterised by widespread organic-rich deposition (Schlanger and Jenkyns 1976; Arthur and Schlanger 1979). These were considered to be evidence of oceanic anoxic events (OAEs) of which three have been identified: OAE1, a prolonged Apto-Albian event; OAE2, a much shorter event at the Cenomanian–Turonian stage boundary; and OAE3, during the Coniacian–Santonian interval. As shown below, only the second of these has been convincingly linked with a mass extinction.

Aptian mass extinction: a missing event

Raup and Sepkoski's extinction curves have consistently predicted a mass extinction in the Early Cretaceous. This is placed in the mid-Aptian although, using the newer time-scale of Gradstein *et al.* (1994), it could equally well be predicted to occur in the Barremian (Fig. 1.5). The Early Cretaceous is in fact a time of relatively low extinction rates (Benton 1995), and the only tentative evidence for an extinction event occurs in the Aptian.

The Cretaceous is characterised by the rise of the rudists, and their domination of reef niches at the expense of scleractinians. The displacement began at the start of the Aptian and accelerated in the mid-Aptian (Scott 1995). However, the rudists were temporarily halted by a crisis in the late Aptian in which several genera were lost in several regions of the tropics (Ross and Skelton 1993). Recumbent forms in particular disappeared, and this morphotype did not reappear again until the Albian. However, this event was not enough to influence or halt the continual Apto-Albian diversity increase noted in Jones and Nichol's (1986) global generic data set for rudists and it does not figure in the intra-Cretaceous reef crises of Johnson *et al.* (1996). Calcareous algae and benthic foraminifers also apparently suffered a late Aptian crisis, at least in the carbonate facies of southern Europe (Masse 1989).

These modest losses of a few tropical taxa are not matched by any extinctions reported from elsewhere in the marine realm. This is not to say that this was a quiet time in life's history for the composition of planktonic populations underwent a rapid change in the Aptian. Open ocean-dwelling planktonic foraminifera initially became established in the Barremian when they consisted of simple, surface-dwelling morphotypes. Gradually, through the Cretaceous, morphotypes characteristic of progressively deeper-dwelling forms appeared (Hart 1980; Caron and Homewood 1983; Leckie 1989). The first minor planktonic foraminifera crisis occurred around the Apto-Albian boundary and involved the loss of a few radially elongated planispiral species (Leckie 1989). A second, equally minor event in the late Albian removed a few species of *Ticinella* (Caron and Homewood 1983). Neither event is comparable in magnitude to the end-Cenomanian crisis and especially not to the end-Cretaceous catastrophe.

Despite the lack of evidence for mass extinction, there is an embarrassing number of potential extinction-inducing events in the Apto-Albian interval. Higher-resolution study of OAE1 has revealed that it can be resolved into three distinct, short-lived events in the early Aptian, early Albian and early late Albian (Bralower *et al.* 1994). Only the early Aptian event appears to have been globally widespread, while the other two are best considered as intervals in which organic-rich deposition was widespread without being ocean-wide. None of these events is recorded as a positive spike in pelagic $\delta^{13}C$ carbonate curves, although the voluminous eruptions of the Ontong-Java Plateau on to the Pacific ocean floor at this time may have contributed light CO_2 to the oceans and thus 'masked' any organic carbon burial event (Bralower *et al.* 1994). Whatever the reason, none of the subevents of OAE1 caused an extinction. Bralower *et al.* (1994) have argued that because deep-dwelling planktonic foraminifers did not evolve until the Albian they were not available to become victims of low oxygen levels in the lower levels of the photic zone – a situation that was to change fundamentally in the Late Cretaceous. Although Apto-Albian black shales are common in many epicontinental areas (see, for example, Tyson and Funnell 1987), there is no single interval when the entire shelf area became anoxic; there were always habitable shelf areas. This presumably explains why shelf benthos passed through OAE1 unscathed.

Rather than anoxic events, the demise of certain rudists appears related to climatic changes. Humidity increase in the late Aptian has been proffered to explain the extensive progradation of clastics at this time (Weissert 1989), with the consequent destruction of carbonate shelves – the preferred rudist habitat. However, like the comparable Carnian episode of the Triassic (p. 147), this event was essentially confined to Europe and cannot be considered indicative of a global crisis.

The late Cenomanian extinction event

Claims for an Aptian mass extinction are much exaggerated, but the late Cenomanian event appears to be based on more substantial evidence. Although only a modest event, with 7% of marine families and 26% of marine genera eliminated (Harries 1993), it is among the best studied of any mass extinction. This is because of a combination of factors including the presence of numerous excellent and easily accessible sections in western Europe and the North American Western Interior and the well-established, high-resolution biostratigraphy.

Global correlation

The Upper Cretaceous is uniquely blessed with the joint occurrence of two of the premier biostratigraphic groups, ammonites and planktonic foraminifera. Larger benthic foraminifera and several bivalve groups, notably rudists and inoceramids, also have biostratigraphic utility in the Late Cretaceous. Ammonite zones provide the finest-scale resolution (Kennedy 1984 and Fig. 8.1), but a degree of provinciality in the Cenomanian somewhat hampers interregional correlation. Planktonic foraminifera are less affected by provinciality, although the Western Interior seaway had its own indigenous taxa (Leckie *et al.* 1991). Around the Cenomanian–Turonian (C–T) boundary the most distinctive faunal change is a turnover of inoceramid populations, with *Inoceramus* spp. being replaced by *Mytiloides* spp. (Fig. 8.2). This apparently isochronous event is widely used as the *de facto* C–T boundary (Fig. 8.1).

Significant perturbations in both the carbon and oxygen isotope curves have also been widely used in both corroboration and definition of C–T stratigraphy. The $\delta^{13}C_{carbonate}$ curve undergoes a positive swing from 2.5‰ to 4.5‰ at the top of the *Rotalipora cushmani* Zone and returns rapidly to its original value at the base of the Turonian (see, for example, Scholle and Arthur 1980; Hilbrecht and Hoefs 1986; Jarvis *et al.* 1988; Gale *et al.* 1993). Substantial changes of oxygen isotope values also occur around the same time, with a prolonged Cenomanian negative swing of $\delta^{18}O_{carbonate}$ values reaching a

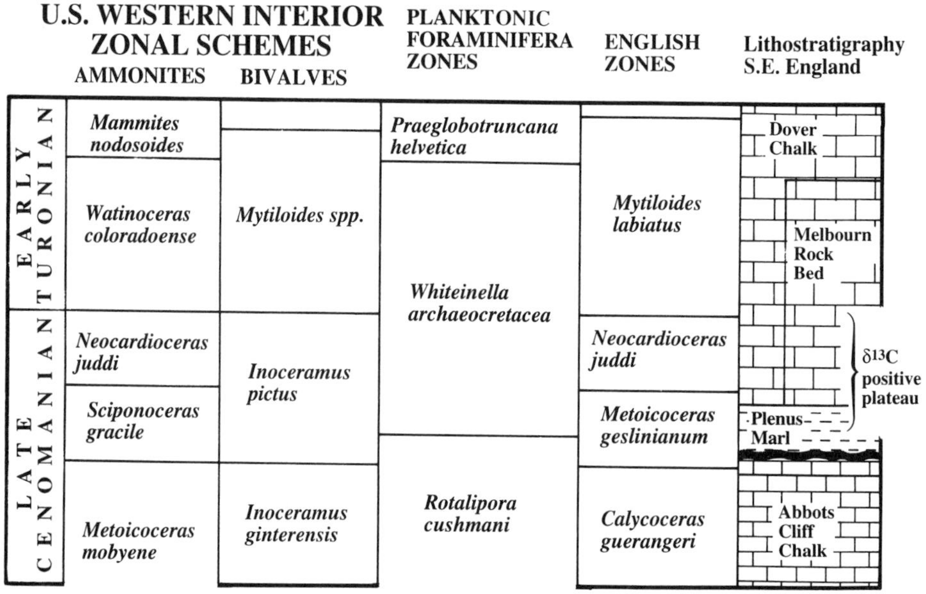

U.S. WESTERN INTERIOR ZONAL SCHEMES		PLANKTONIC FORAMINIFERA ZONES	ENGLISH ZONES	Lithostratigraphy S.E. England
AMMONITES	BIVALVES			
EARLY TURONIAN — *Mammites nodosoides*		*Praeglobotruncana helvetica*		Dover Chalk
EARLY TURONIAN — *Watinoceras coloradoense*	*Mytiloides* spp.		*Mytiloides labiatus*	Melbourn Rock Bed
		Whiteinella archaeocretacea		
LATE CENOMANIAN — *Neocardioceras juddi*	*Inoceramus pictus*		*Neocardioceras juddi*	$\delta^{13}C$ positive plateau
LATE CENOMANIAN — *Sciponoceras gracile*			*Metoicoceras geslinianum*	Plenus Marl
LATE CENOMANIAN — *Metoicoceras mobyene*	*Inoceramus ginterensis*	*Rotalipora cushmani*	*Calycoceras guerangeri*	Abbots Cliff Chalk

Fig 8.1 Correlation chart for the Cenomanian–Turonian interval comparing ammonite biostratigraphic schemes of the North American Western Interior and Europe with planktonic foraminifera and inoceramid bivalve zonations.

Inoceramus pictus
LATE CENOMANIAN

Mytiloides labiatus
EARLY TURONIAN

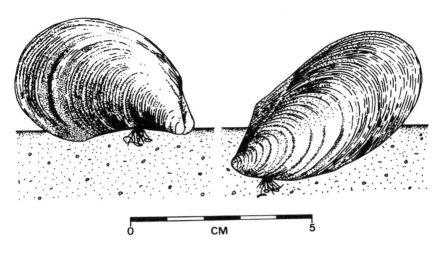

0   CM   5

Fig 8.2   Characteristic inoceramid bivalves of the late Cenomanian and early Turonian.

plateau around the C–T boundary before swinging back to heavier values in the remainder of the Late Cretaceous. However, diagenetic effects readily alter $\delta^{18}O_{carbonate}$ values, rendering them rather unreliable stratigraphic indicators (see, for example, Jenkyns *et al.* 1994).

Bralower (1988) and Jeans *et al.* (1991) have also argued that the $\delta^{13}C$ positive spike is non-synchronous from basin to basin and that it is therefore of little use for high-resolution correlations. However, the non-synchrony may stem from the miscorrelation of the multiple positive peaks within the plateau of generally high values (Gale *et al.* 1993; Jenkyns *et al.* 1994; Paul and Mitchell 1994). Thus, in condensed or incomplete sections, the $\delta^{13}C$ values may only reveal a single positive excursion, while only in the more expanded, complete sections is the true complexity of the carbon isotope curve revealed. In summary, the positive $\delta^{13}C$ excursion is a good indicator of late Cenomanian age but more detailed correlation is best addressed using fossils.

Extinction and recovery

In northern Europe the Late Cretaceous record is characterised by the extensive deposition of coccolith-rich micrites, the Chalk. This prolonged interval of carbonate deposition spanned the 34 m.y. from the Cenomanian to the Maastrichtian. The only significant interruption occurred during a short-lived phase of marly and sometimes organic-rich deposition around the C–T boundary (see, for example, Peryt *et al.* 1994). One of the best-studied but relatively condensed boundary sections occurs at Shakespeare Cliff, near Dover in south-east England (Jefferies 1962; Jarvis *et al.* 1988). Here the mid–late Cenomanian Abbots Cliff Chalk Formation is capped by a burrowed omission surface and overlain by the Plenus Marls, a thin formation ranging from 1.5 to 5 m thick, which consists of alternating marls and limestones: these have been traditionally divided into eight widely traceable beds (Jefferies 1962). The topmost marl is, in turn, overlain by nodular chalk of the Dover Chalk Formation. The C–T boundary is considered to occur in the basal few metres of the Dover Chalk where the typical Cenomanian species *Inoceramus pictus* disappears and several species of *Mytiloides* become common (Jarvis *et al.* 1988). However, the majority of extinctions occur beneath this level within the Plenus Marl (Fig. 8.3).

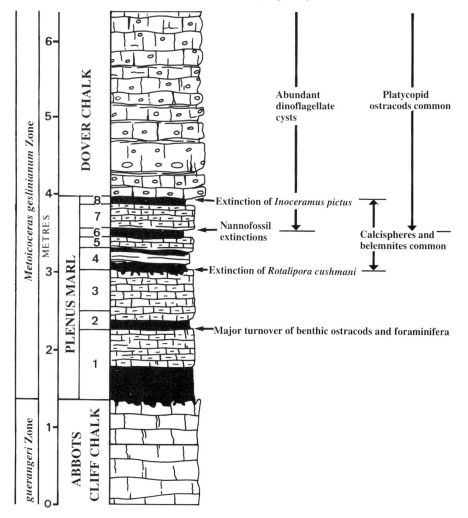

Fig 8.3 Summary lithostratigraphic column for the Shakespeare Cliff section, Dover, south-east England, showing the principal extinction steps (adapted from Jarvis *et al.* 1988). Bed numbers after Jefferies (1962).

The fauna of Bed 1 at the base of the Plenus Marl consists of a diverse assemblage of typical Cenomanian taxa. The first phase of faunal turnover among the macrofauna occurs at the Bed 1–2 boundary where 19 species are lost and 17 new ones appear. Some benthic species disappear higher in the Plenus Marls but the next significant extinction step is the loss of several benthic foraminifera and the planktonic foraminifera *Rotalipora cushmani* at the top of Bed 3 (Leary 1989; Peryt *et al.* 1994; Fig. 8.3). This keeled form is considered to have lived at a considerable depth in the water column (Hart 1980; Jarvis *et al.* 1988; Leckie 1989). The disappearance of *R. cushmani* marks the extinction of the genus and the onset of a crisis in the mid-water column of the world's seas and oceans (Kauffman and Hart 1995). Two other deep-dwelling foraminiferan genera, *Praeglobotruncana* and *Dicarinella*, appeared in the late Cenomanian but they both disappeared around the same level as *Rotalipora*. However, in their case, they reappeared in the early Turonian after the crisis (Leckie 1989).

In the Shakespeare Cliff section the remaining beds of the Plenus Marls (Beds 4–8) are characterised by a distinct change of lithology; calcispheres and inoceramid prisms become common. The Boreal belemnite that gives its name to the formation, *Actinocamax plenus*, also appears at this level. The final wave of extinctions occurs at the Bed 6–7 junction where four species of calcareous nanofossil disappear and ostracod populations become dominated by a few species of the dysoxia-tolerant platycopids – particularly *Cytherella ovata*, which can constitute up to 80% of some samples (Whatley 1991). Thus, the extinction at Dover is confined to the *Metoicoceras geslinianum* Zone.

Many of the faunal events recorded at Dover have been recognised elsewhere in the world; the calcisphere 'event' is a global end-Cenomanian phenomenon, for example (Hart 1991). The fauna of the Dover Chalk is highly comminuted and thus contains a poor record of the Turonian recovery. For this, we must turn our attention to the numerous sections of Colorado, Arizona, Kansas, and adjacent states. As in the Chalk sections, these Western Interior localities record a protracted, stepped extinction crisis spread over the last half million years of the Cenomanian. The number of steps identified varies from author to author, with the record going to Kauffman and Hart (1995) who recognize no less than 11 such events. However, as they note, many of these steps grade into one another when error bars are added to species ranges.

The crisis began late in the *Sciponoceras gracile* Zone with the disappearance of many endemic ammonites (Elder 1989) and the planktonic foraminifera *Rotalipora* and *Guembelitria* (Leckie *et al.* 1991). Transgression in the early *juddi* Zone introduced a distinct fauna with southern affinities into the Western Interior region. Cosmopolitan and subtropical ammonites and inoceramids became common (Elder 1991; Kirkland 1991), while the planktonic foraminifer assemblages are dominated by the eurytopic genus *Heterohelix* (Leckie *et al.* 1991). Benthic diversity is high at the base of the *juddi* Zone, with suspension-feeding bivalves dominating (Elder 1987). However, within one to two metres in most sections, numerous taxa rapidly begin to disappear until, at the top of the zone, only a few species remain. Deposit feeders (particularly the aporrhaid gastropod *Drepanochilus*) and a few eurytopic, long-ranging bivalve genera (*Astarte* and *Corbula*) dominate the late *juddi* Zone assemblages. These occur in marly sediments which are notably organic-rich. This combination of faunal and lithological features led Elder (1989, 1991) to suggest that both soft substrates and bottom-water dysoxia may have been responsible for the diversity decline.

After a temporary recovery of ammonite fortunes at the base of the *juddi* Zone, further extinctions reduced their diversity, particularly of the more cosmopolitan taxa. At the C-T stage boundary all surviving species became very rare (Elder 1989; Harries 1993). Heteromorph ammonites weathered this crisis relatively better than the more 'normal' planispiral varieties. Using Batt's (1989) ecological analysis of Cretaceous ammonite morphotypes, Harries (1993) was able to infer that these extinctions indicate preferential removal of nektobenthic relative to pelagic ammonites.

By the end of the Cenomanian the main extinction phase was over and approximately 50% of the molluscan species had been lost from the Western Interior seaway (Elder 1989). The cosmopolitan, epifaunal inoceramids were particularly severely affected, while the more endemic, infaunal bivalves only suffered a modest 27% species extinction in comparison. The infaunal bivalve genera are notably more species-poor than the inoceramids. Thus, the extinction can be said to have been worst for widespread, speciose genera (Elder 1989). Interestingly, this is the opposite pattern to that observed by Jablonski (1986a, b, 1989, 1991) for end-Cretaceous molluscan extinctions. However, perhaps the most salient aspect of these preferential extinctions is that inoceramids dominated offshore communities of the Cenomanian, while infaunal taxa were more commonly found inshore. The extinction was therefore more effective in deeper-water sites where, incidentally, the majority of nektobenthic ammonites were also encountered.

The early Turonian encompasses the recovery in distinct steps in the Western Interior (Harries and Kauffman 1990; Elder 1991). Initially, a series of low-diversity assemblages were dominant in the *Watinoceras* Zone – first *Mytiloides* spp. and the oyster *Pycnodonte* and then *Discinisca* spp. (Harries and Kauffman 1990). *Mytiloides* is a classic progenitor taxon (cf. p. 14) as it made its first rare appearance in the extinction interval and then speciated rapidly in the early stages of the recovery. Notably, and in sharp contrast to the late *juddi* Zone assemblages, the recovery faunas are dominated by epifaunal taxa. The ammonite faunas were among the last to regain their original diversity but, by the *Mammites nodosoides* Zone, all communities had reached their pre-extinction diversity levels. The recovery was therefore effected in the duration of the *Watinoceras* Zone – approximately half a million years.

The Western Interior and English sections provide interesting comparisons. Both provide evidence for an extinction crisis that preferentially affected benthic and lower water-column groups while leaving surface-dwelling taxa relatively unaffected. However, there is some variability in the details of the extinction timings and the groups affected. At Dover the main extinction interval was in the *geslinianum* Zone; in the Western Interior only ammonites appear to have been affected at this level, while the main phase of benthic extinctions occurred later in the *juddi* Zone. The extinction crisis has been widely recognised in other regions and in other palaeoenvironmental settings. These tend to suggest that the Western Interior extinctions probably took place slightly later than those in other regions of the world.

Rudists suffered a notable extinction in the late Cenomanian followed by a prolonged early Turonian outage for most survivors. Aragonitic-shelled and recumbent rudists were the most severely affected and two families, the caprinids and ichthyosarcolitids, went extinct (Ross and Skelton 1993). In the Pacific, the widespread demise of rudist–coral reefs caused numerous atolls to be irrevocably converted into guyots (Schlager 1981). In both the Pacific region and southern Europe this extinction of some rudists (and the temporary disappearance of the remaining ones) appears to have occurred shortly after the demise of *R. cushmani* and thus closely correlates with the benthic extinctions in England. In contrast, Johnson and Kauffman (1990) have argued that the rudist extinction occurred up to a million years earlier in the Caribbean. However, more recently, Kauffman and

Hart (1995: p. 298) have noted; 'Data for this [much earlier regional] extinction event, is drawn mainly from Mexican localities, [which are] still imprecisely dated'.

The true severity of the rudist mass extinction is difficult to judge; Philip and Airaud-Crumière (1991) noted the loss of 35 out of 44 genera in the late Cenomanian of southern Europe, but Jones and Nichols's (1986) global compilation indicated that it was only a very modest extinction that barely halted the major Late Cretaceous proliferation of this group. Perhaps the most significant aspect of the crisis is the approximately million-year Lazarus interval for most rudists in the early Turonian.

The Late Cenomanian extinction not only affected shallow-water taxa, such as the rudists, but also caused a crisis for many deeper-water groups. The benthic foraminifera from the bathyal sections of Japan exhibit two late Cenomanian extinction steps that removed 29% of species – a higher extinction percentage than that seen at the end of the Cretaceous for this hardy group (Kaiho and Hasegawa 1994). Both the Cenomanian extinction steps correspond to the development of dysaerobic strata. The influence of dysoxic conditions is also evident in the composition of post-extinction benthic foraminifera in abyssal sections of the north Atlantic (Kuhnt 1992). The extinction interval of this region consists of unfossiliferous late Cenomanian black shales but the earliest Turonian sediments contain the agglutinating foraminiferans *Haplophragmoides* and *Glomospira*. These are long-ranging, dysaerobic genera which rapidly proliferated to produce numerous new species in the early Turonian.

A similar link between faunal crisis and dysoxia is seen in the C–T boundary sections of Niger (Mathey *et al.* 1995). A diverse, endemic carbonate ramp biota dominates the Cenomanian of this country but, during the boundary interval, interbedded limestones and dysaerobic shales are developed. The limestones contain a low-diversity community dominated by cosmopolitan oysters and gastropods that are encountered throughout Africa and Europe, while the shales contain a few agglutinating foraminifers and the bivalve *Aucellina*. The 50% species extinction associated with this faunal and facies change is closely comparable to that of the Western Interior.

The link between dysoxia and extinction appears a good one in these regional studies, but a recent compilation of the global fortunes of benthic foraminifera during the Cenomanian and Turonian suggests otherwise. Banerjee and Boyajian (1996) recorded 17% generic extinction (37 of 219 taxa) in the Cenomanian, but when broken down into geographical and ecological categories this revealed a remarkable selectivity. Thus, most extinctions occurred in equatorial latitudes, while taxa encountered above 60°N and below 60°S suffered virtually no losses. In low-latitude areas, the benthic foraminifera of the Western Interior had the lowest extinction rates despite the occurrence of the best evidence for dysoxic deposition in this region. Even more surprising, the highest extinction rates occurred among dysoxia-tolerant agglutinating foraminifera (28% generic extinction) and flattened discoidal forms. The highest extinction rates (48%) of all were therefore experienced by discoidal agglutinating foraminifera, the quintessential dysaerobic taxa. These data are an average of extinctions from throughout the Cenomanian Stage rather than from the C–T boundary interval but, at the very least, they the call into question the dysoxia–extinction nexus (Banerjee and Boyajian 1996).

The late Cenomanian was also a crisis for the top predators of the oceans, namely the sharks, plesiosaurs, and ichthyosaurs (Cappetta 1987; Benton 1988; Bakker 1993). Precise timing of the extinction of such rare taxa is difficult to determine, but it is clear that all these groups suffered generic-level losses in the Cenomanian. Dating of terrestrial verebrate extinctions is even more problematic due to the difficulties of correlation with marine sections. None the less, the compilation of Benton (1988) suggests a modest Cenomanian extinction followed by a major Turonian radiation – particularly of mammals (Benton 1989). However, much more detailed, stratigraphically controlled collecting of the terrestrial record is needed to confirm that the late Cenomanian mass extinction was not just restricted to the marine realm.

Sea-level changes and facies

The Cenomanian–Turonian interval is widely recognised as one of the all-time high-stands of sea level (see, for example, Hancock and Kauffman 1979; Haq *et al.* 1987; Hallam 1992), but the details of the changes during the extinction interval have proved more contentious. Hancock and Kauffman (1979) suggested peak transgression at the C–T boundary, while, in direct contrast, other authors have argued for peak regression (Leary 1989; Jeans *et al.* 1991). However, by far the most popular alternative for C–T sea-level changes, and the only one supported by detailed sequence stratigraphic analysis (see, for example, Robaszynski *et al.* 1993), places the stage boundary within a transgressive phase (Haq *et al.* 1988; Koutsoukos *et al.* 1990; Kirkland 1991; Leckie *et al.* 1991; Philip and Airaud-Crumière 1991). The transgression appears to have been initiated following a minor sea-level fall, immediately prior to the *juddi* Zone, and to have continued into the the early Turonian when maximum flooding conditions were achieved.

In many tropical and mid-latitude shelf sections, the sediments of the C–T interval consist of marls and/or marl-limestone couplets that are distinct from the purer carbonates both above and beneath the extinction interval (see, for example, Jarvis *et al.* 1988; Eicher and Diner 1991: Peryt *et al.* 1994; Mathey *et al.* 1995). It is within these marly sediments that evidence for dysoxic conditions, such as organic enrichment and fine lamination, can commonly be found.

Geochemical evidence

As already noted (pp. 172–3), the C–T interval is characterised by distinctive carbon and oxygen isotope excursions. Other geochemical evidence pertaining to the extinction interval has been collected by Carl Orth and colleagues, initially from the Western Interior and subsequently from sections around the world (Orth *et al.* 1988, 1993). Among the most discussed results of Orth's analysis is the discovery of two Ir 'peaks' of 0.11 ppb which stand out from average background values of 0.017 ppb. The lowest peak occurs within the *Sciponoceras* Zone and corresponds to the onset of the main phase of benthic extinctions outside the Western Interior. The second peak is found near the top of the same zone – a time marked by the demise of numerous subtropical ammonite taxa. The Ir peaks are also associated with enrichment of other trace metals (such as V, Cr, Ni, Pt, and Au) but, as Orth *et al.* (1993) observed, these features are only seen in sections adjacent to the Atlantic, namely the Western Interior and westernmost Europe. Sections further afield, such as those of Poland, show no trace-element enrichment signature. The Atlantic link is also supported by trace element ratios which are very similar to those of Mid Atlantic ridge basalts. Orth *et al.* (1993) therefore suggested a phase of rapid Mid Atlantic ridge formation in the latest Cenomanian to explain the trace-element signature – a suggestion that also helps to explains the origin of the rapid eustatic sea-level rise noted above.

Extinction models

The Late Cenomanian extinctions are closely associated in time with OAE2, the best-known and most convincingly demonstrated of the oceanic anoxic phases. The most frequently invoked extinction mechanism is therefore, inevitably, that of habitat destruction caused by the expanding area of oxygen-poor waters (see, for example, Schlager 1981; Jarvis *et al.* 1988; Elder 1989; Leckie 1989; Kaiho and Hasegawa 1994; Rougerie and Fagerstrom 1994). The associated change to very soft substrates in the Western Interior seaway may also have contributed to the decline of some epifaunal benthic taxa (Elder 1989). The evidence for OAE2 is widespread and comes from combined sedimentological, palaeoecological, and geochemical evidence. Organic enrichment, one of the more characteristic features of oxygen-poor deposition, is seen in oceanic sections and the deeper parts of epicontinental seas (Jenkyns 1985). The positive carbon isotope excursion is widely considered to be further evidence of a major, global, organic carbon burial event (Schlanger *et al.* 1987; Jarvis *et al.* 1988).

While the existence of widespread anoxia is rarely doubted, its ultimate cause is hotly contested. Much of this argument is, however, beyond the remit of this book: Wignall (1994) contains a recent review. Suffice to say, the debate is polarised around high productivity versus enhanced preservation models; the models encapsulate the fundamental dichotomy that underpins much of the more general debate on petroleum source-rock formation mechanisms. Direct evidence for either model is not apparent from fossil evidence, although it has been argued that the widespread abundance of calcispheres in the *juddi* Zone indicates fertile but highly unstable surface-water nutrient levels (Peryt *et al.* 1994).

The stepped waves of extinction in the late Cenomanian have been interpreted as evidence for an upward-expanding zone of oxygen depletion in the lower water column (Jarvis *et al.* 1988). Thus, in the Dover section, first benthic and then deep-dwelling planktonic foraminifers go extinct. The extinction crisis finally culminates in the loss of further benthic species and the development of low diversity assemblages with common platycopid ostracods (Fig. 8.4). In the Western Interior the same bottom-up pattern of extinction is not so clearly developed; nektobenthic ammonites, rather than true benthos, are the first group to suffer. However, the preferential extinction of deeper-water taxa in this region is in accord with an extinction model involving the development of oxygen deficiency in the deeper parts of the region. These sites were the preferred habitats for the *Inoceramus* species that were among the most notable victims of the crisis (Fig. 8.2). The extinction of such widespread species points to the global nature of the crisis among deep-shelf benthos.

The loss of shallow-water groups in the late Cenomanian, particularly the rudists, is not so immediately reconcilable with an extinction attributed solely to the development of anoxia in deeper waters. Indeed, some authors (such as S.M. Stanley 1988) have argued that reef taxa in general should be immune from anoxic events because of their shallow-water, persistently wave-agitated life-sites. None the less, Schlager (1981) considered anoxia to be one of the few feasible mechanisms capable of causing reef destruction; the problem lies in getting relatively deep, oxygen-poor water into the

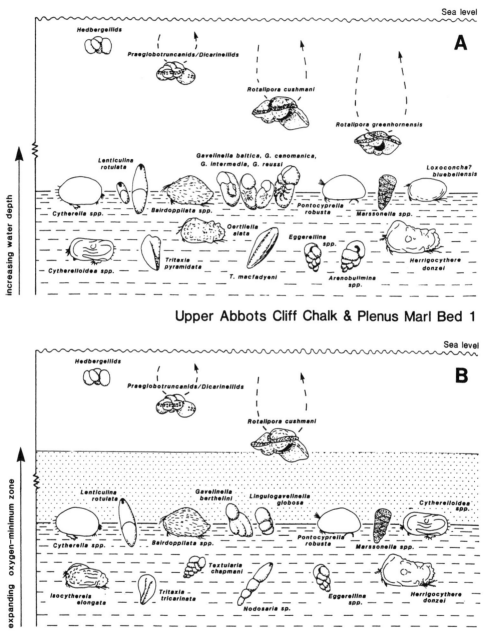

Upper Abbots Cliff Chalk & Plenus Marl Bed 1

Plenus Marl Bed 2

Fig 8.4 Interpretation of the late Cenomanian extinction event based on evidence from Shakespeare Cliff (see Fig. 8.3). A. Pre-extinction, a diverse foraminifera and ostracod assemblage populates the sea floor. B. The rapid development of dysoxic bottom waters during the latest Cenomanian transgression selectively removes those benthic taxa unable to tolerate lowered oxygen levels. C. The height of the water column affected by dysoxic conditions increases during the course of the

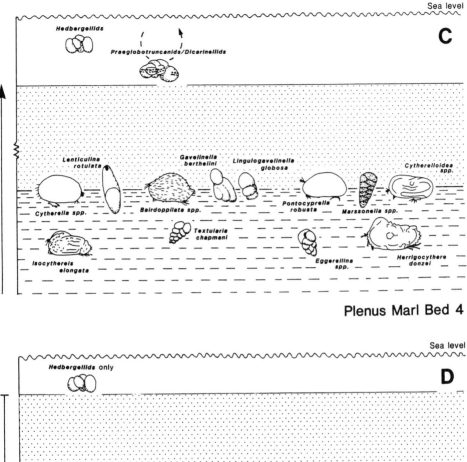

Plenus Marl Bed 4

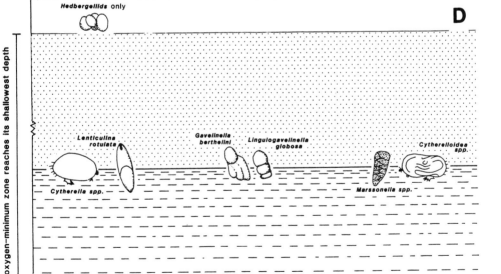

Plenus Marl Beds 6–8 & basal Melbourn Rock

transgression and causes the demise of the deepest-dwelling planktonic foraminifera *Rotalipora cushmani*. D. The dysoxia increases both in intensity and in the height of water column affected, thereby causing all but the surface-dwelling foraminifers to be eliminated. The benthos is reduced to a few dysoxia-tolerant taxa by this stage, notably platycopid ostracods (*Cytherella* spp.) and agglutinating benthic foraminifera. After Jarvis *et al.* (1988).

shallow depths occupied by rudists. A potential solution to this dilemma, proposed by Vogt (1989), involves catastrophic buoyant upwelling of oxygen-minimum zone waters into surface waters triggered by the voluminous submarine volcanism seen in both the Atlantic and Pacific oceans during the C–T boundary interval. Vogt argued that, if such fatal incursions of oxygen-poor waters occurred on a decadal scale or less, there would have been insufficient time for reef recovery between events. In an alternative scenario, Rougerie and Fagerstrom (1994) proposed, for the Pacific atolls, that intermediate water depth, oxygen-poor waters may have been drawn to the surface of the atolls through the weathered, fractured summit of such volcanic edifices – a phenomenon they term 'endo-upwelling' – thereby exterminating the reef fauna. Endo-upwelling is a reasonable mechanism for linking the fortunes of very shallow-water benthos with deeper-water conditions but it cannot plausibly be invoked for the demise of rudist reefs within late Cenomanian Tethyan platform carbonates.

The general lack of shallow-water dysaerobic strata in the late Cenomanian constitutes the main argument against the death-by-anoxia kill mechanism (see, for example, Jeans *et al.* 1991; Harries 1993; Paul and Mitchell 1994). At the Dover locality, for example, the faunal evidence is the best guide to the dysoxic conditions, while there is little if any supporting sedimentological or geochemical evidence. It could be argued that only upper dysaerobic facies are seen at this shallow-water site: palaeoecological criteria are generally more sensitive than other criteria to modestly lowered oxygen levels (Wignall 1994). However, such an argument carries an inevitable hint of special pleading. In the Western Interior, the timing of some of the earliest extinction steps is rather too early for the anoxic event. Thus, the initial ammonite extinction at the top of the *Sciponoceras gracile* Zone substantially pre-dates the main phase of anoxia in the region, which did not occur until the *juddi* Zone. The recent compilation of benthic foraminifera extinction rates adds little support to the anoxia extinction mechanism (Banerjee and Boyajian 1996).

Anoxia is clearly not the only component of the late Cenomanian mass extinction, although it was probably one of the main causes. Another potential contributory factor may have been the productivity decline proposed recently by Chris Paul and his co-workers (Paul and Mitchell 1994; Paul *et al.* 1994), based on coccolith accumulation rates. These calculated rates are based on the assumption that the marl–limestone couplets seen in many sections are a response to an orbital periodicity of 21 000 years, thereby allowing absolute time to be included in the equation. Paul considers the marly sediments of the C–T boundary interval to record severely decreased coccolith productivity and thus an overall decline in primary productivity. This hypothesis contrasts with many geochemical studies (and the calcisphere event) which have suggested that OAE2 records increased productivity (see, for example, Farrimond *et al.* 1990). Paul's calculations may indeed be valid, but the interpretation of productivity decline is based on the further, unsubstantiated assumption that coccolith productivity can be used as a proxy for total primary productivity. A more likely scenario might be that these calcite-secreting autotrophs did not fare as well as purely organic phytoplankton during the OAE, thus triggering the change from almost pure carbonates to more marly sediments. Productivity changes cannot be excluded as a factor in late Cenomanian extinctions, but their failure to explain the patterns of disappearance suggests that they are, at most, of indirect or subsidiary importance.

The only other serious contender for the causative agent of the extinction, global cooling, derives much of its evidence from the fossil record. Jefferies (1962) was the first to note that the faunal turnover within the Plenus Marls marks an invasion of Russian species, typified by *A. plenus*, and the loss of western European species. This faunal event is widely recognised (see, for example, Hilbrecht and Hoefs 1986) and both Jefferies and subsequent authors have attributed it to a cooling event (Leary 1989; Jeans *et al.* 1991). However, additional supporting evidence is scant. Jeans *et al.* (1991) have noted the presence of very rare outsized, exotic clasts in the Chalk of southern England which they interpret as iceberg dropstones. Quite apart from the fact that such rare lonestones could equally well be gastroliths or dropstones from tree roots associated with driftwood, it is inconceivable that icebergs could have occurred at such low-latitude sites: England was around 35–40°N in the Late Cretaceous. The absence of evidence for glaciation at higher latitudes further argues against an iceberg dropstone model. Jeans *et al.* (1991) also suggested that the onset of a prolonged positive shift of oxygen isotope ratios was further evidence for cooling. While this interpretation is reasonable, the cooling trend did not begin until the Turonian (that is, considerably after the mass extinction) and it only really got under way in the Campanian, three stages later (Huber *et al.* 1995). So, how strong is the faunal evidence for cooling? Although the invading Russian species have always been considered cool-water taxa, they are actually derived from areas that were no further north than sites in Europe. The event is therefore better considered as an east-to-west migration. This might have been facilitated by the improved marine connections during the C–T transgression. Overrall, the faunal change is best considered as an expression of the increased cosmopolitanism of faunas caused by migrations in the late Cenomanian (see, for example, Mathey *et al.* 1995). In the Western Interior, for example, the extinction crisis in the *gracile* Zone was followed by an invasion of subtropical species from the south.

Cooling figures again in a recently proposed mechanism for the decline of rudist reefs during the C–T interval. It is argued that the sea-level rise at this time facilitated the flow of warm tropical waters into higher latitudes as the connectivity of the epicontinental seaways improved (Johnson *et al.* 1996). With more efficient heat transport the tropics cooled, leading to extinctions among equatorial rudists. The preferential loss of benthic foraminifera from the same latitudes (Banerjee and Boyajian 1996) further hints at equatorial cooling. This mechanism clearly goes some way towards explaining the faunal changes during the C–T interval but, as with the preceding models, the lack of oxygen isotope evidence casts some doubt on its validity.

Tenuous as the evidence is for late Cenomanian cooling, it appears substantial when compared to the evidence for the third proposed kill mechanism – multiple cometary impact (Hut *et al.* 1987). The stepped extinction and the two horizons of Ir enrichment detected by Orth *et al.* (1988), have been preferred as evidence for this extinction mechanism. This is despite the explicit assertion of Orth *et al.* (1988), subsequently reiterated by Orth *et al.* (1993), that the geochemical signature of the enriched horizons rules out an extraterrestrial origin and strongly suggests a mid-Atlantic basalt source. No other supportive evidence, such as shocked minerals, has been forthcoming and there is therefore little reason to favour a cometary impact model. For the potentially catastrophic effects of bolide impact we must turn to the next great crisis in the world's history, the end-Cretaceous mass extinction.

9 Death at the Cretaceous–Tertiary boundary

It is unusual in science for a major advance to be traceable to the publication of a single paper, but without question the study of mass extinctions received a huge stimulus, and considerable media interest, from the seminal paper in *Science* by Luis Alvarez, his son Walter and two nuclear chemist colleagues (Alvarez *et al.* 1980). Although phenomena from outer space had been invoked on a few previous occasions to account for mass extinctions, they were disregarded because of lack of evidence in support. This situation was changed radically with the discovery of a significant positive anomaly of the platinum-group trace metal iridium at the Cretaceous–Tertiary (K–T) boundary in Italy and elsewhere. This was held by Alvarez *et al.* to be most plausibly accounted for by the impact of an asteroid approximately 10 km in diameter. A few years later the impact hypothesis received strong support from an independent line of evidence, the presence of so-called shocked quartz at the K–T boundary in the North American Western Interior. Shocked quartz possesses distinctive multiple sets of lamellae which have only been recognised in rocks from well-established meteorite impact sites or at nuclear weapons test sites, and appears to signify the passage of shock waves under enormous pressures.

Since that time the impression has become well established among the public at large that the dinosaurs were finally killed off by an asteroid (the media have not bothered to pay much attention to any other organisms). Among informed scientists the matter has been more controversial, and debate continues, but some of the earlier controversy seems now to be in the process of being resolved. The subject has generated an enormous literature, and attention here is concentrated on more recent work. The latest book to deal comprehensively with the K–T extinctions, based on a symposium held in Boston in 1993, is the one edited by MacLeod and Keller (1996); see also MacLeod *et al.* (1997). Reviews of the older literature can be found in Hallam (1989b) and Glen (1994), while the numerous papers published in the second Snowbird conference proceedings (Sharpton and Ward 1990) provide a comprehensive treatment of the subject as it developed during the decade following publication of the Alvarez paper. Since that time by far the most important developments have concerned phenomena located around the Gulf of Mexico and Caribbean (Ryder *et al.* 1996), and these must be reviewed briefly before mass extinctions are addressed.

Recent developments

Although claims for shocked quartz at the K–T boundary have been made for a number of sites across the world, the only really impressive evidence, in terms of number and size of grains, is from the North American Western Interior, notably the Raton Basin of New Mexico. This same region has also yielded stishovite in association with the shocked quartz (McHone *et al.* 1989). Stishovite is a high-pressure polymorph of silica, until recently found only at a few meteorite craters and only in minute quantities. Over a wide area, both in Europe and beneath the oceans, the K–T occurrence of highly oxidised magnetic spinels with high nickel concentrations also supports impact (Robin *et al.* 1992), as do nanometre-sized diamonds in Alberta. Their carbon isotope ratio is close to that of interstellar dust and incompatible with a terrestrial origin (Carlisle 1992).

Although a number of oceanic sites have been proposed for the expected impact crater, the shocked quartz evidence points to one somewhere on the North American continent. The size of the asteroid postulated by Alvarez *et al.* (1980) was estimated by calculation from a formula that links the projectile size to the size of the crater it forms and the size of the dust-cloud plume. In order to create a plume that would encircle the globe, and hence produce a world-wide iridium anomaly, the projectile would have to be about 10 km in diameter, with a crater of up to 150 km in diameter. Claims of K–T boundary tsunami deposits in the Gulf Coast–Caribbean region, held to signify the results of a shock wave nearby, stimulated a search for a local impact crater. The key localities that must be discussed are; the Yucatan Peninsula of Mexico; Haiti; the Brazos River, Texas; Braggs, Alabama; and north-east Mexico (Fig. 9.1).

The Yucatan Peninsula

Geophysical surveys undertaken some years ago in the Yucatan Peninsula revealed circular anomalies in both gravity and magnetic fields centred on the Chicxulub locality. The association of andesitic rocks discovered in boreholes led to alternative interpretations, that it marked either a volcanic centre or an impact crater. More recent work favours the latter interpretation (Hildebrandt *et al.* 1991; Sharpton *et al.* 1992). Thus andesites occur only in the circular zone and occur in association with breccias and shocked quartz and feldspar. The presumed melt rocks contain anomalously high iridium contents, up to 13.5 ppt. The proportion of shocked quartz grains, 27–31%, is within the range of 12–47% reported from 12 K–T boundary sites by Izett (1990). Argon–argon dates of the glassy melt rock beneath massive breccias give a date of almost exactly 65 Ma, virtually indistinguishable from those obtained from purported K–T boundary tektites in Haiti and north-east Mexico (Swisher *et al.* 1992).

There has been some dispute about the diameter of the Chicxulub crater, with estimates ranging between 170 and 300 km. The latest work, based on a study of horizontal gravity gradients, suggests a value of 180 km (Hildebrandt *et al.* 1995).

An intriguing study has been made of the uranium–lead age of zircon crystals from Chicxulub breccias, as well as from the K–T boundary in Haiti, Colorado, and Saskatchewan. They all give a target basement age of about 545 m.y., suggesting strongly that Chicxulub is the source of K–T boundary ejecta, with Palaeozoic silicate basement rocks ejected during impact; there is no need to invoke any other impact site (Krogh *et al.* 1993; Kamo and Krogh 1995).

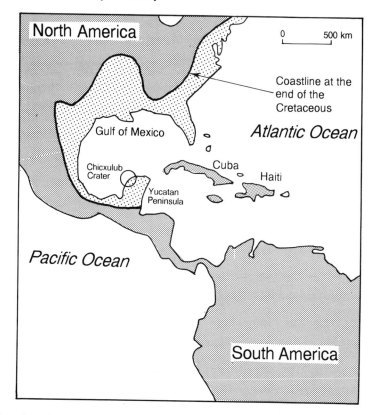

Fig 9.1 Map of the Gulf of Mexico–Caribbean region showing the location of the purported Chicxulub crater and the approximate location of the coastline at the end of the Cretaceous.

While the evidence cited so far suggests an overwhelming case for an impact event at the end of the Cretaceous, a dissenting opinion has been expressed by Meyerhoff *et al.* (1994), who report on a previously unpublished well log (Yucatan no. 6). The borehole is claimed to have penetrated a few hundred metres of Upper Cretaceous carbonates with a Campanian–Maastrichtian microfauna overlying a volcanic sequence; fossiliferous Upper Cretaceous strata lie both above and within the volcanic sequence. The well bottomed in Cretaceous anhydrite. Meyerhoff was the consulting geologist to Petróleos Mexicanos at the time the well was drilled and was closely involved in biostratigraphic correlation. Even if the andesites at depth are indeed impact melt rocks, they would thus be much too old to have any bearing on a K–T boundary event, and the authors propose that all the published radiometric dates suggesting otherwise are unreliable. However, a reexamination of samples recently released by the oil company suggests that the most likely age for sediments immediately above the critical igneous rocks in the Yucatan no. 6 well is in fact Danian (Marin *et al.* 1994).

A more comprehensive study of subsurface stratigraphy in the Yucatan Peninsula has recently been undertaken by Ward *et al.* (1995). An improved stratigraphic context for interpreting the effects of a Chicxulub impact is provided by this integration of data from new biostratigraphic analyses of core samples with data from geophysical logs of deep wells. A thick interval of breccia is recognised throughout the northern Yucatan. It contains clasts of dolomite, limestone, and anhydrite, together with shocked quartz and feldspar, basement rock, and altered melt rock; of various possible origins, impact is considered to be the most plausible. Using planktonic foraminifera, approximately 18 m of Maastrichtian marl overlies this breccia. This is held by the authors to be consistent with the stratigraphy of this interval in sections in Mexico and Haiti. They therefore caution that a scenario declaring Chicxulub as the site of a K–T boundary impact crater is impossible to substantiate on the basis of present biostratigraphic control.

Haiti

Various workers have concluded that black glass particles from the K–T boundary at Beloc, Haiti, are tektites, and that smectised brown spherules are their pseudomorphs. Sigurdsson *et al.* (1992) indicate that they include high-calcium glasses with up to 1% SO_3 by weight suggesting they formed by fusion of anhydrite- or gypsum-rich evaporitic sediments in the presence of high-silica melts. Geochemical evidence from the glasses was consistent with derivation from Chicxulub, where a thick evaporite succession is known at depth. This impact interpretation was challenged by other workers. Lyons and Officer (1992) believed that the glass particles are of volcanic origin, and Jehanno *et al.* (1992) considered that the Beloc glasses are markedly different from tektites, with a structure and composition suggesting that they are not impact-generated. Koeberl (1993) also challenged, on compositional grounds, the supposed link between Haitian glasses and Chicxulub rocks. Subsequently, however, the French workers changed their minds, and believe that the Beloc particles are indeed impact-derived products (Leroux *et al.* 1995). Furthermore, Blum and Chamberlain (1992) have undertaken an oxygen-isotope study that is considered effectively to rule out a volcanic origin. Further support for a proximal impact event is provided by the discovery of large and abundant grains of shocked quartz in the tektite-bearing sediments (Kring *et al.* 1994).

The Brazos River and Braggs sections

Considerable interest was generated by the claim by Bourgeois *et al.* (1988) that an unusual coarse-grained sediment layer in K–T boundary sections in the Brazos River, Texas, was a tsunami deposit provoked by a nearby impact. This interpretation was challenged by Keller (1989; see also the discussion in the *Bulletin of the Geological Society of America*, **103**, 434–6 (1991), who maintained on the basis of plantonic foraminiferal evidence that the layer in question was slightly older than the K–T boundary. Keller considered that it directly overlies an irregular scoured surface overlain by a coarse glauconitic sand of variable thickness, and drew a comparison with a condensed section and sequence boundary at Braggs, Alabama (see below). Thus the purported tsunami bed could represent the infilling of incised topography during the lowstand of a late Maastrichtian sea-level fall. A different age interpretation, also inconsistent with a K–T boundary impact scenario, was given by Montgomery *et al.* (1992). They claimed that rare planktonic foraminifera, coccoliths and dinoflagellates in the purported tsunami bed are of Danian age, and therefore that the bed is too young to signify a K–T impact event.

The Clayton Sands of Braggs and neighbouring sections of Alabama are earliest Danian in age and rest unconformably on the Maastrichtian Prairie Bluff Chalk. They have been variously interpreted as a tsunami deposit or as a non-catastrophic transgressive infilling of incised valleys cut during a previous sea-level lowstand. Savrda's (1993) ichnosedimentological study decisively supports the latter interpretation, with an extended period of deposition being indicated by multiple phases of burrowing.

North-east Mexico

Another K–T boundary tsunami deposit has been claimed at Arroyo el Mimbral in north-east Mexico (Smit *et al.* 1992). A 3 m clastic unit occurring in a biostratigraphically complete deep-water pelagic marl succession can, according to these authors, be located precisely at the K–T boundary on the basis of the planktonic foraminifera. A basal 'spherule bed' contains, it is claimed, both tektite glass and quartz grains with probable shock features. It is overlain by graded laminated beds with intraclasts and plants, with a rippled unit of fine sand at the top, signifying oscillatory currents, maybe a seiche.

A completely different, non-catastrophic, interpretation is given by Stinnesbeck *et al.* (1993), who investigated sections including Mimbral. These authors recognise three successive units. The oldest is a spherulitic bed rich in calcite. Quartz with multiple sets of lamellae is rare. More commonly, the lamellae are curved and irregularly spaced, a characteristic of tectonic origin; there are also very rare glass shards. Overlying this is a massive laminated sandstone and then a cross-bedded sandstone topped by a sandy limestone with ripple marks. This in turn is overlain by a K–T boundary clay unit. The three units can be traced over at least 300 m and give evidence for several discrete events, with the glasses deriving from an event preceding the K–T boundary. The silty limestone–mudstone directly above the clastic layer is thought to represent continuation of normal hemipelagic sedimentation with *in situ* latest Maastrichtian planktonic foraminifera.

The interpretation of Smit *et al.* (1992; see also Smit *et al.* 1996) is, however, supported by the analysis of calcareous nanofossils by Pospichal (1996), which indicates biostratigraphically complete sections at Mimbral and Mulato, with the clastic unit placed at the K–T boundary. Reworking above this unit is supported by correlation of abundance and diversity of uppermost Maastrichtian nanofossils with obviously reworked Campanian–Lower Maastrichtian specimens. Pospichal concludes that it is reasonable to suggest that the nanofossil and foraminiferal extinctions occurred some time during the deposition of the clastic unit and overlying silty limestone, and hence in association with the Chicxulub impact.

According to Keller *et al.* (1993a), of 16 deep-sea core sections in the Caribbean, Gulf of Mexico, Mexico, and the western Atlantic, only Mimbral appears to exhibit relatively continuous sedimentation across the K–T boundary. All other sections have major boundary hiatuses and none has current-bedded clastics as at Mimbral and in Haiti. The hiatus throughout the Caribbean and Gulf of Mexico is thought to be related to intensified current circulation associated with a sea-level low stand.

Thus dispute continues in all the regions considered, although evidently a majority of a large field party associated with the Lunar and Planetary Institute conference at Houston (1994), including several 'neutral' sedimentologists, was persuaded that the coarse clastic deposits at Mimbral were deposited rapidly, in a manner apparently consistent with an impact scenario (Ward 1995). It would at this stage be fair to say that a very strong case has been made for a major impact event in the Yucatan Peninsula at the end of the Cretaceous, and that neither multiple simultaneous impacts nor cometary showers need be invoked for global events. An event of this size, indicated by the estimated diameter of the crater, should have had devastating environmental consequences on a global scale. The extent to which they can be related to the extinction record can only be discussed after this record has been treated in some detail. Initially some remarks are required on stratigraphic correlation.

Stratigraphy

It has been widely assumed by those who invoke a global catastrophe at the end of the Cretaceous that the dinosaurs went extinct at the same time as the oceanic mass extinction of calcareous plankton. To what extent is this assumption justified? Correlation of marine and non-marine strata generally poses problems, and the Cretaceous and Palaeogene are not exceptional in this respect. It became well established some years ago that both the final dinosaur extinctions in North America and the plankton mass extinction fell within the magnetozone 29R, but this does not provide a time resolution of much less than about half a million years. However, the last of the American dinosaurs are claimed to have disappeared at the level of the so-called 'fern spike' established by palynologists, marking evidently a drastic diminution in the angiosperm population. An equivalent fern spike has been recognised in marine deposits in Japan containing calcareous plankton, which can be used to locate the K–T boundary (Saito *et al.* 1986). Evidently the spike coincides precisely with the era boundary as established by the plankton, and hence supports the coincidence in time of terrestrial and marine extinctions.

Planktonic foraminifera provide the key fossils for establishing the K–T boundary in marine strata, with the *Abathomophalus mayaroensis* Zone marking the youngest Maastrichtian and the *Parvularuglobigerina eugubina* Zone the oldest Danian. However, Smit (1982) erected between these two the biozone Po, characterised by *Guembelina cretacica*, which he believed to be the sole survivor of the Maastrichtian fauna. There has indeed been some confusion about whether to place the era boundary at the level of mass disappearance of Late Cretaceous species or the first occurrence of Palaeocene species. The International Commission on Stratigraphy has now accepted as the global stratotype the section at El Kef, Tunisia. Here, the base of the boundary clay coincides with the abrupt decline in abundance of Late Cretaceous species and a significant increase in the survivor *G. cretacica*. Olsson and Liu (1993) consider that mass disappearance is the better criterion for the boundary because the first occurrence of Palaeocene species can be very variable and unreliable as a consequence of downward reworking, extreme rarity, and poor preservation in most sections. Furthermore, there are no first occurrences of Palaeocene species in the boundary clay at El Kef. Based upon sections in Mexico, Longoria and Gamper (1995) suggest that the first appearance of cancellate wall texture in planktonic foraminifera is a phylogenetically significant feature which would make a better criterion for defining the top of the Cretaceous.

The question of how continuous K–T boundary sections are has been the subject of some discussion. MacLeod and Keller (1991) utilised the graphic correlation method to demonstrate that all deep-sea sections have a hiatus at the boundary, but Olsson and Liu (1993) question the reliability of their data sets. Olsson and Liu believe that some of the disputes concerning precise location of the boundary in a number of key sections, especially relatively shallow-water ones in the US Gulf Coast region, can be resolved if the sedimentary setting is taken into account (Fig. 9.2). In deep-sea sites, the K–T transition is marked by an abrupt decline in sedimentation rate, whereas in outer-shelf and upper-slope sections there is a lithological change from carbonate-dominated rocks to black clay, together with evidence of a decrease in sedimentation rate. In the middle and inner shelf there is a possible disconformity. There is a low carbonate content in all environments. In pelagic environments this relates to the mass extinction of calcareous plankton, in shelf environments to the increase in terrigenous input related to sea-level lowstand. The 'tsunami bed' at Brazos is basal Palaeocene in age; one cannot rule out decisively either a sea-level fall or tsunami event.

Extinctions in the oceans
Planktonic foraminifera

No one disputes that planktonic foraminifera underwent a dramatic turnover across the K–T boundary, the most spectacular in the history of the group, but argument persists about how catastrophic this change was. The first important study relevant to this subject was by Smit (1982), who maintained that all species but one went extinct at the boundary, and that the catastrophic nature of this extinction was consistent with the Alvarez impact scenario. Keller (1988) undertook a detailed study of the extinction record at El Kef and claimed that in fact they occurred sequentially, starting below the boundary and with at least 30% of species surviving into the Palaeocene. She maintained that earlier work had missed many small-sized species. According to Olsson and Liu (1993), such conflicting results may be due to discrepancies in placement of the boundary, sampling biases or variations in methods and taxonomic concepts. They examined species from several K–T sections and concluded that all but three species became extinct at the boundary, with no apparent stepwise or sequential extinctions; species reported by Keller to have gone extinct below the boundary at El Kef continue up to the boundary elsewhere.

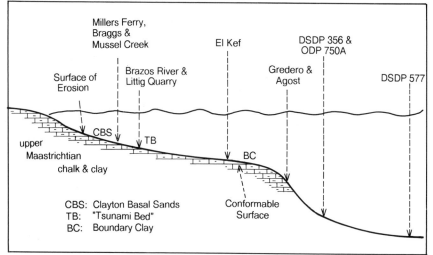

Fig 9.2 Stratigraphic settings of key **K–T** boundary section localities. Afer Olsson and Liu (1993).

The extinction of planktonic foraminifera at El Kef was subjected to a 'blind test' of samples collected in 1993 in the presence of both Smit and Keller. The results of analyses by four other micropalaeontologists were reported at the Lunar and Planetary Institute conference in Houston (1994). The stratigraphic ranges of the identified fossils indicated that a minority of species (2–21%) disappeared before the K–T boundary, but at least half disappeared exactly at the boundary (Ward 1995). This result is something of a compromise between the positions of Smit and Keller, but is certainly consistent with a catastrophic event according to a reasonable usage of the term, although a geological 'instant' could last hundreds or even thousands of years.

Their study of faunas in Mexican basinal sections led Longoria and Gamper (1995) to propose that biological transformations in the form of heterochrony could provide an alternative to account for at least some of the extinctions. Thus four major groups of Hedbergellacea were involved in the K–T evolutionary change. Abathomphalids and globotruncanids underwent size reduction and had stepwise extinction. Other groups, for example the guembelitrids, are ancestors of Tertiary Globigerinacea. The duration of biological transformation, from the extinction of *Abathamphalus mayaroensis* to the first appearance of cancellate wall structure, is difficult or impossible to determine at present. Therefore the controversy about whether the faunal change is instantaneous (catastrophic) or extended (gradual) remains unresolved because there is no unequivocal procedure to estimate the time involved in the K–T boundary bio-event.

No sudden mass extinction of planktonic foraminifera has been recognised at high palaeolatitudes. Indeed, at Nye Kløv in Denmark nearly all the species are claimed by Keller *et al.* (1993b) to have survived and thrived well into the Tertiary, when they gradually disappeared. At high latitudes, according to these authors, only unspecialised small cosmopolitan and presumably eurytopic species were able to thrive. The species that disappeared in low latitudes were large, specialised forms, and the evidence is thought to suggest that the effects of the K–T boundary event may have been more severe at low latitudes.

The contrasting views concerning the degree to which the K–T event was catastrophic depend to a considerable extent on the proportion of survivors. The more catastrophist micropalaeontologists have argued that 'Cretaceous' fossils in basal Tertiary sediments are reworked, but the isotopic study of Barrera and Keller (1990) showed that the carbon and oxygen isotope ratios of post-K–T *Heterohelix globulosa* populations differed substantially from pre-K–T populations at the same locality; if reworked, they should be similar. Isotopic analyses are not, however, feasible for the vast majority of potential K–T survivor species, but MacLeod and Keller (1994) have promoted a biogeographic method. If the occurrence of 'Cretaceous' species in Danian strata accurately reflects the survivors, then both these and indigenous Danian populations should have similar biogeographic distributions, but if they are reworked, then two separate biogeographic signals are likely.

MacLeod and Keller's study is based on 21 sections across the world, including deep-sea cores. They claim that their raw data show that in the most biostratigraphically complete sections or cores, elevated numbers of species extinctions are not tied to the K–T boundary with evidence of impact. Relative to preceding and succeeding faunas, zone Po faunas represent a depauperate assemblage of highly cosmopolitan species dominated by 'Cretaceous' taxa. Application of high-resolution sampling techniques that enhance faunal recovery has resulted in the widespread recognition of Cretaceous morphotypes in the Danian; the reworking alternative is dismissed as implausible. MacLeod and Keller argue that a 'catastrophic' turnover occurs only in stratigraphically incomplete sections, for example in the deep sea. There was a significant reduction in species richness in the uppermost Maastrichtian *P. deformis* Zone.

A more comprehensive global biogeographic analysis of K–T planktonic foraminifera has been undertaken by MacLeod (1995), who argues for a faunal turnover that began in the mid-Maastrichtian and continued through the Po and P1 a–c biozones of the Palaeocene. This must have taken at least 1–2 m.y., with the transition taking place at different rates in different provinces. Of 70 species that occur within the uppermost Cretaceous *P. deformis* Zone, 39 (56%) became extinct within this zone. MacLeod claims that the overwhelming majority of extinctions did not occur synchronously, involved species with very low abundance, and are not associated with independent evidence of bolide impact. Rather, they are spread out over the entire upper Maastrichtian and predominantly involve large, highly ornamented, presumed deep-dwelling species.

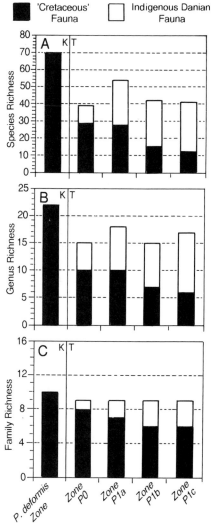

Fig 9.3 The influence of taxonomic scale dependency on estimates of the magnitude of the K–T planktonic foraminiferal turnover. After MacLeod (1995).

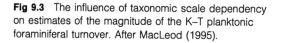

The smaller, morphologically simpler, more abundant presumed surface-dwelling faunas, which should have been devastated according to various catastrophic impact scenarios, were almost entirely unaffected by this initial phase of faunal transition. Extinctions were largely confined to the Tethyan Realm. In comparison, none of 12 cosmopolitan species went extinct. Carbon isotopic analyses show that low- and mid-latitude surface-water productivity began a marked decline in the upper part of the *deformis* Zone, and this is thought to be the likely causal factor for the initiation of the turnover. The potential effects of taxonomic scale dependence are demonstrated in Fig. 9.3. Species-level tabulations show a substantial decrease in biozone-averaged species richness between the uppermost Cretaceous and lowermost Tertiary. Genus-level tabulation shows a severe decline in the perceived magnitude of the K–T turnover, and family-level tabulation shows virtually no change across the boundary.

Benthic foraminifera

In contrast to the planktonic representatives of the group, benthic foraminifera appear to have suffered relatively little extinction at the K–T boundary. This is apparent from studies of fossils extracted from sediments in shallow (Hansen *et al.* 1987), intermediate (Kaiho 1992) and deep water (Thomas 1990). From his studies in Hokkaido, northern Japan, Kaiho (1992) documents a 10% extinction rate for calcareous benthic foraminifera inferred to have lived at intermediate water depths, below about 150 m, compared with an 80% extinction rate within the photic zone. This is consistent with a scenario involving a catastrophe affecting surface but not deeper waters.

From her studies of material from deep-sea cores in the Southern Ocean, Thomas (1990) recognised not an extinction event but a period of increased turnover of species starting probably several hundred thousand years before the era boundary; bioturbational mixing is thought unlikely because samples were taken from the least bioturbated intervals. Inferred epifaunal species were dominant in the Cretaceous, with a pronounced peak just above the boundary, after which the relative abundance of infaunal species recovered. The increase in relative abundance at the boundary of epifaunal species, which are thought to signify low nutrient conditions, and/or high oxygen content of the bottom waters, can plausibly be related to mass extinction in the plankton, and the sharp productivity fall recorded by carbon isotope data.

A detailed, very high-resolution analysis by Coccioni and Galeotti (1994), the first of its kind for benthic foraminifera, has led to rather different conclusions. These authors studied the section at Caravaca, in southern Spain, considered by MacLeod and Keller (1991) to be one of the six best land-based sections that show the most complete record across the boundary. Forty samples were collected from 80 cm below to 120 cm above the boundary clay, probably representing approximately the last 20 000 years of the Cretaceous and first 60 000 years of the Palaeocene. On the basis of depth zonation of planktonic foraminifera, it is believed that the deposits were laid down at depths of 600–1000 m, in a middle bathyal environment. At the K–T boundary there is a sudden and marked decrease in generic richness and what is termed the dissolved oxygen index, associated with a concurrent increase in faunal density and increase in the relative abundance of infaunal morphotypes. This is held to indicate a sudden and exceptionally large flux of organic matter to the sea floor. Oxidation of this would be responsible for the decrease in dissolved oxygen index. Other indications have been reported of anoxic or hypoxic conditions at middle to upper bathyal depths on continental margins (Brinkhuis and Zachariasse 1988, Kaiho 1992;), but there is apparently no evidence for oxygen deficiency at deep oceanic sites. The low-oxygen conditions at the sea floor are considered by Coccioni and Galeotti to be the consequence of the accumulation of a 'nutrient soup' resulting from mass mortality at the sea surface.

A sudden and widespread oxygen deficiency is also inferred by Speijer (1994) for intermediate southern Tethyan waters, embracing the northern margins of Africa and Israel. In particular, the exceptionally well-studied section at El Kef shows no sign of gradual change towards the K–T boundary; instead there was an abrupt extinction followed by a gradual recovery. The sudden drop in diversity at the boundary is associated with the proliferation of opportunistic low-oxygen tolerant shallow-water taxa, with the oxygen deficiency being most severe just above the boundary. More generally, many species with higher nutrient demands, in particular those with elongate ('endobenthic') test shapes, suffered extinction, whereas most oligotrophic deep-sea species were hardly affected. Speijer explained this extinction selectivity by the enormous decrease in nutrient supply to the sea floor, following the collapse of pelagic primary productivity. The contrast with Coccioni and Galetti's 'nutrient soup' explanation could hardly be more marked, but at least both interpretations agree on the subject of oxygen deficiency.

Calcareous nanoplankton

'Calcareous nanofossils' is the term used for small calcareous fossils (2–20 μm), most of which are assumed to be derived from coccolithophorid algae that lived in the plankton. It has been widely recognised since the pioneering work of Bramlette and Martini (1964) that a massive change took place in coccolith assemblages at the K–T boundary. Most Cretaceous species, perhaps some 85%, became extinct at or shortly after the boundary, but at least 15–20 species survived (Perch-Nielsen 1986; Hallam and Perch-Nielsen 1990; Habib *et al.* 1992). A detailed study of the classic El Kef section by Pospichal (1994) revealed that more than 100 species became extinct across the K–T boundary and only 15–20 survived the event. Pospichal's results demonstrate no stepwise or sequential decline in diversity below the boundary, as claimed for planktonic foraminifera by Keller (1988). The marked fall in species richness at the boundary is exceeded by an even more dramatic fall in individual abundance. Pospichal claims that results from other K–T boundary sections also indicate no change before the boundary.

Other microorganisms

The organic-walled marine phytoplankton seem to have survived the K–T event better than calcareous phytoplankton, with only a minority of species going extinct (Hallam and Perch-Nielsen 1990). Such changes as can be recognised through K–T boundary sections seem to relate largely to facies changes associated with sea-level rises and falls (Brinkhuis and Zachariasse 1988; Habib *et al.* 1992). It should be remembered that dinoflagellates have a cyst resting stage in the benthos, so that the group could survive disturbances in surface waters. The same phenomenon could account for the high survival of diatoms at the boundary, because resting spores occur at one stage in the group's life history (Kitchell *et al.* 1986).

In comparison with the amount of evidence available for calcareous micro- and nanofossil groups, there are few localities at which the change from the Mesozoic to the Cenozoic radiolarian fauna can be traced; no complete K–T boundary section containing radiolaria has yet been described. The available evidence has appeared to indicate a profound decline in radiolarian diversity at the K–T boundary. According to Sanfilippo *et al.* (1985), a high proportion of Mesozoic genera and families became extinct, only a few surviving into the early Palaeocene. This interpretation differs, however, from that of Hollis (1996) for New Zealand faunas. He perceives no significant turnover at the K–T boundary, but instead an early mid-Palaeocene decline.

Molluscs

Some of the most characteristic and abundant macroinvertebrates of the Cretaceous were the inoceramid and rudist bivalves, the ammonites, and belemnites. All had disappeared by the end of the period, and until the relatively recent recognition of a mass extinction event among the calcareous micro- and nanoplankton this was one of the clearest means of distinguishing the Cretaceous from the Tertiary. All the macrofossil groups in question had declined before the end of the period. In the case of the inoceramids, ammonites, and belemnites the decline in diversity was prolonged over millions of years, with an accelerated decline through the Maastrichtian (Fig. 9.4). The rudists underwent a significant radiation in the Late-Cretaceous but had all but disappeared by the beginning of the late Maastrichtian, a few million years before the end of the period (Kauffman 1984). With regard to the belemnites, in the late Campanian and early Maastrichtian *Belemnitella* and *Belemnella* were dominant in the

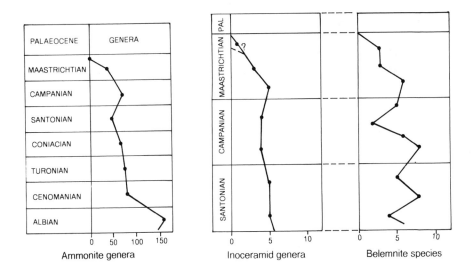

Fig 9.4 Decline in molluscan diversity in the Late Cretaceous. After Hallam and Perch-Nielsen (1990).

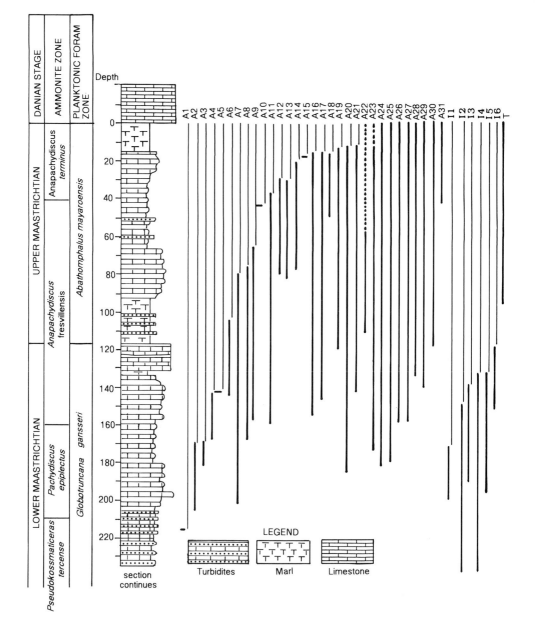

Fig 9.5 Composite range chart of ammonites and inoceramid bivalves for Bay of Biscay stratigraphic sections. A = ammonite species, I = inoceramid species, T = *Tenuiptera*. Simplified from Ward *et al.* (1991).

Boreal realm. At the end of the Cretaceous only rare representatives of these genera remained alongside the extremely rare *Fusiteuthis* (Christensen 1976).

Only in recent years have detailed, fine-grained studies been made of ammonite and bivalve extinctions at or close to the K–T boundary. Some the the best sections suitable for this purpose occur in several excellent coastal exposures in the Basque country of France and Spain adjacent to the Bay of Biscay; they are among the most complete of all known land-based K–T sections. The boundary, as determined by planktonic foraminifera and calcareous nanofossils, is marked by a distinctive dark clay-rich marlstone that separates red or brown Maastrichtian marlstones from pink Danian limestones; an iridium anomaly occurs at the base of the boundary marlstone. Ward *et al.* (1991) and MacLeod (1994) have studied the stratigraphic distribution of the two most abundant macrofossil groups, the ammonites and inoceramids. The ammonites maintained a species richness of 8–16 through the late Maastrichtian and then disappeared simultaneously at or less than 20 cm below the K–T boundary (probably equivalent to less than 10 000 years). The gap between the boundary and the last-appearance datums of eight species, within the uppermost 1 m, may be a sampling artefact (the Signor–Lipps effect). In contrast, the inoceramids declined gradually in abundance and disappeared at the beginning of the late Maastrichtian from a minimum of six species in the early Maastrichtian. The enigmatic genus *Tenuiptera*, a pteriomorph related to but not congeneric with *Inoceramus*, extends to 1 m below the boundary (Fig. 9.5). The mid-Maastrichtian extinction of inoceramids apparently reflects a global pattern (MacLeod 1994). It is believed that they might have disappeared because the influx of oxygen-rich Antarctic bottom waters largely eliminated the low-oxygen soft bottom conditions in the deep ocean.

The only other area in Europe where good molluscan data are available is in Denmark (Fig. 9.6). At the top of the well-known Cretaceous section at Stevns Klint, immediately below the Fish Clay, seven ammonite genera are recorded (Birkelund and Hakansson 1982). However, a well-known hardground directly above the last ammonite clearly indicates a stratigraphic hiatus. Following Heinberg's (1979) detailed work on the topmost Cretaceous hardground fauna at this locality, where preservation of originally aragonitic bivalve shells is unusually good, a dozen genera hitherto believed to be Cenozoic are now known to occur also in the latest Maastrichtian. Evidently a radiation event took place before the end of the Cretaceous and was little affected by the extinction event.

Rich molluscan (mainly bivalve) faunas are also known from the American Gulf Coast region. In both the Brazos River (east Texas) and Braggs (Alabama) sections there appears to be no significant change through the upper Maastrichtian before the K–T boundary; a species decline in diversity at the Brazos River is attributable to local facies change (Hansen *et al.* 1987). More recent work does suggest, however, that diversity decreased 60 cm below an event bed associated with the K–T boundary, but that a more abrupt drop in species diversity and a major ecological shift to deposit feeders occurred at the K–T boundary (Hansen *et al.* 1993). Bryan and Jones (1989) undertook a particularly detailed study of the molluscan fauna in the Braggs section. These authors find no persuasive evidence of either gradual or stepwise extinctions through the upper Maastrichtian Prairie Bluff Formation. Time resolution ranges up to 0.8 m.y. but could be much less. Thus the extinctions may have been cataclysmic or could have extended over a period of several hundred thousand years. A suspected truncation of the boundary, relatively high time resolution, comparable preservation, similar lithologies

Fig 9.6 K–T boundary section at Stevns Klint. Denmark showing 1, latest Maastrichtian chalk (capped by a hardground); 2, the Fish Clay at the boundary, and 3, Danian bryozoan calcarenites. Thomas Wignall for scale.

and persistently high Cretaceous and low Tertiary diversities all argue for a primary, not artefactual, abrupt faunal decline. A detected Lazarus effect results from local extinction and (or migration), not taphonomic or collection difficulties.

Using the molluscan data of Bryan and Jones (1989), Koch (1991) has undertaken a statistical study to determine whether an apparent stepwise extinction pattern in the Prairie Bluff Formation could be due to a sampling effect. The relative abundance of taxa fits well the log-normal frequency distribution. Because the occurrence frequency and relative abundance of each taxon is known, it is possible to predict the proportion of Bryan and Jones species that range to the top of the lower, middle, and upper Prairie Bluff divisions. The predicted pattern is not significantly different from the observed pattern (Fig. 9.7) and thus the stepwise pattern is attributable to a sampling effect. This result is in accord with what Bryan and Jones inferred.

The only locality in the southern hemisphere where there is good information across the K–T boundary for a number of fossil groups is Seymour Island, west Antarctica (Feldman and Woodburne 1988). Zinsmeister *et al.* (1989) inferred for molluscs a gradual, not catastrophic, change through the succession, but some doubt is thrown upon this conclusion by the statistical work of Marshall (1995), using well-recorded ammonite data. Elliott *et al.* (1994) report a substantial iridium anomaly ~9.5 m above the last recorded ammonite occurrence. and this interval is the one thought most likely to contain the K–T boundary. In Marshall's opinion the ammonite data are compatible with a mass extinction that coincides with the iridium anomaly but is also compatible

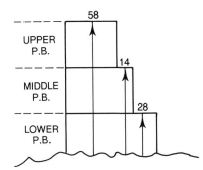

BRAGGS, ALA. (PER 100 SPP.)

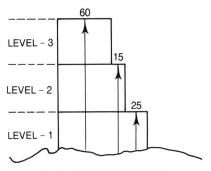

PREDICTED – BASED ON ACTUAL
SAMPLE SIZE (PER 100 SPP.)

Fig 9.7 Stepwise molluscan species extinction pattern towards the Cretaceous–Tertiary boundary at Braggs, Alabama (upper diagram) compared with expected stepwise extinction pattern for 100 species ranges towards a sudden mass extinction (lower diagram). After Koch (1991).

with an earlier mass extinction or gradual decline. They are not, however, compatible with a gradual disappearance over more than 20 m of section.

Other macroinvertebrates

Other groups of macroinvertebrates are generally much less common than the molluscs and information is correspondingly more limited. The best information comes from Denmark.

An interesting study was performed on the Maastrichtian and Danian micromorphic brachiopods of Nye Kløv, the most complete Danish section, by Surlyk and Johansen (1984). Whereas the Maastrichtian is dominated by cancellothyrids, the Danian is dominated by *Argyrotheca*. Of 30 species recognised in the Lower Danian, only 6 occur also in the Maastrichtian, and there was indeed an abrupt extinction at the very end of the Cretaceous of most of the Maastrichtian fauna, coincident with the plankton mass extinction, which was not preceded by any decrease in either diversity or individual abundance (Fig. 9.8). The brachiopod extinction is interpreted as being due to the

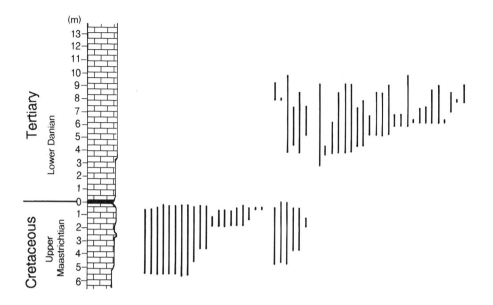

Fig 9.8 Range chart of brachiopod species in the Cretaceous–Tertiary boundary sequence of Nye Kløv, Denmark. Simplified from Surlyk and Johannsen (1984).

destruction of the chalky substrate habitat of this specialised group as a consequence of the mass destruction of calcareous plankton. Restoration of chalky substrates in the early Danian led to a return of the brachiopods, but the species were almost wholly new.

The Maastrichtian and Danian bryozoans of Denmark exhibit an immense diversity, with some 500 species being recognised, and they have received detailed attention from the viewpoint of end-Cretaceous extinction by Hakansson and Thomsen (1979). The two major groups, cyclostomes and cheilostomes, exhibit different responses. Seventy-five per cent of species of the low-diversity, evolutionarily conservative cyclostomes occur in both the Maastrichtian and Danian, whereas less than 20% of the Maastrichtian cheilostome species persisted into the Danian. In the Nye Kløv section, there are 60 cheilostome species restricted to the Danian, with only 11 common to both stages. There was an extreme diversity reduction in the basal Danian followed by a rapid rise to a stage maximum 6 m above the K–T boundary. This corresponds to the time of significant increase in brachiopod diversity (Fig. 9.8).

According to Birkelund and Hakansson (1982) there was virtually no change across the K–T boundary in the genera of crinoids, asteroids, and ophiuroids, but there was substantial change at the species level, with little overlap between the Maastrichtian and Danian. The basal Danian fauna is dominated by crinoids, especially *Bourgueticrinus*, and is marked by a complete absence of cyclostomes, brachiopods, and calcitic bivalves.

It is interpreted by Birkelund and Hakaansson as a pioneer community capable of directly colonising a soft bottom.

Reptiles

Several important groups of marine reptiles characteristic of the Mesozoic had disappeared by the close of the era. Whereas the ichthyosaurs and plesiosaurs had been in slow decline through the Cretaceous, the mosasaurs underwent a major Late Cretaceous radiation, and indeed did not exist before this time. They apparently evolved from small amphibious aigialosaurs and comprised up to 70 species, many of which appeared in the late Campanian and Maastrichtian, exhibiting a global distribution, with representatives in high-palaeolatitude regions including Australia, northern Canada, and Sweden. Their dramatic disappearance is consistent with a catastrophic extinction event (Lingham-Soliar 1994).

Extinctions on the continents

The only groups for which a detailed record of change has been established are terrestrial and aquatic vertebrates and plants, to which discussion must accordingly be confined. However, it is worth noting that for the extremely important and diverse insect faunas, for which the fossil record has improved considerably in recent years, there are no indications of any significant change across the K–T boundary (Labandeira and Sepkoski 1993). Nor is there any good evidence of a extinction event among birds (Chiappe 1995).

Vertebrates

Although they attract the greatest popular interest, dinosaurs are one of the least satisfactory groups for this kind of study, because of the paucity of suitable stratal sections and the comparative scarcity of fossil material. Virtually all the conclusions that have been drawn about the final dinosaur extinction episode derive from a few sections in the North American Western Interior, arguably the only complete succession of vertebrate-bearing strata across the K–T boundary, with the best sections being in eastern Montana. For all we know, the group might well have gone extinct in other parts of the world before the end of the Cretaceous, or even locally have persisted into the Palaeocene. In any case, too much has been made of the end-Cretaceous dinosaur mass extinction as a unique event. In fact, as Padian and Clemens (1985) have pointed out, the dinosaur generic turnover rate was exceptionally high throughout the group's history, and the most unusual feature of the end-Cretaceous event was the failure of a new replacive group of dinosaurs to emerge. The implication of the high generic turnover rate is that dinosaurs were always relatively vulnerable to extinction throughout their long history, and that no environmental event of exceptional magnitude need necessarily be invoked.

It now seems to be generally accepted that the last dinosaur bones disappear about a metre below the iridium-enriched layer generally taken as the K–T boundary. An earlier argument took place about the statistical probability of finding dinosaurs within this short interval, on the assumption that the group went extinct as a direct consequence of bolide impact, as signified by the iridium anomaly. Fastovsky's (1987) work on the depositional environments of the Maastrichtian Hell Creek Formation and Danian Tullock Formation suggests that the unfossiliferous metre could well be the product of leaching of fossils from the uppermost Hell Creek Formation by acidic groundwaters derived from widespread swamps, because the inception of deposition of the Tullock Formation reflects a shift to a much wetter environment, with high water tables.

Whereas Russell (1979) has claimed that the dinosaurs were cut short in their prime, with the group exhibiting no decline from a high Late Cretaceous diversity level at the end of the period, other workers have expressed a different view. Carpenter and Breithaupt (1986) studied the latest occurrence of nodosaurid ankylosaurs in Wyoming and Montana, using the relative abundance of teeth as a good measure of species abundance, and inferred a real decrease in population levels during the late Maastrichtian, with the group going extinct well before the end of the stage. That this pattern could be true of dinosaurs in general is suggested by the more comprehensive work of Sloan *et al.* (1986). According to these authors, who take into account the discovery of articulated bones to eliminate the possibility of reworking, dinosaur extinction in Wyoming, Montana, and Alberta was a gradual affair, beginning about 7 m.y. before the end of of the Cretaceous and accelerating rapidly in the last 0.3 m.y. This decline up the succession cannot apparently be dismissed as an artefact; there is more outcrop available for examination in the top 30 m of the Cretaceous in the Montana section than an equivalent thickness of strata below, where more dinosaur remains have been found. The rapid reduction through time of both diversity and abundance is attributed to a combination of environmental deterioration and more tentatively to competition from immigrant ungulate mammals. Sloan *et al.* also make the more controversial claim (see the discussion in *Science*, **234**, 1170) that dinosaur 'genera' (best interpreted as species) persisted into the Early Palaeocene in Montana, with only four disappearing at the end of the Maastrichtian (see also Rigby *et al.* 1987). The fossil teeth on which this claim is based occur in undoubted Palaeocene strata, above the local iridium anomaly and correlated palynological change, but in stream channel sediments, which raises the possibility of derivation by reworking from Maastrichtian strata. Sloan *et al.* discount this on two grounds: first that the teeth bear no signs of abrasion; and second, that dinosaurs are less common than mammals in the Maastrichtian sediments through which the channel is cut, yet no mammals have been found with dinosaurs in the channel deposits.

These interpretations of gradual dinosaur extinction have been challenged by Sheehan *et al.* (1991) who conducted a very thorough team study of Hell Creek Formation outcrops in Montana and North Dakota. They subdivided the deposits, 70–90m thick, into lower, middle, and upper units, and distinguished three facies, thalweg, point bar, and floodplain; the first two were also analysed together as channel deposits. Large-scale standardised survey procedures were undertaken by many trained workers over several field seasons; eight dinosaur families were recorded, all of which range into the upper unit. Ecological diversity was also studied because it measures changing abundances of taxa rather than just presence or absence. Sheehan *et al.* (1991) find no statistically meaningful fall in ecological diversity through the Hell Creek Formation, at least at family level, a result which is held to be consistent with a catastrophic impact-related scenario. However, Archibald (1992a, b) argues that the persistence of all eight families through the Hell Creek Formation provides only weak evidence against a 'deterioration' of the dinosaur fauna, as such familial persistence could have been accompanied by a 43% decline in the total number of genera.

A more comprehensive study has been undertaken by Archibald and Bryant (1990) of all 111 species of non-marine vertebrates in the most analysed part of Montana. Extinction was evidently not simply size-related, because both immense dinosaurs and diminutive mammals were severely affected while very large crocodiles and very small amphibians were not. There is a reasonably good correlation between terrestrial and aquatic habitats, with animals in the latter suffering a much lower extinction rate, but the wholly aquatic elasmobranchs disappeared completely. The rapid species turnover of mammals at the close of the Cretaceous gives the impression that the group suffered nearly total extinction. However, allowance must be made for 'pseudo-extinctions', specifically anagenetic and cladogenetic speciation events (Archibald 1993). This underlines the importance of phylogenetic analysis of individual clades (cf. Patterson and Smith 1987). Several mammal species apparently extinct in the study area occur elsewhere in post-Cretaceous sediments. When only common species are utilised, approximately 53% survived the K–T event. Allowing for pseudo-extinctions, the total survivorship approaches 65%. Archibald and Bryant suggest that the K–T boundary marks a time of accelerated turnover for mammals rather than a catastrophic mass extinction.

Sheehan and Fastovsky (1992) made some criticisms of Archibald and Bryant's method of analysis, but concurred in accepting that the elimination of all rare species removes the most important bias of the data base, namely the sampling error of very rare species. They discriminate two assemblages, freshwater (fish, champsosaurs, crocodiles, turtles) and land-dwelling (salamanders, frogs, dinosaurs, lizards, mammals). When Archibald and Bryant's data are corrected for the sampling effect of rare taxa, there can be inferred a 90% survival of the freshwater and 12% survival of the land-dwelling species.

Most recently, Hurlbert and Archibald (1995) have undertaken a close examination of the two diversity indices used by Sheehan *et al.* (1991), the Shannon index and expected species richness, and endeavour to demonstrate that they are not capable of answering the questions posed by the Hell Creek study. They find no support for the contention that they tell us something about extinction rates prior to the K–T boundary. Hurlbert and Archibald conclude that the poor quality of the dinosaur record precludes us from determining whether or not there was gradual or abrupt extinction.

In view of the considerable geographic differentiation among Late Cretaceous tetrapods global-scale generalisations from western North America are unwise, but unfortunately very little adequate information is as yet available from elsewhere. One of the best-studied regions is in Languedoc, southern France (Le Loeuff *et al.* 1994). Dinosaurs from upper Maastrichtian red beds in the Corbières region demonstrate a change from a sauropod-dominated to a hadrosaur-dominated community, thought probably to be associated with the late Maastrichtian regression and climatic change, since the palynology indicates the replacement of a tropical-subtropical flora by a more temperate one. These changes cannot, however, provide a convincing explanation for the later extinction of the dinosaurs.

Plants

It has been established for some time that the Aquilapollenites pollen province of western North America and north-eastern Asia suffered up to 75% species extinction across the K–T boundary (Hickey 1981, 1984) but change elsewhere was less pronounced, such as in the Normapolles Province of eastern North America and Europe (Collinson 1986). In the southern hemisphere, in Antarctica and New Zealand, little or no change is recognisable (Askin 1988, 1992; Johnson 1993).

One of the more striking discoveries in the years directly following publication of the Alvarez bolide impact hypothesis was that of the so-called fern spike in the North American Western Interior, coincident with an iridium anomaly and a horizon of shocked quartz (Tschudy *et al.* 1984; Tschudy and Tschudy 1986). This coincidence has since been traced from New Mexico to Alberta, and marks an abrupt change independent of lithofacies. Extinction levels varied among angiosperms, gymnosperms, and pteridophytes but simultaneously affected different plant communities throughout the region. The extinctions were most drastic among endemic angiosperms, less so among conifers and other gymnosperms and least among ferns, other pteridophytes, and mosses, hence the fern spike (Nichols and Fleming 1990). In their work in the Canadian part of the Western Interior, Sweet *et al.* (1990) recognise five sequential changes in the palynofloras of the uppermost Maastrichtian and lowermost Palaeocene. As these changes either pre-date or post-date the K–T boundary, marked by a geochemical anomaly and palynological extinction event, they cannot plausibly be explained by an impact-related scenario.

Plant macrofossils offer greater taxonomic resolution than microfossils and are not subject to reworking into younger strata, but there has been only one thorough study of the record across the K–T boundary. This is by Johnson and Hickey (1990), who investigated the northern Great Plains and Rocky Mountains region embracing Montana, Wyoming, and the Dakotas. The best data come from the latest Maastrichtian and earliest Palaeocene of North Dakota. Their results are presented in Fig. 9.9. There was a substantial floral turnover through the Maastrichtian Hell Creek Formation and Palaeocene Fort Union Formation, but by far the most marked change is at the K–T boundary, with only 21% of the flora persisting into the Palaeocene. There was an abrupt disappearance of most dominant taxa, including nearly all dicotyledonous angiosperms. The basal Palaeocene flora is composed of taxa that were extremely rare in the latest Cretaceous or that lived in environments not represented by deposits. The palynological and macrofloral record for the remainder of the Western Interior basins from New Mexico to the Yukon is considered by Johnson and Hickey to be consistent with this scenario.

Extinction selectivity

It has been argued by Stanley (1986a) that tropical organisms in the marine realm are relatively vulnerable to extinction and this has been maintained specifically for Cretaceous bivalves by Kauffman (1984). However, in their analysis of the geography of end-Cretaceous marine bivalve extinctions, Raup and Jablonski (1993) failed to find any evidence to support this, if the rudists, which were confined to the tropics, are excluded.

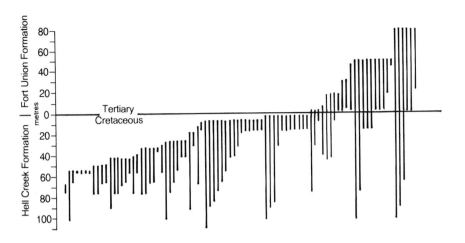

Fig 9.9 Biostratigraphic range chart for all megafloral taxa that occur at more than one stratigraphic level in the composite section at Marmarth, North Dakota. Simplified from Johnson and Hickey (1990).

With regard to the plankonic foraminifera, Keller *et al.* (1993b) have also claimed that end-Cretaceous extinctions were more severe among the low-latitude representatives. Among terrestrial plants, extinctions in the tropics were evidently less severe than in the northern hemisphere, most notably in the Aquilapollenites Province (Hickey 1984). As already noted, the relative extinction vulnerability in this province of the angiosperms and to a lesser extent the gymnosperms led to the production of the so-called fern spike. It is generally believed that plants are more resistant to extinction than animals, because of the resistance to short-term environmental disturbance conferred by seeds and rhizome systems, and indeed there was a rapid recovery of angiosperms in the North American Western Interior following the disturbance indicated by the fern spike (Tschudy and Tschudy 1986).

The fossil record is very clear about a mass extinction of the plankton at the end of the Cretaceous, implying a catastrophic environmental deterioration of ocean surface waters. As pointed out earlier, groups that survived relatively unscathed, such as the diatoms and dinoflagellates, could have done so because of having benthic resting stages or the possibility of encystment during times of unfavourable conditions. The implications of plankton mass extinction for the benthos were explored in an important pioneering paper by Sheehan and Hansen (1986). Filter-feeding organisms such as bryozoans, crinoids, corals, articulate brachiopods, and filter-feeding bivalves would have been very vulnerable to disturbance of their plankonic food supply and all these groups experienced substantial extinctions. On the other hand, detritus feeders, carnivores, and scavengers would have been relatively resistant and, as exemplified by detailed studies in the Brazos River sections, experienced comparatively low extinction rates.

Using data from the Gulf and Atlantic coastal plains of the United States, Jablonski and Raup (1995) confirmed for bivalves that suspension feeders suffered more than deposit feeders, but pointed out that interpretation of the latter group was more complicated than had been assumed. Thus tellinids experienced significantly higher losses than nuculoids and lucinoids, although they have a greater flexibility between suspension and deposit feeding, so their higher extinction rate is puzzling. Furthermore, the greater extinction resistance of nuculoids and lucinoids cannot be linked to broader geographic ranges (cf. Jablonski 1986a). Turning to benthic foraminifera, their relative resistance to K–T extinction is in apparent agreement with the Sheehan and Hansen hypothesis that detritus feeding offered a buffer (Thomas 1990).

Primary productivity decline was seen as an important selective controlling factor for some benthos by Rhodes and Thayer (1991). Available evidence indicates that brachiopods have much lower energy budgets than bivalves. Within the bivalves, swimmers and burrowers should starve sooner than inactive forms attached by a byssus, cemented or reclining passively in recesses. Many bivalves as well as articulate brachiopods survive starvation for longer than the 2–3 months of darkness hypothesised by some impact supporters for the K–T event. Rhodes and Thayer maintain that bivalve genera relatively susceptible to starvation underwent significantly greater extinction than more starvation-resistant genera.

Based on his studies of the US coastal plain deposits, Gallagher (1991) offered an intriguing additional explanation for comparative resistance to extinction of benthic or nektic organisms. It is very likely that typically Cretaceous infaunal and semi-faunal bivalves possessed planktotrophic larvae that would have been drastically affected by mass extinction of the plankton. The common Danian survivors probably exhibited a variety of reproductive strategies independent of the plankonic food supply, such as asexual reproduction of corals and sponges. By and large, lophophorates had lecithotrophic larval stages, freeing them from dependence on plankton. Thus an important determinant of survival across the K–T boundary could, according to Gallagher, have been reproductive strategy. Ammonites probably produced many small embryos that spent more time in the plankton, whereas nautiloid embryos were large-yolked and hatched as nektobenthic swimmers. The smaller ammonite embryos probably starved during the plankton population crash, and this could account for their final dramatic extinction, as recorded, for instance, by Ward *et al.* (1991). Gallagher questions Jablonski's (1986a) generalisation that K–T survivability depended on widespread geographic distribution, because it is inconsistent with the extinction of ammonites and exogyrid bivalves, both of which had widespread ranges. He broadly supports the conclusions of Sheehan and Hansen, however, pointing out that the food chain could have continued up to durophagous predators, such as mollusc-eating fish, and in turn the predators of these, such as mosasaurs, which also went extinct rather dramatically at the end of the Cretaceous.

Sheehan and Hansen (1986) also applied their model of detritus feeding as an extinction buffer to continental environments (see also Sheehan and Fastovsky 1992). Today's niche of mammalian browsers was occupied in the Late Cretaceous by herbivorous dinosaurs, on which the minority of carnivorous dinosaurs would have been dependent for food. Damage to the plant world by inhibition of photosynthesis, as envisaged in the Alvarez scenario, would have inevitably had adverse consequences for the dinosaurs (cf. Buffetaut 1990). On the other hand, many if not most of the small-sized mammals would have been insectivorous. This is clearest for the eutherians and detatheridians, and these groups had the lowest extinction rates. The high survival rate of turtles, crocodiles, and champsosaurs is explicable, according to Sheehan and Hansen, by the fact that freshwater ecosystems are based on land-derived detritus.

Geochemical changes across the boundary

Platinum-group and other trace elements

As is widely appreciated, the first serious evidence for bolide impact came from the discovery by Alvarez *et al.* (1980) of a large anomaly of the platinum-group element iridium in the K–T boundary clay at Gubbio, Italy. Since that time an iridium anomaly at this level has been found at numerous localities across the world. In their analyses of thousands of samples throughout the Phanerozoic record, Orth *et al.* (1990) found the K–T anomaly to be far greater than at any other horizon, strongly suggesting to them a unique event associated with bolide impact.

Osmium is another platinum-group element, and Turekian (1982) suggested that the $^{187}Os/^{186}Os$ ratio in K–T boundary clays supported an extraterrestrial origin, because it compares more closely with that found in meteorites than in rocks of the continental crust. He conceded, however, that there was likely to be no difference from mantle material. The more recent work on K–T boundary beds in Turkmenistan by Meisel *et al.* (1995) confirms that the osmium isotope ratios are not plausibly accounted for by derivation from volcanic rocks but are consistent with an extraterrestrial or mantle source.

Considering the platinum-group elements as a whole, Tredoux *et al.* (1989) found elevated concentrations in K–T boundary sections which show marked differences in both hemispheres. They argued that these and other differences are not easily explained by impact and that mantle sources should also be considered. Further complications to interpretation have emerged from detailed work on the classic boundary clays at Stevns Klint, Denmark, and Caravaca, Spain. Earlier work had identified a layer enriched in noble metals as fallout from a world-embracing dust cloud indicated by impact, but both Schmitz (1988) and Elliott (1993) have convincingly demonstrated that the Stevns Klint clay is locally derived, water-transported material. Furthermore, even if the metals are ultimately derived from extraterrestrial matter, their enrichment in boundary clays in various localities was probably due to organic and inorganic precipitation from seawater. The abundant pyrite at Stevns Klint was produced in early diagenesis by bacterial activity in anoxic conditions, and is the major carrier of Ni, Co, As, Sb, and Zn, together with rare earth elements (Schmitz *et al.* 1988).

Carbon and oxygen isotopes

In modern oceans the calcareous skeletons of plankton are characterised by positive $\delta^{13}C$ values because the dissolved bicarbonate in surface water is relatively depleted in ^{12}C, due to photosynthesis. On the contrary, bottom waters are relatively enriched in ^{12}C due to respiration of isotopically light organic carbon, and benthic foraminifera are characterised by this. Thus a surface- to bottom-water isotopic gradient exists. Numerous detailed studies of sediments and their contained microfossils across the K–T boundary show a distinct anomaly exactly at the boundary, marked by a negative shift of several per mille (Fig. 9.10). This is generally held to signify a collapse of the gradient due to a productivity fall consequent on mass destruction of the plankton (Hsü and McKenzie 1990).

One of the best studies of this phenomenon is by Zachos *et al.* (1989) who analysed the sediment core of Deep Sea Drilling Project site 577 on the Shatsky Rise in the Pacific.

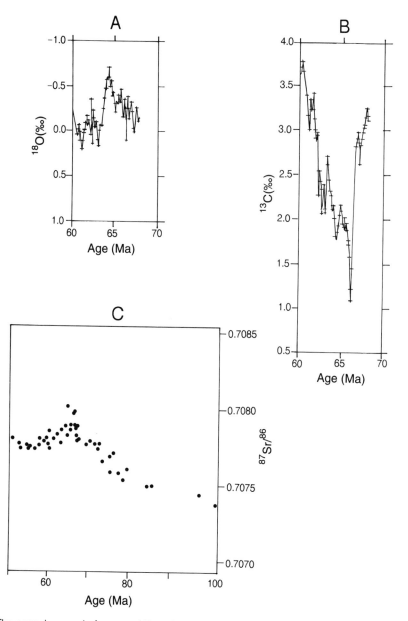

Fig 9.10 The oceanic record of oxygen (A), carbon (B), and strontium isotope (C) changes across the Cretaceous–Tertiary boundary. Adapted from Shackleton (1986) and Elderfield (1986).

This site has yielded the first continuous undisturbed succession of pure nanofossil ooze across the K–T boundary, and the calcareous microfossils have not undergone significant recrystallisation due to burial. The carbon isotope data indicate, as elsewhere, a rapid and complete breakdown of the biologically mediated gradient from surface to deep water. The implied substantial reduction of oceanic primary productivity persisted for about half a million years before the carbon isotope gradient was re-established, a phenomenon not readily explicable by bolide impact or anything else. The environmental change is thought to have begun at least 200 000 years before the K–T boundary. Carbonate accumulation rates decreased by a factor of four in all pelagic marine successions across the boundary examined. The extinction of the plankton and disappearance of the carbon isotope gradient was geologically abrupt, taking place in less than 10 000 years.

According to Keller *et al.* (1993b), the $\delta^{13}C$ shift was largest in the palaeotropics, about 3‰, but was greatly reduced in high palaeolatitudes in Denmark and absent in Antarctica. This suggests that the surface- to deep-water gradient remained virtually unchanged in high latitudes, indicating that surface productivity remained stable. The implication is that Hsü and McKenzie's (1985) Strangelove Ocean was not global in extent. One of the collaborators in this study has also found major $\delta^{13}C$ variations across the lower–upper Maastrichtian boundary, suggesting either important regional or global changes at this time (Barrera 1994). She speculates that erosion of carbonate and organic carbon from extensive Maastrichtian epicontinental marine deposits exposed during regression could have led to transfer of low $\delta^{13}C$ material to 'carbon pools' in the oceans. During the late Maastrichtian sea-level highstand organic carbon accumulations could be recorded as more positive $\delta^{13}C$ pelagic carbonate.

The oxygen isotope record shows no such sharp change across the K–T boundary (Fig. 9.10). The short-term oscillations of the curve in this figure are not to be trusted as environmental signals, because oxygen isotopes are notoriously vulnerable to postdepositional diagenetic changes, even in the deep-sea core record. Maastrichtian data suggest a cooling trend through the stage in high southern latitudes marked by distinct intervals of temperature decrease spanning a million years or less (Barrera and Huber 1990). The mid-latitude site 528 in the south Atlantic has a Maastrichtian record indicative of long-term cooling largely resulting from geologically rapid thermal 'steps', at least in the southern oceans (D'Hondt and Lindinger 1994). The kinds of temperature change envisaged in various K–T impact scenarios would have been far too brief to leave a trace in the oxygen isotope record.

Strontium isotopes

The pattern of $^{87}Sr/^{86}Sr$ change across the K–T boundary is shown in Fig. 9.10. A rising trend through the Late Cretaceous appears to accelerate to a peak at the K–T boundary and then fall into the Palaeocene. Two independent groups of workers, Hess *et al.* (1986) and Koepnick *et al.* (1985), recognised this K–T peak. It was interpreted by Hallam (1987b) as resulting from an increase in riverine delivery of radiogenic strontium to the oceans, derived from continental weathering and runoff associated with a marked fall of sea level and consequent expansion of continental area. The subsequent fall was related to later sea-level rise. An alternative explanation linked with bolide impact was preferred by Macdougall (1988) and Martin and Macdougall (1991), with acid rain provoking an increased rate of continental weathering. These workers based their results not on bulk carbonate but on picked microfossils, and maintained that the positive excursion coincides closely in time with the K–T boundary. The admittedly cruder results of other workers suggests that the sharp rise leading to the peak began before the boundary. If this is true it rules out the impact scenario. More research is clearly needed to resolve the issue decisively.

Possible causal factors

Before discussing possible causal factors for the end-Cretaceous extinctions, it is desirable to evaluate how catastrophic they were, taking into account the evidence reviewed earlier in this chapter. It needs to be remembered that most geologists, who are used to dealing with events lasting for millions of years, would consider a dramatic event lasting a few thousand or even tens of thousands of years something which it would not be an exaggeration to describe as catastrophic. Bolide impact scenarios require something much briefer, however, usually no longer than a few months or years at most. Unfortunately, the stratigraphic record is such that it is unlikely ever to be resolved as finely as this. Therefore the best one can hope to do is establish a pattern which is consistent with the sort of geological instant required by impact.

The best evidence comes from the record of calcareous plankton. There is no dispute among micropalaeontologists that a catastrophic mass extinction took place at the end of the Cretaceous, but no general agreement yet whether it was geologically instantaneous or preceded by thousands of years of gradual or stepwise extinctions. Opinion seems to be moving, however, towards the former interpretation and, taking into account the Signor–Lipps effect, this is perhaps the more probable.

Extinctions among the benthos and nekton are less critical, because the organisms involved would have been dependent ultimately on phytoplankton as the base of the food chain, with a variety of cascading effects possible. So whether it involved starvation of suspension feeders, adverse effects on organisms with planktotrophic larvae, or nektic predators higher in the food chain, it is unsurprising that more recent work has tended to support some kind of end-Cretaceous catastrophic scenario. It should not be overlooked, however, that the important bivalve groups, the inoceramids and rudists, apparently disappeared in the mid-Maastrichtian, several million years before the end of the era, and that other major molluscan groups had been in decline earlier than this.

The evidence for non-marine vertebrates is much poorer, and in fact quite good only for one region of the world, the northern part of the North American Western Interior. In this region also there has been a conflict of views about whether the dinosaur extinctions were gradual or catastrophic. This issue has not yet been decisively resolved, but the extensive and highly focused work of Sheehan *et al.* (1991), favouring the latter interpretation, cannot be lightly disregarded. As with the marine benthos, dinosaurs would ultimately have been dependent on other organisms, namely the plants, and indeed the plant record in the Western Interior appears to be consistent with a catastrophe. This was evidently not global in extent, however, yet dinosaurs disappeared for ever.

Bolide impact

The environmental effects invoked by supporters of bolide impact involve severe reductions of light levels, changes of temperature, acid rain, and wildfires.

In the original scenario of Alvarez *et al.* (1980) a world-embracing dust cloud, composed of expelled crustal rock laced with meteoritic material, severely reduced light levels, thereby inhibiting photosynthesis and hence adversely affecting the base of the food chain both on the continents and in the oceans. Theoretical and experimental evidence suggests that significantly reduced light levels for as little as 9–104 days would eliminate 99% of the zooplankton (Paul and Mitchell 1994). These may be underestimates if massive wildfires are also invoked, because soot absorbs sunlight more effectively than rock dust (Wolbach *et al.* 1990).

Clemens and Nelms (1993) and Rich *et al.* (1988) have argued that impact-induced darkness would not necessarily have killed off the dinosaurs, because they occur in both polar regions. Johnson (1993) points out, however, that their interpretations fail to account for the extreme seasonality of ancient polar climates. Dinosaurs could have survived a polar winter that followed a highly productive polar summer. If, however, the summer light was removed by impact dust the effect on the animals could have been very severe.

With regard to short-term temperature change, an early suggestion by Emiliani *et al.* (1981) was that water vapour in the atmosphere would induce a global greenhouse effect. The dominance of $CaCO_3$ and $CaSO_4$ minerals in the upper 3 km of the Chicxulub section has led to active research on the effects of devolatilisation. Experiments have suggested that expulsion of CO_2 from the limestone target could have given rise to a global warming of 2–10°C (O'Keefe and Ahrens 1989). On the other hand, sulphate aerosols could have led to several years' cooling of the atmosphere (Sigurdsson *et al.* 1992). Would these events have taken place in succession or simultaneously? In the latter case they could conceivably have cancelled each other out.

A Chicxulub impact model was explored further by Pope *et al.* (1994). In their model greenhouse global warming due to CO_2 release from vaporised carbonates would have led to a 4°C warming. This is regarded as a maximum and would be less if there were more CO_2 in the terminal Cretaceous atmosphere than today, as widely assumed. The effect would indeed be much less than cooling due to impact-induced aerosols (up to ~ 100°C) but the actual temperature fall would be buffered by the heat released from the oceans for several years. Significant cooling could have lasted for a few years, followed by a more prolonged period of moderate warming, because of the longer residence time of CO_2 in the atmosphere. Such a time-scale is, however, far too short to leave an oxygen isotope record in the strata. Pope and his colleagues hypothesised that continental biota were severely stressed by impact as a consequence of widespread freezing. The experiments and calculations of Chen *et al.* (1994) led the authors to conclude, however, that shock devolatilisation of sulphates, resultant production of H_2SO_4 aerosols, and temporary decrease of solar radiation could not have been a major K–T extinction mechanism.

As regards the more widely discussed view of expelled dust and wildfire soot promoting darkness and cooling, account should be taken of the recent discovery of dinosaurs in high palaeolatitudes. Since the North Slope of Alaska was near to the Cretaceous North Pole, the occurrence there of Campanian and Maastrichtian dinosaurs (Parrish *et al.* 1987) implies that some could either tolerate three months of winter darkness or were capable of seasonal migrations across a considerable latitudinal distance. Dinosaurs of Valanginian–Albian age have also been found in south-eastern Australia, a part of the world where palaeogeographic reconstructions indicate a high-latitude position at that time, about 80°S. A cool, seasonal climate is indicated by fossil plants and oxygen isotope results (Rich *et al.* 1988). The fact that dinosaurs coped with high latitudes for at least 65 m.y. through the Cretaceous suggests that cold and darkness may not have been the prime factors causing their extinction, as some have maintained. Archibald (1994) points out that the vertebrates most likely to be affected would have been ectothermic reptiles and amphibians. However, apart from a decline in lizards, Archibald's North American data do not accord well with a temperature-decrease scenario (cf. Buffetaut 1990). He notes the lack of ectotherms in the Upper Cretaceous of Alaska, whereas endothermic mammals and probably endothermic dinosaurs are present (Clemens and Nelms 1993). It is apparent that the Pope *et al.* (1994) scenario fails to account for the observed pattern of vertebrate extinctions.

Impact-induced acid rain on a massive scale has also been invoked to account for destruction of calcareous plankton, by lowering the pH of surface waters. For most estimates of massive nitrogen oxide production and SO_2 volatilisation, atmospheric oxidation to nitric and sulphuric acid would not have created enough acid rain to affect significantly global surface ocean chemistry. However, rapid, globally uniform rain of masses corresponding to the highest estimates would have destroyed the carbonate-buffering capacity of ocean surface waters and led to a catastrophic reduction of pH. Nevertheless, modern plankton culturing studies and taxonomic patterns of K–T survival are not readily compatible with scenarios that rely on acid rain as the primary cause of extinction. For example, it has been assumed that carbonate-secreting organisms would be more heavily damaged than silica-secreting forms, but planktonic foraminifera and coccolithophorids can survive relatively acidic environments, albeit in an uncalcified state (D'Hondt *et al.* 1994). There is also a problem with the high survival rate of freshwater vertebrates, because modern lakes suffering from acid rain are substantially stripped of life (Archibald 1994).

An analysis of soot concentration in the purported impact fallout boundary clay layers in Denmark, Spain, and New Zealand by Wolbach *et al.* (1985) revealed a striking enrichment not hitherto found anywhere, and led to the hypothesis of a global wildfire induced by impact. This initial study was subject to the criticism that it was assumed that all the clay was deposited within the geological instant associated with impact. If this is not the case, and the clay is a normal one, locally derived (see, for example, Schmitz 1988), then such a concentration is not significantly different from sediments stratigraphically above and below. Wildfires on at least a modest scale were no doubt commonplace in geological history and the mere occurrence of soot is of no great significance; the amount of unusually large enrichment is crucial. More recent work by Wolbach *et al.* (1990) made some allowance for this problem, and major global wildfires were still insisted upon.

The wildfire hypothesis nevertheless suffers from several difficulties. In the first place, so little work has been done using a rather specialised technique that there is an almost total absence of control, in marked contrast to the situation with iridium anomalies. In effect, hardly anything is known about the concentration of soot in the stratigraphic record as a whole. Furthermore, it is surprising that no abnormal concentration of soot have been found, despite intensive search, in one of the most obvious places to look, the North American Western Interior. Finally, the effects of the claimed global wildfires would have been far too drastic and the hypothesis fails to account for the large proportion of survivors in the continental record.

Volcanism

It has been established in recent years that the extensive and thick flood basalts of the Indian subcontinent, known as the Deccan Traps, were erupted at a time effectively coincident with the K–T boundary, which has led to their being invoked as an alternative causal agent for the mass extinction (Courtillot 1990). For example, some of the predicted effects of large-scale basaltic eruptions, based on modern examples, include cooling of the atmosphere by sulphur oxide aerosols and acid rain, similar to predictions about bolide impact (Officer *et al.* 1987). Accepting both impact and this volcanism, could they be merely coincidental, as, for example, Sutherland (1994) argues, or could a large impact have triggered the volcanism through pressure-relief melting in the asthenosphere, as proposed by Alt *et al.* (1988)? One obvious difficulty with this interpretation is that the only well-authenticated impact site, in Mexico, is a long way from India. A more crucial one is that the Deccan volcanism evidently began well before the K–T boundary, and could indeed have reached peak activity ∼ 2 m.y. before the end of the Cretaceous (Venkatesen *et al.* 1993).

Support for this comes from the first record of an iridium anomaly in the Deccan sequence, in an intertrappean sediment succession in Kutch (Bhandari *et al.* 1995). It occurs just above the uppermost horizon containing dinosaurs, with a concentration increasing abruptly to a peak of 1271 pg/g. The high iridium, osmium and nearly meteoritic Os/Ir ratio are characteristic of the K–T boundary layer globally, and taking into account its age of ∼ 65 m.y., a good case is made for it being the same layer. The iridium is accompanied by an unusually high concentration of siderophile and chalcophile elements. These elements cannot plausibly be derived from seawater, as for marine sediments, and are more likely to be derived from the weathering of surface sediments (cf. Schmitz 1992). The key issue is that the iridium-enriched layer is in intertrappean sediments – in other words, it was *preceded* by volcanism.

During the 1980s, as well reviewed by Glen (1994), there was an active debate between supporters of the impact and volcanism scenarios. The latter could argue that many of the postulated environmental effects were the same, and that recent eruptions in Kilauea had demonstrated that mantle-derived iridium could be enormously enriched in aerosols associated with volcanism; eruptions on a sufficient scale could conceivably result in a global iridium-enriched sedimentary layer. The vulcanists failed, however, to account satisfactorily for the occurrence of shocked quartz, and by the end of the decade, even before the new research in the Caribbean region, the debate seemed to be substantially resolved in favour of the impact supporters. It would be a mistake, however, to consider that these are the only serious contenders in accounting for the end-Cretaceous extinctions, because several other Earth-bound causes must also be discussed.

Climatic changes

The kinds of temperature change invoked under various impact scenarios are essentially short-term, in some cases almost brief enough to be considered as 'weather'. It is widely accepted, however, that the latest Cretaceous marked a time of significant longer-term climatic change. Oxygen isotope studies of planktonic and benthic foraminifera generally indicate long-term cooling of surface and/or deep marine waters during the Campanian and Maastrichtian (Douglas and Savin 1973). At least in southern latitudes and presumably therefore elsewhere, the cooling seems to have been accomplished in a series of geologically rapid 'steps' (D'Hondt and Lindinger 1994). One such step was in the mid-Maastrichtian (Barrera 1994), which might have a bearing on the disappearance of inoceramids and rudists at that time. At least in the mid-Pacific an important cooling trend began at least 200 000 years before the K–T boundary (Zachos *et al.* 1989). A similar water-cooling trend, from the Turonian to the Maastrichtian, has been inferred from oxygen isotope analyses of belemnites and brachiopods from the Chalk of north-west Europe, as well as whole-rock analyses of the English Chalk. Although it is only possible to show relative temperature trends from $\delta^{18}O$ analysis of limestone, data from the Bottacione Gorge in Italy also show a cooling from the Turonian to the Maas-trichtian, but with a rise in the mid-Maastrichtian and in the late Maastrichtian, which are counter to the general oceanic trend (Spicer and Corfield 1992, Fig. 9.11).

The best information on air temperatures comes from the study of angiosperm leaf physiognomy in North America. A Late Cretaceous cooling trend is also recognised, but the work of Wolfe and Upchurch (1987) in the North American Western Interior fails to indicate any significant temperature change through the Maastrichtian, and indeed shows an immediate pre-K–T boundary warming. Such warming is not evident in north Alaska (Spicer and Parrish 1990; Fig. 9.11), although this result may change with higher-resolution work. There are in fact conflicting results for north Alaska. Fredericksen (1989) infers a late Maastrichtian cooling based primarily on diversity change in pollen assemblages, whereas Spicer and Parrish (1990; Fig. 9.11) deduce from leaf physiog-nomy no such cooling over this time interval. These diverging interpretations clearly need to be resolved, but evidently significant temperature change is likely to have been an important factor in controlling high floral turnover rates in the Late Cretaceous, independent of any end-Cretaceous catastrophe. Johnson (1993), indeed, recognises an important mid-Maastrichtian event marked by major extinctions of terrestrial flora, which may possibly correlate precisely with the inoceramid and rudist extinctions in the oceans. On a broader time-scale, the diversity of dinosaurs in the North American Western Interior, especially the herbivorous ornithischians, decreased from the Cam-panian to the Maastrichtian (Clemens 1986). Very likely the causal factors of dinosaur extinctions are bound up with interactions of both long-term and short-term environ-mental changes affecting complexly integrated ecosystems.

Marine regression

That the end of the Cretaceous was marked by a major regression has been well established for many years, and it is quite natural to invoke it as a possible factor in the mass extinction (Hallam 1987b, 1992). The K–T boundary in epicontinental marine sections is characteristically marked by a major hiatus or sequence boundary, indicating a significant fall of sea level followed by a rise. The amount and rapidity of the sea-level fall, and its precise timing in relation to the boundary, remain uncertain. Three detailed studies of extinction events in relation to sea-level change are instructive and illuminat-ing in this respect.

Three sections in Alabama, including the well-known one at Braggs, were studied by Habib *et al.* (1992). The boundary between the Prairie Bluff Chalk and basal Clayton Formation marks a type 1 sequence boundary, according to Donovan *et al.* (1988). The lowstand systems tract comprising incised valley-fill carbonaceous, estuarine sands

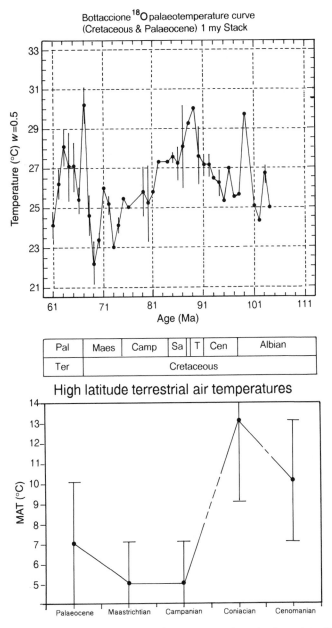

Bottaccione ^{18}O palaeotemperature curve
(Cretaceous & Palaeocene) 1 my Stack

Pal	Maes	Camp	Sa	T	Cen	Albian
Ter	Cretaceous					

High latitude terrestrial air temperatures

Fig 9.11 Oxygen isotope palaeotemperatures from the Bottacione Gorge, Italy (upper diagram) and mean annual air temperatures from North Alaska based on leaf physiognomy (lower diagram) for the Late Cretaceous and earliest Tertiary. After Spicer and Corfield (1992).

exposed at Mussel Creek is not present at Braggs, where the lowest bed of the transgressive systems tract consists of limestone. The sequence boundary corresponds with a mass extinction event among the nanofossil species, as also among the molluscs (Bryan and Jones 1989). Sharp fluctuations in dinoflagellate diversity also relate closely to inferred sea-level change in the late Maastrichtian and Danian, but the extinction rate was much lower; diversity increase relates to deeper-water, more marine conditions.

According to Keller *et al.* (1993b), who cite many references in support, the latest Cretaceous sea-level fall took place immediately before the end of the period, and was rapidly followed by a rise. Long-term oceanic instability is evident in sea-level fluctuations for a period of time that is estimated to last from $\sim 100\,000$ years before to $\sim 300\,000$ years after the K–T boundary. As exemplified in the well-studied Danish sections, the maximum lowstand occurs at ~ 1–2 m below the boundary, followed by a sea-level rise for 20–40 cm (5–10 000) years below the boundary. The well-known Fish Clay of the basal Danian was deposited during a sea-level highstand. Two hiatuses or periods of non-deposition mark sea-level lowstands at the top of zones Po and P_{1a}, $\sim 50\,000$ and 230 000 years after the boundary. Thereafter sea level rose and normal conditions were re-established.

Brinkhuis and Zachariasse (1988) made a study of dinoflagellates and planktonic foraminifera in relation to sea-level change in the K–T boundary succession at El Haria, Tunisia. The associations of dinoflagellate cysts, proportions of spores and pollen and quantity of derived organic matter are inferred jointly to reflect a rapidly falling sea level during the final 17 000 years of the Cretaceous, which culminates at the K–T boundary. This steep sea-level fall at K–T time is held to represent a peak regressive pulse at the end of the well-documented latest Cretaceous regressive trend. For those who have related the extinction of neritic benthos to reduction in habitat area of epicontinental seas, following the original model of Newell (1967), it has remained a puzzle how regression could account for the mass extinction of plankton. It is therefore intriguing that Rohling *et al.* (1991), expanding on the earlier work of Brinkhuis and Zachariasse, attempt to account for plankton extinctions related to the end-Cretaceous regression in an original oceanographic model.

Deep-water production at the present day is highly localised in small, high-latitude areas, due to cooling and/or freezing processes. In the more equable world of the Cretaceous, deep-water production was probably induced by excess evaporation in shallow marginal and epicontinental seas at low to middle latitudes (Brass *et al.* 1982). The upward flux of nutrients is dominated by general deep-ocean ventilation via an upflow process. The vertical advection of nutrients is related to the intensity of upflow and therefore the intensity of deep-water formation. Due to recycling within the mixed layer the 'shallow' system is much less vulnerable to decreases of vertical nutrient advection than the 'deep' system. The latest Cretaceous regression would have reduced the extent of areas producing deep waters, namely shallow epicontinental seas. Therefore less deep water would have been formed, and the related upflow intensity and vertical nutrient advection reduced, giving rise to a gradual diversity decline as phytoplankton productivity was reduced. The end-Cretaceous regressive pulse finally curtailed deep-water formation and minimised vertical nutrient advection, leading to rapid destruction of the 'deep' phytoplankton community and dependent organisms. The 'shallow' community was affected at the very end of the period. Rohling *et al.*'s model, illustrated in Fig. 9.12, clearly depends on the plankton extinction event being a gradual affair, albeit restricted to a few thousand years, rather than a geologically instantaneous catastrophe not preceded by any significant change.

Keller *et al.* (1996) have made a detailed study of the neighbouring K–T boundary global stratotype section at El Kef. They infer that the biotic record in this shallow-shelf section was significantly influenced by local conditions which, combined with the latest

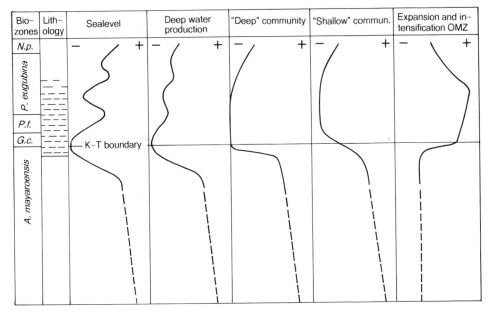

Fig 9.12 Schematic overview of the effects of sea-level changes around the K–T boundary. Horizontal hatching indicates 'boundary clay'. The 'shallow' community consists of phytoplankton inhabiting the mixed layer. The 'deep' community consists of phytoplankton thriving below the mixed layer at the base of the euphotic zone. *N.p. = Neogloboquadrina pseudrobulloides; P.f. = Parvularuglobigerina fringa; G.c. = Guembelina cretacea;* OMZ = oxygen-minimum zone, − = a fall or decrease; + = a rise or increase. After Rohling *et al.* (1991).

Maastrichtian sea-level fall and subsequent rise, resulted in shallowing of the oxygen-minimum zone relative to the sea surface. This led to selective disappearance of benthic faunas and may have adversely affected the surviving photic zone members. The selective nature of species extinctions, however, appears to be related partly to long-term global oceanographic changes which were accelerated at the K–T boundary, possibly by a bolide impact.

Pelagic successions may also indicate evidence of a regressive interval at the K–T boundary in the form of a sharp increase in the proportion of continent-derived kaolinite to other clay minerals. This is the case, for example, in the classic Italian section at Gubbio. Making reasonable estimates based on sedimentation rate, a time interval of a few hundred thousand years is indicated for the regressive pulse (Hallam 1987b).

Whatever the plausibility of invoking regression as a prime cause of extinction of marine organisms, how could it cause the extinction of the dinosaurs? One possibility is to invoke an increase in 'continentality' of the climate on land, in other words an increase in seasonal extremes of temperature, and perhaps also increased aridity of continental interiors (Hallam 1987b). A more direct influence is proposed by Archibald (1993b) for the North American Western Interior. He argues that the departure of epicontinental seaways would have increased the length of rivers, increasing the diversity of freshwater aquatic habitats which, as we have seen, experienced relatively few extinctions. There would have been no such buffering in the terrestrial realm. Though the total land area increased, the size and variety of low coastal plain habitats declined drastically, as with shallow coastal habitats. Habitat fragmentation would have further stressed the surviving populations. Sheehan and Fastovsky (1992) are, however, unconvinced by Archibald's regression model, failing to see how more coastal lowland could have been exposed via marine regression, yet the total amount of coastal lowland plains have decreased. They maintain that terrestrial palaeoenvironmental changes of the sort claimed have never been documented in any facies study. Retallack (1994b) has made a study of the latest Cretaceous to earliest Tertiary palaeosols in eastern Montana. He points out that the striking extinctions of vertebrates and plants recorded from this region seem out of all proportion to the slight environmental changes across the K–T boundary indicated by the palaeosols. In his recently published book, Archibald (1996) maintains a major role for regression but accepts some role also for impact in the dinosaurs' final demise, probably because of the mass destruction recorded among plants which would have served as the base of the food chain.

Anoxia

Due to the decrease in oxygen solubility with increased temperature, the flux of oxygen to the deep ocean in the Cretaceous may have been much less than today, and this reduced oxygen supply would have promoted the occurrence of anoxia (Southam *et al.* 1982). Rohling *et al.* (1991) speculate that mass mortality at the end of the Cretaceous caused the consumption of oxygen by oxidation of dead organic matter, leading to an expanded oxygen-minimum zone (Fig. 9.12), with dysaerobic conditions expanding across the shelf edge. There is indeed evidence of black shale or anoxic equivalent classic K–T boundary sections across the world, including Stevns Klint, Denmark, Caravaca, Spain, Woodside Creek, New Zealand, and El Kef, Tunisia (Schmitz 1988). The thickness of the Fish Clay at Stevns Klint (Fig. 9.6), however, a classic kerogen-rich laminated shale signifying bottom-water anoxia persisting well into the Danian, is too great to be plausibly accounted for by the geologically transient effect of catastrophic mortality at the end of the Cretaceous. Since this and equivalent deposits elsewhere form part of a transgressive systems tract associated with rising sea level, they seem instead to be further examples of a common phenomenon in the stratigraphic record, few of which are associated with mass extinction on the end-Cretaceous scale, and especially not of phytoplankton (Hallam 1989a). As mentioned earlier, anoxic or dysoxic conditions have been invoked to account for changes in benthic foraminiferal populations at the K–T boundary.

Changes of sulphur isotope ratios in whole-rock sulphide from a K–T boundary section in Japan have been used to argue for a short-term anoxic event of around 70 000 years' duration (Kajiwara and Kaiho 1992). However, the large positive excursion in the boundary beds suggests a brief of interval of closed-system sulphate reduction, typical of oxic environments.

In summary, evaluation of possible causal factors is as yet inadequate because in general there is insufficient evidence to resolve competing claims, but considerable progress has nevertheless been made since the original Alvarez hypothesis was put forward. There has, for instance, been a certain amount of shift towards a more catastrophist position and an increasing sympathy among palaeontologists towards some kind of impact scenario. There remains, however, no consensus about which factors would have been the most significant, and some scenarios seem to imply environmental deterioration too devastating to account for the high survival rate of many groups of terrestrial organisms. There is much to be said for the original 'lights out' scenario involving inhibition of photosynthesis by a global-embracing dust cloud, and selective extinctions of the sort proposed by Sheehan and Hansen (1986).

Even if one accepts the likelihood of end-Cretaceous impact, it cannot be ignored that the latest Cretaceous was a time of considerable environmental change involving both climate and sea level. Regression could possibly have affected plankton as well as benthos, though anoxia associated with subsequent transgression was probably not as important an extinction mechanism as for other major extinction episodes. Although a Late Cretaceous cooling trend seems to be well established for both air and water temperatures, there remains no clear picture of what was happening to the climate in the time immediately preceding the end of the period. The fluctuating environment of the Maastrichtian is likely to have been the prime cause of the high species turnover rate of many groups or organisms, and the extinction of at least some. A compound scenario involving both gradual extinctions followed by a catastrophic *coup de grâce* seems to be the one best fitted to the facts as we know them at present (Hallam 1987b, 1988). One suspects, furthermore, that even without a bolide impact there might well have been a mass extinction recorded at the end of the Mesozoic era.

Recovery after the extinction

Compared with the intensive research on the mass extinction event, there have been as yet only a few detailed studies of recovery of organisms in the immediate aftermath, and these are confined to marine microfossils and molluscs.

The fullest data come from planktonic foraminifera, where two contrasting views have been expressed. The first view is that there was a devastating mass extinction of species at the end of the Cretaceous, leaving a mere three survivors. There followed a dramatic radiation in the early Palaeocene of spinose globigerinids, which appears to indicate an environment radically altered and/or vacated by the extinctions (Olsson and Liu 1993). The spines are thought to have been an adaptation that enhanced the ability of these organisms to capture actively moving organisms in the water column. This 'catastrophist' view implies that many 'Cretaceous' morphotypes occurring in the earliest Palaeocene strata are reworked specimens, an interpretation that has been challenged on a number of grounds, including stable isotope studies, morphometrics, preservational state, and biogeography (MacLeod 1995).

The analysis of MacLeod and Keller (1994) indicates that a substantial proportion of the youngest Maastrichtian faunas survived the K–T event and went on to dominate in the earliest Tertiary, being eventually replaced by two successive and geographically distinct waves of indigenous Danian speciation. Based on abundant and continuous occurrence through the lowermost Danian in over 20 K–T boundary sections, they recognise no fewer than 31 Cretaceous survivor species, including the widely recognised *Guembelitria cretacea*. 'Cretaceous' faunas make up averages of 89% and 94% species richness of biozone Po faunas from the Brazos River and Ocean Drilling Program Site 738. In P1a the corresponding figures are 59% and 79%. The indigenous Danian species richness values increase, and the putative survivor fauna decrease, monotonically through Po to P1c. Relative to the preceding and succeeding faunas the Po faunas represent a depauperate assemblages of highly cosmopolitan species dominated by 'Cretaceous' taxa. Zones Po and P1a have high survivor fauna species richness values in low to middle latitudes and low values in deep-sea sites and high latitudes. This pattern progressively reverses itself through P1b and P1c, for which there is a high survivor species richness in high latitudes and low values in low to middle latitudes. As rates of speciation rose in P1a a large number of new Danian species entered low- to middle-latitude habitats and displaced many formerly cosmopolitan survivor species to high latitudes via the extinction of low- to middle latitude populations, as well as driving all 'Cretaceous' forms restricted to these low to middle latitudes completely extinct.

This extended temporal interval over which the faunal turnover took place, from the mid-Maastrichtian to P1c, along with its characteristic biogeographic structure, fails, according to MacLeod and Keller (1994), to support a causal link with the direct effects of bolide impact.

In their high-resolution study of deep-sea benthic foraminifera in the section at Caravaca, Spain, Coccioni and Galeotti (1994) inferred a rapid recovery, achieved within about 7000 years after the K–T event, with the reappearance of polytaxic assemblages with a complex trophic structure. They recognised a distinct four-step pattern.

1. Decimation and survival (500–600 years). Only infaunal deposit-feeding *Spiroplec-tammina* and *Bolivina*, indicative of low-oxygen conditions, survived and blossomed opportunistically.

2. Between 600 and 1200 years after the K–T event some epifaunal, deposit-feeding taxa reappeared, in a dysaerobic to quasianaerobic facies.

3. About 1500 years after the K–T event, *Bathysiphon*, a presumed epifaunal suspension feeder, reappeared, together with some epifaunal and infaunal deposit feeders.

4. Just above the boundary clay layer epifaunal, attached suspension feeders reappeared (*Rhabdammina*, *Rhizammina*). From here upwards the biofacies again became aerobic, the nutrient levels oligotrophic, and the fossil assemblages K-selected and polytaxic.

At Caravaca the benthic foraminifera recovered much more rapidly than the planktonic foraminifera, probably because the former migrated away from the location of the section whereas the latter underwent global decimation and needed more time to recover. The benthic recolonisation is seen as a stepped process related to the return of more oxygenated bottom waters, a wider spectrum of utilisable nutrient resources on and within the sediment, and greater environmental stability.

With regard to the calcareous nanofossils, at least 15–20 species survived the K–T boundary event. The $\delta^{13}C$ values of the fine fraction, mainly coccoliths, show a characteristic change to more negative values just at the K–T boundary despite the fact that the calcareous nanofossils present are mainly ($\pm$ 90%) vanishing Cretaceous species, which would be expected to carry a 'Cretaceous' signal if they were reworked. The environment returned rapidly to conditions favourable for the evolution and preservation of new species, most of which are very small and easily overlooked among the vanishing and persistent species which are larger. A succession of environmental changes over several hundred thousand years led to a sequence of successive blooms of survivors in high latitudes and to new species in the Tethyan realm (Hallam and Perch-Nielsen, 1990).

The most detailed research on molluscan faunas, mainly bivalves, is that on the United States Gulf Coast deposits by Hansen and colleagues (Hansen 1988; Hansen *et al.* 1987, 1993; Kelley and Hansen 1996). The early Palaeocene faunas are composed almost entirely of new species of genera that originated in the Cretaceous. While the number of species dropped by ~80% in most molluscan families from the Late Cretaceous to the Early Tertiary, a few families either declined slightly or actually diversified, namely the Ostreidae, Carditae, Cuculleidae, and Turritellidae (Hansen

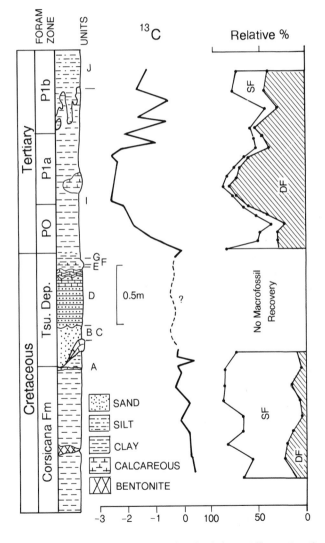

Fig 9.13 Composite stratigraphic section for K–T boundary beds in east Texas, together with carbon isotope curve, showing changes in the molluscan fauna. Tsu. Dep. = tsunami deposit of Bourgeois *et al.* (1988); DF = deposit feeders; SF = suspension feeders. After Hansen *et al.* (1993).

1988). It is not known for sure if these suffered massive extinction but radiated more rapidly, or were less affected by extinction. That there are virtually no species in common between the Upper Cretaceous and lower Palaeocene after P1b suggests that the four families suffered large extinctions and rapid radiation. They abruptly declined in species richness after the early Palaeocene. Hansen considers them to be equivalent to an opportunistic 'bloom' fauna, characterised by rapid maturation, short life-spans and wide dispersal capabilities to take advantage of post-disaster niches. Their replacement later by other groups suggests that they were not successful competitors. The overall ecological structure was essentially rebuilt by P1b, producing a fauna relatively similar in proportions of feeding types and life habits to that of the Late Cretaceous. After extinction of the 'bloom' species the fauna took on a more typical Tertiary aspect, but it took nearly 25 m.y. for the Palaeogene diversity to build up to near the Late Cretaceous level.

The best results come from intensive investigation of sections in the Brazos River, east Texas (Hansen *et al.*, 1987, 1993). The Upper Cretaceous part of the section is dominated by suspension feeders ($\sim 60\%$), with minorities of deposit feeders ($\sim 15\%$) and carnivores ($\sim 25\%$). These basic proportions remained relatively constant throughout the 1 m.y. interval preceding the K–T boundary, in spite of changes in taxonomic composition. Immediately above the boundary suspension feeders drop to less than 20% of the second macrofossiliferous sample above the boundary and to less than 10% further upsection. The changes in proportions of suspension to deposit feeders are mirrored by changes in $\delta^{13}C$ (Fig. 9.13). That some suspension feeders survived the extinction, but did not flourish, argues against a delayed recovery due to an evolutionary lag ($\sim 100\,000$ years for Po and P1a). It seems more likely that the lack of suspension feeders relates to a loss of primary productivity, because they eat phytoplankton. Deposit-feeding molluscs feed on bacteria-rich detritus and faecal pellets and are therefore more independent of primary productivity. It is puzzling, however, that even a deposit-feeding community could be maintained in the absence of primary production of the order of hundreds of thousands of years. Possibly organic matter was derived from continental runoff.

10 *Cenozoic extinctions*

For the Cenozoic era there are a number of well-documented extinction events both in the marine and terrestrial realms, a few of which qualify as mass extinctions. Our knowledge of Cenozoic environments, considerably enhanced within the last few decades by an extensive programme of deep-sea drilling, is greater than that for earlier eras. A major contributor to this is the use of actualistic comparisons in palaeoecology, because of the relatively close affinities to extant taxa of the fossilised organisms. Accordingly it is to be expected that there should be a better prospect of understanding causal factors.

Rather than postponing a discussion of possible causes of extinction until after a presentation of the key data, it is desirable to acknowledge from the outset that a wealth of data indicates that the dominant environmental *leitmotif* of the Cenozoic is climatic change, with a more or less progressive and stepwise decline of global temperatures from a climatic optimum in the early Eocene (Frakes *et al.* 1992). This is especially well demonstrated for the Palaeogene by oxygen isotope curves, based on analysing the shells of benthic and planktonic foraminifera, for oceanic bottom and surface waters. Figure 10.1 shows that both curves have a similar trend, signifying a decline in temperature from the early Eocene which was greatly accelerated across the Eocene–Oligocene boundary. Also shown on the figure is the carbon isotope curve for bulk sediment, where the most striking changes are between the latest Cretaceous and early Eocene. A contemporary change of air temperatures, to a considerably higher range of mean annual temperature, across the Eocene–Oligocene boundary can be inferred from the terrestrial plant record, the best data coming from Washington State, USA, with the climatic inferences being based on the Climate-Leaf Analysis Multivariate Program (CLAMP) (Wolfe 1992). The climatic deterioration in the Middle and Late Eocene correlates with the growth of Antarctic ice. Sedimentological and oxygen isotopic evidence supports a model whereby there was a progressive increase in the amount of ice, with extensive ice sheets being established by the early Oligocene (Ehrmann and Mackensen 1992; Miller 1992).

Details of the Neogene climate are very well known (Crowley and North 1991; Frakes *et al.* 1992) and need not concern us in this introductory section.

The end-Palaeocene event in the deep ocean

Benthic foraminifera are the most abundant fossils of deep-water organisms, as known from the study of ocean cores. Whereas they were apparently little affected by the end-

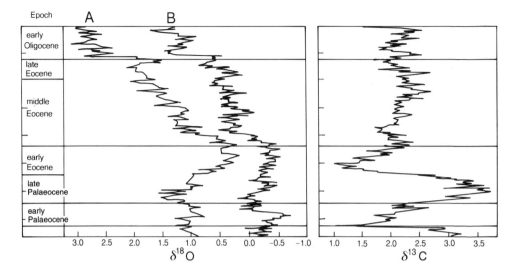

Fig 10.1 Stable isotope ratios of marine microfossils from the Palaeocene to the early Oligocene. A = bottom waters, B = surface waters. Adapted from McGowran (1990).

Cretaceous extinction event, they suffered a striking mass extinction at the Palaeocene – Eocene (P–E) boundary; according to Thomas (1990), some 50% of species went extinct. Nothing comparable is known at other horizons in the Cenozoic.

Some of the best deep-ocean cores for studying this phenomenon come from the Maud Rise, east of the Weddell Sea in the Southern Ocean. In particular, site 690 appears to be the most complete section of this interval available from all Deep Sea Drilling Project and Ocean Drilling Program sites (Thomas 1990; Kennett and Stott 1991). As indicated in Fig. 10.2, there was a precipitous fall in diversity followed by a period of unusually low diversity. Using the best available time-scale biochronological and palaeomagnetic data, Thomas (1990) has estimated that the event was geologically brief, less than 25 000 years, while Kennett and Stott (1991) estimate an even briefer time of less than 3000 years. The fossils are evidently well preserved, with no solution loss, and a lack of bioturbation encompasses the extinction level, the underlying and overlying sediments being clearly bioturbated (Kennett and Stott 1991). According to Thomas (1990), estimated palaeodepths are between 1000 and 2000 m, indicating the lower bathyal zone.

Isotopic values show little change through the cores except at the extinction horizon, which is located within one of the largest negative $\delta^{13}C$ excursions of the whole Cenozoic. The horizon occurs within a long-term negative $\delta^{18}O$ trend (Fig. 10.1), interpreted as reflecting the warming of Antarctic surface waters. An associated peak in kaolinite suggests a warm, humid climate on Antarctica. Deep waters were warmed to temperatures close to those at the ocean surface, temporarily eliminating the vertical temperature gradient. The large $\delta^{13}C$ gradient ($\sim 2‰$) that had previously existed was virtually eliminated before being re-established within about 3000 years, the surface waters warming by 3–4°C. A total warming of subsurface waters of 6°C in 6000 years is inferred (Kennett and Stott 1991).

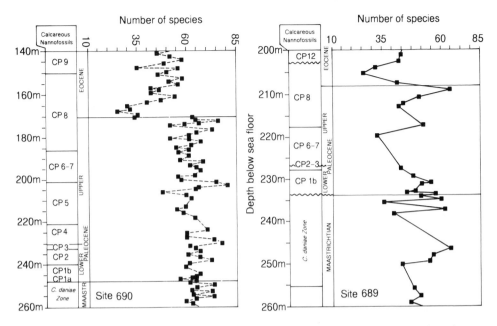

Fig 10.2 Diversity of benthic foraminifera (expressed as number of species per 300 specimens) plotted versus sub-bottom depth for Sites 689 and 690 in the Southern Ocean. Redrawn from Thomas (1990).

The extinction is associated with an assemblage dominated by relatively small and thin-walled uniserial, triserial and other forms characteristic of an infaunal habitat (Thomas 1990) and a marked diminution in abundance and diversity of ostracods, leaving a rare assemblage composed almost exclusively of small, smooth, thin-walled forms (Kennett and Stott 1991). In marked contrast to the benthic foraminifera and ostracods, there was no significant plankton extinction, with the diversity of planktonic foraminifera, calcareous nanofossils, and dinoflagellates actually increasing through the time interval (Thomas 1990).

An exactly contemporary event among benthic foraminifera has been recognised at bathyal – abyssal depths elsewhere in the oceans, such as the Pacific (Douglas and Woodruff 1981), Atlantic (Tjalsma and Lohman 1983), the southern Tethys (Speijer 1994) and in Japan and New Zealand (Kaiho 1988). It is thus clearly global in extent, with the extinct faunas being cosmopolitan (Kaiho 1991). Kaiho's regional study (1988), followed by a global one (1991), is of particular interest. He perceives a relationship between the foraminifera and the degree of oxygenation of bottom waters. Taxa tolerant of stagnant environments, such as *Praeglobobulimina, Pleurostomella, Bulimina,* lagenids, and agglutinated forms, become the main faunal component immediately after the extinction event in New Zealand and Hokkaido, suggesting that oxygen deficiency was the cause of the extinction, a view concurred with by both Thomas (1990) and Kennett and Stott (1991), on the grounds that the proportion of infaunal species increased. This is a view also shared by Speijer (1994), who studied the extinction and recovery patterns in benthic foraminiferal palaeocommunities on the northern margins of Africa and in Israel. He inferred an episode of sudden widespread oxygen deficiency in deep- and intermediate-water masses, which caused the extinction of many cosmopolitan deep-sea species. This dysoxic event affected South Tethyan shallow-water communities as well. However, due to more effective survival strategies against oxygen deficiency, most shallow-water species were able to survive; a few opportunistic species settled in the deeper parts of the basin, temporarily replacing the bathyal community. After some 50 000 years a relatively impoverished Eocene upper bathyal community gradually returned, indicating the re-establishment of more normal conditions.

The most recent study of the extinction event is by Ortiz (1994), who analysed the benthic foraminifera in the classic sections of Caravaca and Zumaya in Spain. He confirms that the mass extinction was rapid, coinciding with a negative $\delta^{13}C$ shift of 2–4‰, the onset of dark grey shale deposition and increased carbonate dissolution. After the extinction event, small, finely agglutinated, and thin-walled epifaunal taxa dominate in the north Atlantic Zumaya section. This assemblage indicates sluggish circulation, bottom waters undersaturated in calcium carbonate and dysoxic conditions. Low-oxygen conditions prevailed for the succeeding 400 000 years. In the western Tethys, Caravaca section infaunal taxa ('buliminids') dominated after the extinction event and prevailed through the succeeding 400 000–450 000 years of decreased $\delta^{13}C$ values, indicating low-oxygen conditions. In the Zumaya section, the extinction event occurred over less than 50 000 years and resulted in a sudden reduction of species richness after the $\delta^{13}C$ shift. In contrast, in the Caravaca section species diversity declined gradually, beginning before and in concert with the gradually declining $\delta^{13}C$ values. This gradual change occurred over a period of 250 000 years and culminated in a 4‰ $\delta^{13}C$ shift. Thus it seems that the P–E oceanographic changes that triggered the benthic foraminiferal mass extinction were first manifested in the Tethys region.

There is a general agreement that the dysoxic (not anoxic) event (cf. Kennett and Stott 1991) was related to a major episode of oceanic warming, associated with a pronounced decrease in the abundance of stratigraphic hiatuses in most ocean basins, implying a decreased vigour of ocean currents (Moore *et al.* 1978). In the absence of polar ice, oceanic circulation was probably completely different from that of today, and the model of Brass *et al.* (1982) is generally favoured (Miller *et al.* 1987; Thomas 1990; Kennett and Stott 1991). Dense, saline bottom waters formed by evaporation in low latitudes would have driven oceanic circulation. Warm waters would be relatively oxygen-depleted especially if formed at low latitudes, and would become even more depleted during their long travel to high latitudes. The cause of warming, which is consistent with the oxygen isotope evidence, is unknown, but Eldholm and Thomas (1993) have suggested that CO_2 output of north Atlantic volcanic province basalts could have triggered rapid environmental change through increased high-latitude surface temperatures, in turn changing deep-sea circulation and productivity. However, an increase in atmospheric CO_2 is not supported by research on coexisting palaeosol organic matter and carbonate from a terrestrial succession in the Paris Basin (Sinha and Stott 1994).

The pronounced negative $\delta^{13}C$ excursion across the Palaeocene–Eocene boundary is consistent with a marked productivity fall, but there is apparently no decrease in biogenic $CaCO_3$ production at this time, in striking contrast to what happened at the Cretaceous–Tertiary boundary, and the benthic extinctions occurred before any large-scale change in surface-water processes that could have affected primary production (Kennett and Stott 1991). Similarly, the increase in relative abundance of infaunal species appears to rule out productivity fall as a causal factor (Thomas 1990).

Extinctions in the late Eocene and early Oligocene

The time interval from the middle Eocene to the early Oligocene is now generally acknowledged to have been one of significantly increased extinction rates both among marine and terrestrial organisms, which is generally related to climatic change (Prothero and Berggren 1992; Prothero 1994). Establishment of the contemporaneity of events in different environments depends, of course, on accurate correlation and dating, and this can pose problems for shallow-marine and terrestrial environments. The best dating is based on the study of planktonic foraminifera and coccoliths in pelagic facies, in conjunction with magnetostratigraphy and interpolations from $^{40}Ar/^{39}Ar$ dates. Thus the Eocene–Oligocene boundary is defined by the last occurrence of the planktonic foraminifer *Hantkenina* (in the magnetozone C13R), with a revised age of 34 m.y.; the stratotype is in the northern Apennines (Berggren and Prothero 1992).

Correlation of continental with marine sections remains problematic, but considerable improvements have been made in recent years, at least in North America, by the use of magnetic stratigraphy in conjunction with $^{40}Ar/^{39}Ar$ dating. The North American successions have traditionally been subdivided into a number of land mammal 'ages', which are not defined with the same rigour as marine stages. The new dating has required some revision to their approximate correlation with marine stages (Table 10.1).

Table 10.1 Correlation of Eocene and Oligocene marine stages and land mammal 'ages'

Epoch	Marine Stage	Land mammal 'age'
Oligocene	Chattian	Arikareean
	Rupelian	Whitneyan Orellan
Eocene	Priabonian	Chadronian
	Bartonian	Duchesnean
	Lutetian	Uintan
	Yypresian	Bridgerian Wasatchian

Source: adapted from Stucky (1992).

The marine realm

Foraminifera

The most comprehensive data for marine faunal turnover came from the study of planktonic foraminifera in deep-sea cores (Keller *et al.* 1992; Fig. 10.3). Among low-latitude taxa there was a major turnover from the late middle Eocene to the early late Oligocene, which involved more than 80% of the species and took place more or less continuously over a period of about 14 m.y. The overwhelming majority of species that went extinct at that time were surface dwellers that were replaced by more cold-tolerant subsurface-dwelling forms. This inference is based on stable isotope results, not morphology. Two intervals stand out as characterised by a brief but markedly intensified turnover, the middle – late Eocene boundary where, according to Boersma

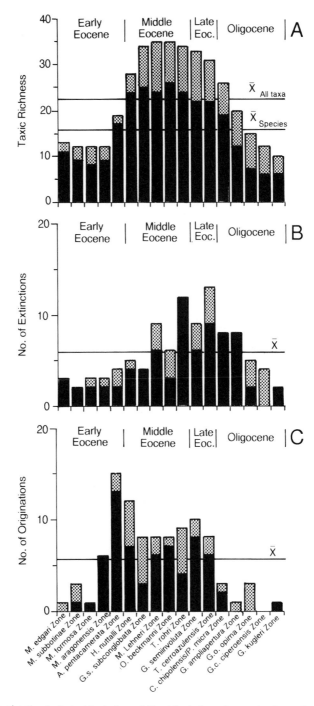

Fig 10.3 Histograms of estimated planktonic foraminiferal taxic (species and subspecies) richness (A), numbers of taxic extinctions (B) and originations (C) in each of the 17 Eocene and Oligocene planktonic foraminiferal biozones. Black bars = species data stippled pattern = subspecies data. After Keller *et al.* (1992).

et al. (1987), there were 18 species extinctions, and the early–late Oligocene boundary. Contrary to previous reports, there was no major faunal change across the Eocene–Oligocene boundary.

A comparable picture is recognised for bathyal benthic foraminifera from Maud Rise cores, with a pattern of more or less gradual stepwise extinctions (Thomas 1992). The first period of change was in the middle middle Eocene, when large, heavily calcified *Bulimina* species and many lagenids disappeared. The second period was in the late Eocene and the third in the late early Oligocene. There is evidently no correlation between the sudden increase in $\delta^{18}O$ values of benthic foraminifera and the faunal change, which occurred earlier. This is clearly brought out in Fig. 10.4. The initiation of benthic foraminiferal changes always pre-dates the isotopic change. The same is true for the planktonic foraminiferal change used to define the Eocene–Oligocene boundary (Miller 1992). In Thomas's (1992) opinion these facts support the hypothesis that the isotope change was at least partly and probably largely caused by ice-volume increase,

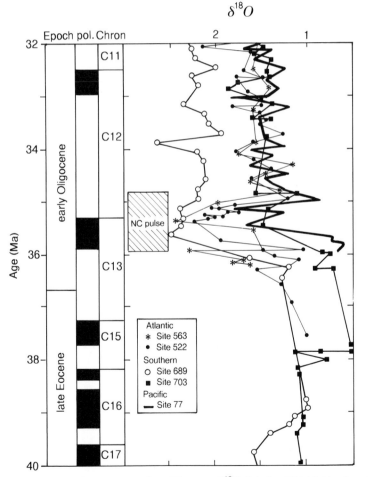

Fig 10.4 Summary of late Eocene to earliest Oligocene $\delta^{18}O$ data for *Cibicidoides* from the Atlantic, Pacific, and Southern oceans. After Miller (1992).

not temperature decrease, and was the result of a threshold effect. There was a gradual cooling, which affected the faunas, and then a sudden increase in ice volume.

According to Thomas (1992), the overall gradual pattern of faunal change established for the Maud Rise is valid world-wide, but this view is challenged for the Austalian region, and also the warm-water neritic benthic foraminifera of the Indo-Pacific and American provinces. Thus McGowran *et al.* (1992) argue that there was a sharp extinction event among benthic foraminifera at the end of the Eocene, which was not merely part of a continuum of change from the middle Eocene onwards.

Calcareous nanoplankton

Deep-sea cores also provide a good record of coccoliths. As with the planktonic foraminifera, a more pronounced species turnover occurred near the middle–upper Eocene boundary than the Eocene–Oligocene boundary; a larger number of extinctions occurred in the early Oligocene than in the late Eocene. Within an approximately 0.5 m.y. interval which started about 1.2 m.y. after the Eocene–Oligocene boundary, mainly within zone NP22, at least 12 species went extinct (Aubry 1992). There was an approximately 70% reduction in diversity between the middle Eocene and early Oligocene. In general, there is close parallelism with planktonic foraminifera. The fact that high-latitude assemblages of both coccoliths and foraminifera invaded low latitudes in the two pulses in the late Eocene and Oligocene (Haq *et al.* 1977) confirms the presumption that cooling climate was the principal agent responsible for the extinctions.

Planktonic foraminiferal records suggest that the Antarctic seas were thermally isolated from warm-water boundary current systems as early as the late middle Eocene (Boersma *et al.* 1987). It seems clear that such isolation was effective in the late early Oligocene as it prevented a nanofossil rebound echoing the contemporary recovery in low latitudes. Very low diversity but high species dominance persisted in high southern latitudes through the rest of the Oligocene (Aubry 1992).

Molluscs

Gastropods and bivalves are the most abundant elements among the neritic faunas, but only for the Gulf Coast region of the United States is there a well-documented record of change through time. According to Hansen (1987, 1988, 1992), extinctions were spread throughout the entire late Eocene, with no clear-cut mass extinction events. Hansen argues for a temperature control on the extinctions, pointing out that there is a good correspondence between diversity peaks, for example the early Eocene, and temperature highs established on other evidence. The percentage of warm-water taxa, as recognised from actualistic comparisons, increases during the temperature highs, and cool-tolerant genera such as the gastropods *Calyptraea*, *Crepidula*, and *Turritella* form a larger percentage of the fauna during low-temperature intervals. Dockery (1986) has put forward the alternative view that the extinctions were caused by reduction in habitat area consequent upon regression, but Hansen points out that there is little correspondence between mollusc diversity and estimated sea level (Dockery and Hansen 1987).

Echinoids

The global diversity of this group shows a peak in the Eocene (nearly 1500 species) bounded by much lower values in the Palaeocene ($\sim$ 100) and Oligocene ($\sim$ 500). The low Palaeocene diversity is thought probably to correspond to a lag phase after the end-

Cretaceous extinctions, and the low Oligocene diversity caused by extinctions due to global cooling (McKinney *et al.* 1992). The general pattern is that the Eocene peak occurred rather early in the middle Eocene, with a small to moderate loss at or close to the Lutetian–Bartonian boundary. A second, much larger, fall in diversity occurred at the end of the Eocene, when massive species reductions of over 50% occurred in all orders. These patterns are recognised in the United States, Caribbean, Europe, and the Indo-Pacific. North Africa is anomalous in that the diversity peak is in the early Eocene. This is thought to be due at least partly to better preservation, so the low global early Eocene diversity may be an artefact of poorer preservation at this time. There is a strong correlation in the south-eastern United States between diversity and temperature. Cold-adapted marsupiate echinoids occur in the late Eocene and Oligocene of Australia. McKinney *et al.* (1992) conclude that temperature is the predominant, though perhaps not the only, controlling factor on diversity and extinction.

Whales

The only marine vertebrate group for which there are well-documented data is the Cetacea (Fordyce 1992). Primitive archaeocete whales appeared at about the Yypresian–Lutetian boundary and diversified in the eastern Tethys in the Lutetian. The lower diversity of the Priabonian may reflect extinctions at the Bartonian–Priabonian boundary. There is no evidence for any major extinctions at the end of the Eocene. The early Oligocene saw some dramatic changes, with the Mysticeti appearing in the early and Odontoceti in the late Rupelian; Chattian whales are diverse and mark an explosive radiation of modern-type Cetacea that continued through much of the Neogene. In Fordyce's opinion food resources, and ultimately the physical structure of the oceans, seem to be the key driving force in cetacean evolution and extinction. Thus in the early Oligocene progressive cooling was allied with new feeding opportunities (food resources) offered by the increasing vertical and horizontal differentiation of the oceans associated with the isolation of Antarctica, with the first toothless (baleen-feeding) mysticetes and odontocetes occurring at the end of the Rupelian.

Tektites

For some time there has been discussion about the significance of the discovery in the upper Eocene part of deep-sea cores of microspherules, including microtektites, generally believed to be terrestrial material melted and ejected by the impact of extraterrestrial bodies. Impact enthusiasts have even suggested that stepwise extinctions in the microfauna could be the consequence of impact by cometary showers (Hut *et al.* 1987).

The upper Eocene impact-generated microspherule layers are widely distributed in the low-latitude Pacific, Indian, and Atlantic oceans. Considerable debate has take place about how many bolide impacts can be inferred for the late Eocene, the number ranging from one to six (Poag and Aubry 1995). Wei (1995) has dated microspherule layers from six critical sites using calcareous nanofossils, which have a common age of about 35 m.y. His careful examination of the literature suggests that proposals for several layers have resulted largely from misinterpretation and miscorrelation, and all the available data are compatible with a model of one couplet of microtektites and microcrystites. Wei finds no support for the Hut *et al.*, hypothesis and concludes that impacts did not cause extinction of species of planktonic and benthic foraminifera, calcareous nannoplankton, or diatoms. Species extinctions coincident with the layers have only been documented for five radiolarians, which account for only 2% of the total radiolarian species at that time, hardly an impressive correlation. The probable site of the impact crater associated with a tektite layer traced across the south-eastern United States and Caribbean has recently been located in Chesapeake Bay (Kerr 1995).

To conclude, there is a strong consensus that the overwhelmingly most important causal factor in the late Eocene–early Oligocene marine extinctions was, either directly or indirectly, lowered water temperature associated with global cooling, with the extinctions being extended through as long as 14 m.y. Only a minority of workers perceive a sharp extinction event at the end of the Eocene, but a major turnover near the middle–upper Eocene boundary is widely recognised. With regard to the planktonic foraminifera, there has been an interesting debate about the role of productivity changes. Hallock *et al.* (1991) interpreted the extinctions in relation to inferred changes in nutrient flux to surface waters. The early Eocene was characterised by K-strategists as indicative of oligotrophic conditions, replaced by less specialised taxa characteristic of deeper and cooler waters as high-latitude cooling intensified. This caused an increased rate of oceanic overturn and return of nutrients to the surface, leading to a condition of eutrophy in the early Oligocene. In contrast, Keller *et al.* (1992) argue that the carbon isotope gradients for both benthic and planktonic foraminifera reflect decreasing surface productivity towards the Oligocene. This interpretation appears, however, to be at variance with the evidence of high nutrient availability in Antarctic waters during the early Oligocene, with high biosilicic productivity (Aubry 1992).

The terrestrial realm

Significant information on organic turnover is confined to vertebrates, principally mammals, and plants.

Vertebrates

The rich record of Palaeogene mammals in the non-marine deposits of western North America allows a more thorough examination of change through time than anywhere else in the world (Stucky 1992). In Stucky's opinion the patterns of diversity change are consistent with global warming near the early–middle Eocene boundary and global cooling and reduction of climatic equability through the late Eocene and Oligocene. There was a concomitant shift from generally subtropical, closed forest habitats to more open, savanna-like habitats in the Western Interior. The faunal turnover in the latter part of the Eocene (Uintan to Chadronian – see Table 10.1) resulted in the extinction of many primitive mammal groups and the appearance of many modern families; there were high rates of extinction and origination in the middle Chadronian. It is not clear in the present state of knowledge how stepwise or catastrophic the extinctions were. By the end of the Eocene, more cursorial adaptations appear at a time coincident with the opening of habitats.

With regard to individual groups, there was a high extinction rate of primates in the middle Bridgerian, and in the late Uintan and Whitneyan a high generic extinction of perissodactyls and artiodactyls; high rodent extinctions took place in the middle Chadronian and Orellan. Obligatory arboreal taxa (primates and dermopterans) declined through the Eocene and were virtually extinct by the end of the Eocene. By

Oligocene times most perissodactyls, artiodactyls, and rodents had higher tooth crowns and greater loph lengths, implying a reliance on browsing together with a tendency to more efficient oral food processing. Overall there was a steady reduction in generic richness from the middle Eocene to the Oligocene, with high extinction rates in the early and middle Bridgerian and middle Chadronian.

Among the reptiles and amphibians of western North America changes initiated in the Uintan and completed by the Orellan–Whitneyan appear to be associated primarily with a general increase in aridity and decrease in the number and even absence of permanent rivers and streams (Hutchison 1992). There may also have been some decrease of mean annual temperature but not so severe as to exterminate tortoises and large-bodied reptiles. The general diversity decline over about 16 m.y. does not seem to be attributable to a single catastrophic event, though the tempo of change appears to have quickened across the Eocene–Oligocene boundary – that is, between the Chadronian and Orellan. Whereas there was a clear decline of aquatic forms (crocodiles, turtles) into the Oligocene, there was no change among the terrestrial turtles (tortoises).

The only other good vertebrate record is from Europe. Early in the twentieth century Stehlin (1909) recognised a major mammalian turnover in the Paris Basin, which he called the Grande Coupure. This is now widely recognised across Europe and has traditionally been placed at the Eocene–Oligocene boundary, but has now been redated as early Oligocene, about 1–2 m.y. younger than the epoch boundary, based on correlation with marine sequences in the Rhine Graben and Mainz basins (Hooker 1992). Figure 10.5 gives the extinction and origination rates for Palaeogene mammals (Legendre and Hartenberger 1992). The Grande Coupure event stands out clearly, with approximately half the genera going extinct within a million years and a similar proportion appearing, making this by far the most important time of turnover in the whole Palaeogene. Thereafter both extinction and origination rates remained uniformly low. Two other notable extinction events are recognisable from the figure, in the earliest Eocene and middle Eocene. During the latter event, most of the large perissodactyls went extinct, replaced by species of *Palaeotherium*, with soft-browsing forms being replaced

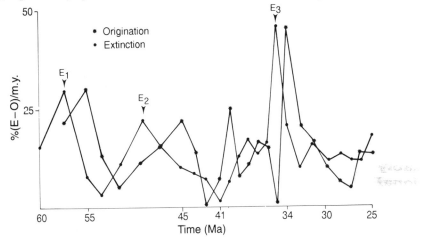

Fig 10.5 Extinction and origination rates (percentage of genera per million years) in the Palaeogene mammalian record. The three main events are noted E_1 E_2 E_3. After Legendre and Hartenberger (1992).

by coarse-browsing forms. There was also a notable reduction in arboreal mammals and insectivores, together with an increase in large ground mammals and browsing herbivores (Hartenberger 1986).

According to Legendre and Hartenberger, the drastic change at or close to the beginning of the Oligocene involved large-sized species becoming rare and medium-sized species absent, indicating, it is thought, a change towards greater aridity and a more open environment. A lower species diversity in each local fauna during the Oligocene can be related to decrease in temperature. Legendre and Hartenberger invoke a drastic change across the epoch boundary from humid, warm, forested conditions to arid, colder, more open savanna-type environments. These changes appear to mirror those in North America, but Berggren and Prothero (1992) maintain that the evidence of drying is much greater in the latter continent; they accept that the Grand Coupure is much more important than the Chadronian–Orellan transition in North America. The post-Grande Coupure faunas are dominated by rabbits, advanced carnivores, artiodactyls, and perissodactyls, with arboreal mammals having disappeared completely and granivorous rodents making their first appearance. Almost all the taxa are immigrants from Asia or North America via Asia (utilising the Bering land bridge). Thus at least to some extent the Grande Coupure is less a climatic event than a sudden immigrant influx comparable to the so-called Great American Biotic Interchange between North and South America in the late Neogene (Berggren and Prothero 1992).

Plants

As for the vertebrates, the best records come from North America and Europe. According to Wolfe (1992), regional extinctions were greatest across the Eocene–Oligocene boundary at high latitudes, and at least moderate extinctions at middle latitudes in North America. However, many lineages at middle latitudes survived environmental deterioration either by migrating from montane to lowland areas or by surviving in southern refugia and thus reappearing in warm intervals. The high-latitude (microthermal) flora of the Oligocene was very depauperate but by the end of the epoch diversity again increased, reaching a maximum in the middle Miocene.

Palynological evidence from the Anglo-Belgian-Paris Basin indicates that the Eocene–Oligocene transition was marked by the incoming of temperate elements, the loss of tropical and subtropical elements and an increase in conifer pollen, with evidence for cooling starting early in the late Eocene. In north-western Bohemia and the Weisselster Basin of Germany there is macrofloral evidence of a change from a dominantly evergreen, subtropical flora in the late Eocene to a mixed evergreen and deciduous flora in the early Oligocene, indicating a warm but temperature-seasonal climate (Collinson 1992). Collinson notes that many taxa in north-west Europe continued from the Eocene to the Oligocene, but Wolfe (1992) points out that all the cited taxa lived in aquatic habitats and most extant taxa are relatively insensitive to climate and have broad geographic ranges.

In summary, evidence from the terrestrial fauna and flora broadly mirrors that of the marine fauna in indicating an extended period of turnover, with no single catastrophic event, either at the Eocene–Oligocene boundary or at any other horizon. The changes seem to be bound up in considerable degree with climatic cooling, but with increase of aridity towards the Oligocene also being important.

Events in the Neogene

Miocene extinctions

A middle Miocene mass extinction among marine faunas was recognised by Raup and Sepkoski (1984) in their familial analysis based on the literature and formed a key data point in their claim of extinction periodicity. This conclusion was subsequently supported by generic analysis (Sepkoski 1986, 1989). There has, however, been remarkably little endorsement from researchers on Cenozoic marine fossils of such an extinction event. The only study worth noting is by Kaiho (1994) who recognised an extinction event among foraminifera in the early middle Miocene (16–13 Ma) in the Pacific and Atlantic oceans. Evidently the intermediate- and deep-water benthic forms were more affected than the planktonic, with estimates varying from 38% to 52% extinction of species, a rate significantly higher than anything for benthic species since the end of the Palaeocene. The planktonic foraminiferal extinction rate, though relatively high, was not much higher than the background extinction rate, due to the high evolutionary/extinction rates among this group. As with the late Eocene, no catastrophic event can be discerned, and the event involved either stepwise of gradual extinctions extended over several million years. Kaiho attributes a similar cause to both events, long-term global cooling.

With regard to the terrestrial realm, important late Miocene extinction events have been recognised among mammals in North America (Webb 1984; Webb and Barnosky 1989). In the early Miocene, mammals reached a global peak of ordinal and familial diversity, with major radiations taking place among North American ungulates, presumably as a response to the expansion of savannah and steppe at the expense of forests. In the late Miocene there began a series of Neogene declines, the first sharp extinction event taking place in North America at the end of the Clarendonian (~9 Ma). The greatest extinction episode, however, was in the late Hemphillian, at or near the close of the epoch, and involved 62 genera, 35 of which were large (Fig. 10.6). Because of inadequate stratigraphic and chronological data, it is not clear how catastrophic this major event was, and it could have extended over a substantial time interval, up to 1.5 m.y. (Webb 1984). Webb considers that climatic deterioration in the late Miocene destroyed the large browsers as extensive midcontinent savannah changed into steppe. During the late Hemphillian event, mixed feeders and even grazers (such as peccaries, horses and rhinos) were heavily affected. There was a net trend of replacing large by small herbivores, the most diverse of which were rodents. The record suggests that generally immigrations tracked extinctions, filling the vacuums created by prior extinctions. These and other Neogene extinctions are generally attributed to intervals of rapid climatic change and decreasing equability (Webb and Barnosky 1989).

Marine Pliocene–Pleistocene extinctions

In general the Pliocene and, especially, the Pleistocene are characterised by a remarkably low extinction rate among marine faunas (Vermeij 1987), but an important regional event among bivalves has been documented by Stanley and Campbell (1981) and Stanley (1986b). This event took place in the western north Atlantic, off the east coast of the United States, where, according to Stanley (1986), two or more extinction pulses in the late Pliocene and Pleistocene (between 3 and 1 Ma) removed about two-thirds of early

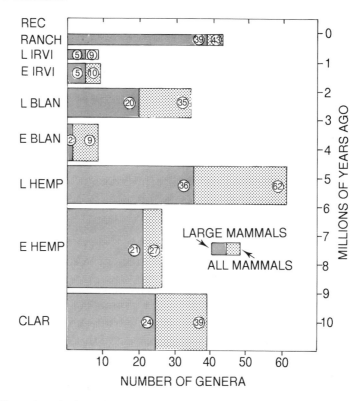

Fig 10.6 Late Cenozoic extinction episodes in North American land mammals. Clar = Clarendonian; Hemp = Hemphillian, Blan = Blancan; Irvi = Irvingtonian; Ranch = Rancholabrean; Rec = Recent. After Webb (1984).

Pliocene species. In marked contrast, there was a high survivorship of bivalve species on the north Pacific borders, in California and Japan.

Stanley attributes the extinctions to the association of climatic cooling and a particular palaeogeographic setting. In the north Pacific, faunas responded to cooling by shifting southwards but the north Atlantic was more heavily influenced by the growth of major ice caps. Whereas in the faunally more stable regions of California and Japan species of siphonate burrowers of small size experienced lower extinction rates than large forms, in the western north Atlantic the opposite is true. This is thought by Stanley to be due to the fact that small species are more endemic and therefore more vulnerable to temperature changes.

According to Jackson *et al.* (1993) the mass extinction event first documented by Stanley for bivalves north of the Caribbean occurred in a vastly greater region around the Caribbean and affected corals as well as molluscs. Jackson *et al.* consider that changes in patterns of upwelling and nutrient distribution may have been more important factors in causing the mass extinction than refrigeration due to the onset of the Pleistocene glaciation, because there was no evident temperature decline in the southern Caribbean after the Pliocene and only a slight decline in Florida. A similar view has been expressed by Allmon (1992) for turritelline gastropods, which exhibited their highest rate of extinction in the western Atlantic in Plio-Pleistocene times, with almost all species going extinct. Allmon considers that nutrients were a more important factor than temperature. The gastropods could have been adversely affected by a collapse and/ or major reorganisation of oceanic productivity associated with the formation of the Panama Isthmus.

Terrestrial late Pleistocene extinctions

Towards the end of the Würm/Wisconsin ice age the continents were much richer in large mammals than today – for example, mammoths, mastodonts, and giant ground sloths in the Americas; woolly mammoths, elephants, rhinos, giant deer, bisons, and hippos in northern Eurasia; and giant marsupials in Australia. Outside of Africa most genera of large mammals, defined as exceeding 44 kg adult weight, disappeared within the last 100 000 years with an increasing number going extinct towards the end, indicating that there was a significant extinction event near the end of the Pleistocene; this is brought out clearly in Fig. 10.7. This event was not contemporary across the world, however. North and South America were strongly affected 10 000–12 000 years ago, but well over 15 000 years ago in Australia and New Guinea. Oceanic islands have lost mammals and birds more recently, 1000–6000 years ago, while Eurasia and Africa have escaped a significant proportion of extinctions (Fig. 10.8). Table 10.2 presents quantitative data on extinctions of large mammal genera in different continents within the last 100 000 years (Martin 1984a). Because of the recency in time of these extinctions, the activities of early *Homo sapiens* have been widely invoked as a causal explanation, while others have preferred an interpretation involving abiotic environmental, essentially climatic, change. As Marshall (1984) points out, it is risky to generalise too freely, and extinctions on each landmass should be viewed as discrete events, each of which may be unique. Before evaluating rival interpretations, it is therefore desirable to outline some of the principal events in different parts of the world that suffered severe extinctions before there are full records of human activity.

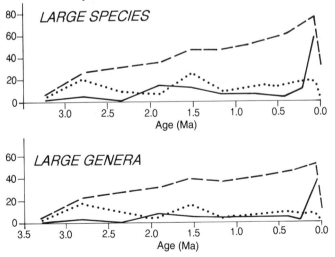

Fig 10.7 Extinction (solid line), origination (dotted line), and standing diversity (dashed line) for large mammals of the Plio-Pleistocene. After Martin (1984a).

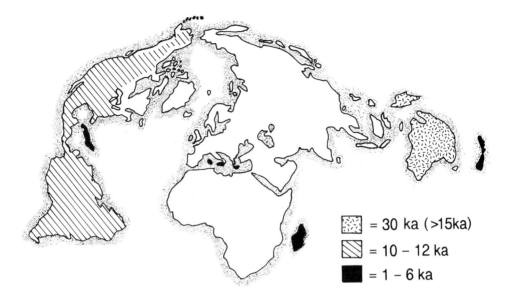

Fig 10.8 Timing of Late Pleistocene extinction of mammals and birds in different parts of the world. After Martin (1984a).

Table 10.2 Extinctions of large mammal general over the last 100 000 years

	Extinctions	Living	Total	Extinctions (%)
Africa	7	42	49	14.3
N. America	33	12	45	73.3
S. America	46	12	58	79.6
Australia	19	3	22	86.4

North and South America

As Table 10.2 clearly indicates, extinctions in both of these continents were highly significant. According to Martin (1984a), 23 genera of large mammals went extinct over a period of 3 m.y. prior to the Wisconsin glaciation. At the end of this glaciation there was a loss of 33 genera including all North American members of seven families and one other, the Proboscidea; equids and cameloids disappeared from the area of their origin. The maximum disappearance of large mammals was in the Rancholabrean (Fig. 10.6). There was no corresponding loss of small mammals. Late Pleistocene extinctions in South America included the last liptopterns and notoungulates, mastodonts, various artiodactyls, large edentates, and a giant rodent, but whereas North America lost its camelids, the tapir and capybara, their relatives survived in South America.

The most reliable radiocarbon dating suggests a relatively sudden event for the extinctions, about 11 000 years ago (Martin 1984b; Mead and Meltzer 1984).

Australia

The large mammals that went extinct include giant wombats, large kangaroos, and the hippo-sized *Diprotodon*, with an earlier date of perhaps as much as 30 000 years ago. No comparable extinctions are known among small marsupials (Martin 1984a; Murray

1984). Extinctions of moas (giant ratites) on New Zealand were much younger, within the last few thousand years (Trotter and McCulloch 1984).

There has been some discussion about whether the mammalian extinction rate in the late Pleistocene was higher than earlier in the Neogene, as implied for instance by Fig. 10.7. Gingerich (1984) disputes this, pointing out that North American origination rates greatly exceeded extinction rates in the early Pleistocene. Given the equilibrium observed between origination and extinction throughout the Cenozoic, the high rate of late Pleistocene generic extinctions could be viewed as the natural sequel to an unusually high rate of origination in the early Pleistocene.

Rival hypotheses

The principal arguments in favour of human beings or climate as the dominant control can now be summarised.

Human activity

There are three reasons for citing humans as the main reason for the late Pleistocene extinctions. First, the extinctions follow the appearance of humans in different parts of the world. Very few megafaunal extinctions of the late Pleistocene can definitely be shown to pre-date human arrival, with a sequence of extinctions following human dispersal, culminating most recently on oceanic islands. Thus there is evidence of human occupation in Australia before about 35 ka, perhaps as much as 40 ka (Murray 1984). The best evidence comes from North America, where Clovis culture sites have been radiocarbon-dated, with little scatter, at 11 ka; the evidence for any pre-Clovis culture remains weak (Martin 1984b). People reached New Zealand about 1000 years ago and there is abundant evidence of moa hunting, mainly as middens (Trotter and McCulloch 1984). In the Mediterranean region nearly 90% of endemic genera went extinct in the Holocene, apparently at the hands of Neolithic invaders (Sondaar 1977).

Second, it was generally only large mammals that went extinct. It is obvious that large animals would make a better target for hunting, and they would also have been more vulnerable to human-induced habitat destruction, because of a combination of smaller population size and lower reproductive rate. In contrast, there was no such discrimination between large and small mammals during the late Hemphillian extinction discussed earlier, which obviously pre-dated the arrival of our species by several million years.

Finally, the only reasonable alternative, involving environmental changes implicating climate in some way either directly or indirectly, is implausible (McDonald 1984). The limiting intensity of most selective forces, such as cooler air temperatures, simplicity of habitats, limited patchiness, low primary productivity, *relaxed* after 18 ka, and the associated environmental changes should not have produced the cluster of extinctions that occurred. The advent of human hunting is the only new factor. The climatic-change hypothesis fails to predict the increasing likelihood of extinctions with increasing body size, greater severity in both North and South America than in Eurasia, the lack of simultaneous extinction in Africa and tropical Asia, and the absence of extinctions at the end of previous Pleistocene glacial periods (Owen-Smith 1987).

However, the human overkill hypothesis has been criticised on four fronts. First, why were there no significant late Pleistocene extinctions in Africa, which has a record of hominids going back several million years? In an attempt to counter this objection,

Martin (1984a) notes that there was an appreciable number of extinctions in the early Pleistocene, unlike elsewhere in the world, followed by a long period of stability. He speculates that the highly diverse and cursorial nature of surviving African mammals, the Pleistocene radiation of bovids, and the lack of slow, ponderous herbivores like ground sloths, glyptodonts, and diprotodonts may reflect a coevolutionary history of humans and other mammals. Some have maintained that the survival in Africa of large mammals does not favour Martin's overkill hypothesis, but he uses this fact as an argument against a combination of climatic and cultural cause of extinction.

Second, hunting-induced extinctions should leave abundant evidence of kill sites, such as the New Zealand middens with butchered moa remains, but, unlike also in Eurasia and Africa, neither American nor Australian archaeological sites have yielded the bones of extinct mammals (Grayson 1984). Martin's (1984a) counterargument is that few such sites would be expected if human impact were truly swift and devastating, a veritable 'blitzkrieg'.

Third, the human predation hypothesis fails to explain the simultaneous extinctions of a number of mammalian or avian species not obviously vulnerable to human overkill. To account for this, Owen-Smith (1987) proposes a 'keystone herbivore' hypothesis. In present-day Africa elephants, rhinos, and hippos can transform tall grasslands into 'lawns' of more nutritious grasses. Elimination of megaherbivores elsewhere in the world by human hunters at the end of the Pleistocene would have promoted reverse changes in vegetation. These could have been detrimental to the distribution and abundance of smaller herbivores dependent upon the nutrient-rich and spatially diverse vegetation created by megaherbivore impact.

Finally, Martin's blitzkrieg hypothesis predicts that the ranges of megafauna in North America were constricted as the semicircular front of hunters moved south-eastwards, hence the extinctions should be time-transgressive from the north-west to the south-east. This prediction was tested by Beck (1996) in three separate analyses, none of which support the hypothesis. In fact all the patterns in the data appear to be in a direction opposite to that predicted.

Climate

There are three arguments in favour of this mechanism. First as indicated in Fig. 10.6, the late Pleistocene (Rancholabrean) event is but one of several North American mammal extinction events in the Neogene, the most important of which was at the end of the Miocene (late Hemphillian). All of these events appear to be associated with episodes of climatic change, involving glacial terminations (Webb 1984). Climate can thus not be ruled out for the youngest event, though Webb concedes that the advent of humans is also likely to be relevant.

Second, Vereshchagin and Baryshnikov (1984) argue that the extremely rapid decline of some non-exploited mammal species in the Quaternary of northern Eurasia points to the supreme importance of external, abiotic factors. The near-total extinction of mammoths and associated large species in tundra and taiga zones, and their partial survival in forest steppe and steppe, confirms, in their opinion, the decisive effect of changes in climate and landscape. The best proof is the ubiquitous transformation of late Pleistocene steppe fauna into a forest or taiga fauna over huge areas in upper and middle latitudes, where human influence was minimal. The destructive activity of humans became important in Eurasia only during the last few thousand years, being greatest in the ancient heartlands of civilisation in the Mediterranean region, west and central Asia, and China, and least in the polar desert.

Finally, it is probably over-simplistic to invoke climatic change on its own while disregarding concomitant biotic reorganisation. Individualistic response of various species to, for instance, loss of equability could have reduced the predictability of the composition and structure of new communities. In coevolved systems these changes could have had a cascading effect, by disrupting the coevolutionary relationships between plants and animals, thus creating a disequilibrium for the system. Competition in herbivore communities could have driven species with reduced fitness to extinction (Graham and Lundelius 1984). While these authors rightly emphasise how complicated the effect of climatic change can be, their work suffers from an excess of vague generalisation and lack of critical detail.

Disregarding the advent of humans, climatic hypotheses have made few predictions and are consequently very difficult to test (Grayson 1984). Why, for instance, did some large mammals, and not others, survive? Thus horses became extinct in North America at the end of the Pleistocene, yet thrived when they were reintroduced by Europeans. Why? On the other hand, Martin's overkill hypothesis has become so resilient that it can survive a succession of criticisms. As with the climatic hypothesis there is a danger here also of special pleading and *ad hoc* explanations such as why some large mammals, but not others, went extinct (as with the climatic hypothesis). On balance, however, it is difficult to deny the likelihood of human involvement in the late Pleistocene extinctions, though climatic factors may indeed also have played a role. Clearly, much remains to be learned.

Knowledge for more recent times is much better, and our species is believed to be responsible for most or nearly all recorded extinctions (Diamond 1984b, 1989). It is now clear that the first arrival of human inhabitants has always precipitated a mass extinction in the island biota. Well-known victims include New Zealand's moas, Madagascar's giant lemurs and elephant birds, and numerous bird species on Hawaii and other tropical Pacific islands. It is interesting to note that, in contrast to moas, there is a lack of elephant-bird hunter sites in Madagascar, because this could have relevance to the Rancholabrean extinctions in North America, where a similar situation holds. Possibly large bones in the latter region could have been dispersed by animal scavengers. As regards the future, consideration of the main mechanisms of human-caused extinctions (overhunting, effects of introduced species, habitat destruction and secondary ripple effects) indicates that the rate of extinction is accelerating. The basic reason is that there are now more humans than ever before, armed with more potent destructive technology, and encroaching on the world's most species-rich habitats, the continental tropical rainforests. Evidently we are in the midst of a new phase of mass extinction, and only a panglossian optimist would believe that this is likely to be checked in the foreseeable future.

11 *The causes of mass extinctions*

Before going on to discuss the possible causes of mass extinctions it is desirable to consider two subjects that have an important bearing on interpretation.

Raup and Sepkoski's claim of a 26 m.y. extinction periodicity since the end of the Palaeozoic, already discussed in Chapter 1, strongly suggests that, if true, there was only one prime causal factor. We have already noted how controversial this claim has proved to be, provoking a variety of criticisms embracing statistical methods used, radiometric dating, and the problems of paraphyly, which should raise doubts about at least the lesser extinction events. But even if, for the sake of argument, all the events are accepted as valid, there remain serious problems. Thus there are significant gaps in the periodicity, with no good evidence available from detailed scrutiny of well-marked extinction events in the Callovian, Tithonian, Aptian, and mid-Miocene. Furthermore the Eocene–Oligocene boundary does not signify a short-term extinction event but is merely the conclusion of an extended period of increased extinction rate lasting several million years. Raup and Sepkoski also missed a clear-cut extinction event in the deep sea at the end of the Palaeocene. They concede that no periodicity is apparent from their data for the Palaeozoic prior to the Late Permian, and Benton's (1995) independent analysis, based on data from the second edition of *The Fossil Record*, fails to find any evidence supporting periodicity. We conclude that Phanerozoic mass extinctions were episodic rather than periodic in character, and could have had a multiplicity of causes.

The second subject concerns the extent to which the mass extinctions were catastrophic, having been accomplished within a mere geological 'instant'. Some impact supporters have expressed the opinion that the early resistance of most palaeontologists and geologists was the consequence of their being indoctrinated as students with Lyellian uniformitarianism, implying gradual and denying catastrophic change. While this may well be true of some, there has been for several decades a strong school of thought dubbed 'neocatastrophism' (see, for example, Ager 1973) and there is no reason why all catastrophes on Earth should have been induced by bolide impact. As yet we know too little about the workings of our planet to dismiss Earth-induced catastrophes, probably bound up ultimately with changes in the mantle. The evidence must decide in particular cases, and it is to this evidence that we now turn.

Among a multiplicity of possible causes only a limited number merit consideration here, as having the potential to cause significant environmental change on a global scale. One should also take due note of the recently discovered phenomenon of **self-organised criticality**, based originally on the study of avalanching sandpiles. The same trigger, at the limit a mere single added grain of sand, can cause avalanches of a whole range of magnitudes obeying a power law, with a few large and many small ones. The relative magnitude of mass extinctions also appears to approximate to such a power law, as do other natural phenomena such as earthquakes and floods (Kauffman 1995). One key implication is that there is not necessarily a simple relationship between the size of a given mass extinction and the size of the causal factor. Furthermore, there may sometimes have been a combination of interacting factors that exceeded some critical threshold, with possibly cascading effects.

Bolide impact

Bolide impact has been a serious contender ever since Alvarez *et al.* (1980) published their paper on the Cretaceous–Tertiary boundary, and the extreme view has been put forward that not just mass extinctions but virtually all extinctions are probably due to this cause (Raup 1991). For this view to be more than an act of faith good evidence must be brought forward in support.

Iridium anomalies provide the most discussed evidence but, as the wide-ranging survey of Orth *et al.* (1990) brings out, only for the Cretaceous–Tertiary boundary can a really convincing case be made out for extraterrestrial impact. The many smaller iridium anomalies in the Phanerozoic record, by no means all located at extinction horizons, can variously be attributed either to extreme condensation or diagenetic changes at redox boundaries, as mentioned in Chapter 1. On the other hand, it needs to be recognised that comet, as opposed to asteroid, impact may leave little to no geochemical trace (Jansa 1993).

Shocked quartz is a good indicator of impact, provided it is not confused with quartz that has undergone Earth-bound tectonic deformation. Once again, the only convincing evidence comes from the Cretaceous–Tertiary boundary. The claim made for the Triassic–Jurassic boundary is decidely equivocal, as pointed out in Chapter 6. A good case can be made for at least two modest-sized meteorite impact events in the Late Devonian, but in neither case do they correspond to mass extinction intervals – on the contrary, they occurred at times of radiation. Microtektites also provide good evidence of impact, provided they are well authenticated. Various claims have been made for glassy spherules, but distinguishing genuine tektites from volcanic products can require sophisticated geochemical analysis of the sort that has been rarely attempted. The best examples are the purported microtektites at the K–T boundary in Haiti (see Chapter 9). The well-authenticated upper Eocene tektites undoubtedly signify bolide impact events but do not seem to coincide with extinction events (Miller *et al.* 1991; Wei 1995). Upper Devonian microtektites are known from a number of horizons in the Famennian, but they significantly post-date the Frasnian–Famennian mass extinction (cf. Chapter 4).

Other evidence from the stratigraphic record is more equivocal. Thus various impact-induced K–T boundary tsunami deposits have been claimed for the circum-Gulf of Mexico region, but the claims remain controversial (Chapter 9). McLaren (1983) and McLaren and Goodfellow (1990) have argued that catastrophic mass killings detected by biomass disappearance at bedding planes are a likely indication of impact. Certainly, depending solely upon taxon counts and ignore individual abundance can lead to underestimating the catastrophic character of biotic change, but acceptance of catastrophe is not in itself evidence for impact, as noted earlier.

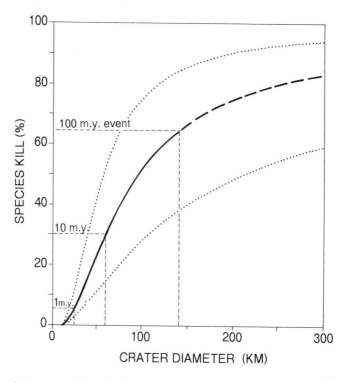

Fig 11.1 Relation between species extinction and impact crater diameter. Dotted curves define the worst-case limits of uncertainty in the placement of the impact–kill curve. After Raup (1992).

The Earth's cratering record also provides pertinent evidence and many well-authenticated Phanerozoic craters are now recognised (Grieve 1987). Raup (1992) has proposed what he calls a 'kill curve' for marine species in order to investigate large-body impact as a cause of species extinction. Current estimates of Phanerozoic impact rates are combined with the kill curve to produce an impact–kill curve, which predicts extinction levels from crater diameter (Fig. 11.1). Raup's model can be tested for some of the best-authenticated impact craters. The Ries crater of Bavaria has a diameter of about 30 km and formed at about 15 Ma, in the middle Miocene. According to the Raup curve about 10% of species should have gone extinct globally, but no species extinctions are recognisable from the region for either mammals (Heissig 1986) or plants (Gregor 1992).

The Montagnais impact structure on the Nova Scotia shelf is 45 km wide and is dated as late early Eocene in age. The Raup curve indicates a 17% species extinction, but there is in fact no recognisable biological change on a local, regional, or global scale (Jansa *et al.* 1990). The two largest craters in the Phanerozoic record, apart from Chicxulub, are the Popigai crater in Siberia and the Manicouagan crater in Quebec, both about 100 km in diameter (Grieve 1987). The Popigai crater is dated as 39 ± 9 m.y. old, and Raup has claimed that it could have been associated with the end-Eocene extinctions, but both marine and continental extinctions extended over several million years and there is no clear-cut end-Eocene event; furthermore, 39 Ma is several million years earlier than the Eocene–Oligocene boundary. The latest dating of the Manicouagan crater by Hodych and Dunning (1992) is 214 $\pm$ 1 Ma. The Gradstein *et al.* (1994) time-scale signifies this as early Norian in age, a time for which no one has recognised any extinctions, either in the marine or continental realm. Yet according to the Raup curve, no fewer than 50% of species should have gone extinct.

It must be concluded that, contrary to Raup's (1991) opinion, bolide impact cannot plausibly be invoked as a general cause of extinctions. Even for the K–T boundary event, it remains by no means clear what role impact had in causing the necessary environmental deterioration, as opposed to other Earth-bound factors. The evidence is good only for one major impact in North America, and the comet-shower extinction hypothesis of Hut *et al.* (1987) is not supported, either for the K–T boundary, which lacks good evidence for multiple impacts extended over time, or for the late Eocene, when multiple impacts evidently had no effect on the biota.

Volcanism

The eruption of vast provinces of continental flood basalts constitutes one of the most spectacular manifestations of igneous activity; many have considered that they may have precipitated mass extinction events (Vogt 1972; Keith 1982; Loper *et al.* 1988; Rampino and Stothers 1988; Stothers 1993). Recent redating of many flood basalt provinces has revealed that the eruption times may be as little as one million years or less, further strengthening the case that volcanism may be capable of causing geologically rapid environmental crises. The most frequently cited link between volcanism and extinction is that between the K–T event and the Deccan Traps (Officer *et al.* 1987; Sutherland 1994). Beyond India, the presence of smectitic boundary clays, such as the Fish Clay at Stevns Klint in Denmark, provide an additional pointer to the occurrence of volcanism during the extinction crisis (Courtillot 1990). However, suggestions that shocked minerals and Ir anomalies can be produced by flood basalt eruption are no longer generally accepted.

Individual flows within the Deccan Traps commonly exceed 10 000 km^2 in area and 1000 km^3 in volume. This far surpasses in magnitude the largest flood basalt eruption witnessed in historical times, the Lakagigar eruptions in Iceland in 1783–84, which only extruded 15 km^3 of lava. None the less, these eruptions killed 75% of the livestock and 25% of the human population of Iceland, probably due to the release of noxious gases (Courtillot 1990). It therefore seems reasonable to assume that the local and regional effects of the Deccan eruptions would have been devastating, but the global effects are more debatable. The most serious consequences may have come from the release of huge amounts of volcanic carbon dioxide and sulphur dioxide (Coffin and Eldholm 1993). The extinction mechanism is therefore variously attributed to the effects of global darkness, due to the presence of sulphate aerosols and/or acid rain (Officer *et al.* 1987; Rampino and Stothers 1988; Courtillot 1990). Such effects are only likely to have lasted a few years after each eruption episode, and other authors have suggested that global warming due to CO_2 release is a more likely cause of extinction (Loper *et al.* 1988). In particular, oceanic changes consequent upon warming may have created oceanic anoxia/dysoxia (Keith 1982; Coffin and Eldholm 1993; Wignall and Twitchett 1996).

Other than the Deccan Traps/K–T connection, the only other convincing links between mass extinctions and flood basalts can be found for the P–T and end-Palaeocene crises and the Siberian and Arctico-British flood basalt provinces respectively. Notably, both events are closely linked with global warming, suggesting that the emission of large volumes of CO_2 may be the most deleterious consequence of volcanism. Claims have been made that all post-Palaeozoic mass extinctions were caused by flood basalt eruptions, initially by Rampino and Stothers (1988) and with more statistical rigour by Stothers (1993). These in turn have been linked to periodic or quasi-periodic instabilities at the core–mantle boundary and the generation of deep-mantle plumes (Loper and McCartney 1986; Loper *et al.* 1988). Such claims have been supported by models of core–mantle boundary behaviour, but these are of little consequence because the original data set fails to stand up to close scrutiny. Stothers (1993) was only able to achieve a statistical link between flood basalts and extinctions by including Pliocene, middle Miocene, Aptian, Tithonian and Bajocian events among his major mass extinctions. As we have shown, these are non-events or at best of only regional significance. Even more perversely, Rampino and Stothers (1988) were only able to achieve a correlation by excluding the P–T event from their plot and yet this is one of the best extinction-flood basalts conjunctions. In fact the failure of many of the major flood basalt provinces to correlate with extinction events is one of the more telling arguments against a link. Thus, the Paraná basalts of eastern South America are one of the greatest of all flood basalt provinces, greater in extent and volume than the Deccan Traps and comparable to the Siberian Traps, and yet their brief eruption around the Barremian–Aptian boundary (120 ± 5 Ma) caused no notable extinction.

That some flood basalts coincide with major mass extinctions is undoubted, but they may be 'only accidental flukes of timing' (Stothers 1993: p. 1399), or perhaps they only served to exacerbate rather than directly trigger environmental crises.

Climate

Climatic changes, either cooling or warming, have been invoked for most mass extinctions; for Stanley (1984), cooling is held as the most frequent cause. Cooling can take two different forms: **high-latitude cooling** associated with a steepening of the equator-to-pole temperature gradient and commonly polar glaciation; and **global cooling**, in which the average global temperature declines but not necessarily the gradients. This is an important distinction but it is rarely made. Three high-latitude cooling/glaciation phases have occurred during the Phanerozoic and, notably, only the first of these, in the Hirnantian Stage of the latest Ordovician, is associated with a major mass extinction. The Permo-Carboniferous glaciation of Gondwana may have caused a minor extinction event at its onset in the mid-Carboniferous, while the Tertiary to Quaternary cooling caused an equally modest crisis. Even for the Hirnantian event the main mass extinction interval probably occurred during the warming phase at the termination of the glaciation (Fortey 1989). Of the earlier Hirnantian extinctions, the benthos may have been affected by the glacioeustatic regression but not by cooling *per se*. Only the pelagic extinctions of graptolites and trilobites may directly relate to cooling, in that oceanographic changes effectively eliminated their habitats (Fig. 3.11). Only for the protracted crisis during the late Eocene and early Oligocene can a good case be made for extinctions directly related to cooling (Chapter 10).

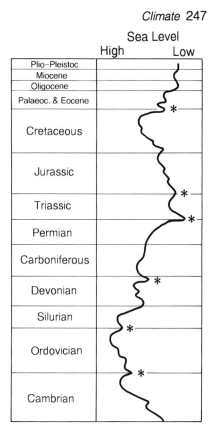

Fig 11.2 Newell's six mass extinction events (asterisks) plotted against Hallam's Phanerozoic sea-level curve. After Hallam (1992).

High-latitude cooling principally causes the wholesale movement of faunas and floras as climatic belts shift. Tropical faunas consequently become restricted to a narrower equatorial belt. This is clearly seen in the latest Ordovician, when the latitudinally restricted Edgewood fauna became the principal repository for much marine diversity at the height of the glaciation. Much the same is seen for brachiopod provinces during the Permo-Carboniferous glaciation (Raymond *et al.* 1989).

Global cooling is required if tropical habitats are to be completely eliminated, as may have happened during the K–T and F–F mass extinctions and possibly also during the trilobite crises at biomere boundaries. As already related, the K–T cooling/darkness interval is considered to be a relatively brief episode caused by impact ejecta, while the longer periods of global cooling, associated with the other events, are attributed to drawdown of atmospheric CO_2 levels. Evidence for global cooling is primarily derived from fossil data such as the preferential elimination of tropical taxa and the appearance of cold-water taxa in tropical areas during the immediate post-mass extinction interval (for example, the hexactinellid sponges in basal Famennian strata and the olenid trilobites above biomere boundaries). However, it is rarely possible to rule out the alternative possibility that such occurrences record the appearance of deep-water taxa during transgressions. Deep-water and cold-water preferences frequently go hand in hand, as is the case for both the hexactinellids and olenids.

Global warming, generally attributed to a rise in concentrations of greenhouse gases, is also an important component in some extinction scenarios. In the terrestrial realm global warming will be at its most severe at high latitudes due to the elimination of polar habitats; the elimination of the glossopterid forests at the end of the Permian appears a particularly good example of such an effect (Retallack 1995a). In the marine realm global warming is rarely held as the proximate cause of death, except for the death-by-cooking scenario proposed for the Frasnian–Famennian extinctions on the basis of some rather dubious oxygen isotope data (Chapter 4). More significant is the effect of warming on oceanic circulation and the decreasing solubility of oxygen as temperature rises (see, for example, Keith 1982). Thus, global warming-induced anoxia/dysoxia is considered a principal contender in the end-Ordovician, end-Permian and end-Palaeocene extinctions (Table 11.1).

Table 11.1 Summary of the proposed causes of the main Phanerozoic mass extinction events

	Bolide impact	Volcanism	Cooling	Warming	Regression	Anoxia/transgression
Late Precambrian						●
Late Early Cambrian						●
Biomere boundaries			○			●
Late Ashgill			●	●	●	●
Frasnian–Famennian			○		○	●
Devonian–Carboniferous			○			●
Late Maokouan					●	
End-Permian		●		●	○	●
End-Triassic					●	○
Early Toarcian						●
Cenomanian–Turonian			○			●
End-Cretaceous	●	●	●		●	○
End-Palaeocene		●		●		●
Late Eocene			●			

● strong link   ○ possible link

Marine regression

Chamberlin (1909) was perhaps the first person to perceive a relationship between regressions and extinctions in the marine realm, and the subject was further pursued with respect to the Palaeozoic of North America by Moore (1954). It was, however, his American compatriot Newell (1967) who first put forward an explicit hypothesis based on his recognition of six major mass extinction events in the Phanerozoic record. He argued that shrinkage of the area of epicontinental sea habitat should have a deleterious effect on neritic organisms and should therefore lead to widespread extinction. Radiation of the survivors would take place during the expansion of habitat area consequent upon a succeeding transgression. A quantitative analysis based on the ecologists' species–area relation was subsequently undertaken by Schopf (1974) and Simberloff (1974) for the end-Permian extinction, and was held to provide support for Newell's claim of a strong relation between regression and extinction.

On the face of it Newell's hypothesis is a reasonable one, because there is nearly always a positive correlation among living organisms in a wide range of habitats between species number and area (Connor and McCoy 1979) and decreasing habitat area is widely perceived as the prime cause of extinction today (Eldredge 1991). There appears, furthermore, to be a strong relationship between Newell's six extinction events and marked falls of sea level (Fig. 11.2), and many lesser extinction events have also been correlated with relatively minor regressions (Hallam 1989a, 1992).

In recent years, however, the hypothesis has been subjected to a number of criticisms. The most obvious one is that not all regressions correlate with extinction events. A good example portrayed in Fig. 11.2 concerns the Early Devonian regression. It is well established that there was a succession of major regressions of shelf seas during the Quaternary, at times of global cooling and ice-volume expansion, but the extinction rate for neritic invertebrates was remarkably low (Valentine and Jablonski 1991). Indeed, Jablonski (1985) argued that in the modern world most marine faunas would persist in undiminished shallow-water regions around oceanic islands at times of Quaternary sea-level fall. Several reservations have been expressed about uncritical application of the species–area relationship, whatever it may signify, to the fossil record. Thus Wyatt's (1987) hypsometric study has demonstrated that the response of continents to sea-level change depends upon their size, and that the global curve of shallow sea area with time that he produced does not vary in a straightforward manner with respect to sea level. Accordingly, predictions of organic diversity change with sea level should take account of their nonlinear relation. A comparable point was made by Martin (1981); the species– area curve is steep for small areas but the slope becomes asymptotic as the area increases.

Regarding the Quaternary, one important factor that should not be overlooked, however, is the comparative rapidity of glacioeustatic sea-level change. Bearing in mind that extinction risk increases with time in a stressful environment without further diminution of area (Diamond 1984a), it could be that the rapid restoration of less stressful conditions in the Quaternary has served to diminish the extinction rate owing to reduction of habitat area through sea-level fall. The slower and less spectacular regressions of much of the Phanerozoic might have been environmentally more significant because they lasted longer, and quite modest falls in very extensive and extremely shallow-water epicontinental seas with no close modern analogue could have had a correspondingly major effect (Hallam 1981b: Chapter 5). Johnson (1974) stressed the importance of organic adaptations to changed circumstances in understanding the likely causes of extinction of neritic invertebrates. During episodes of sustained enlargement of epicontinental seas, organisms become progressively more stenotopic and an equilibrium is established. They are in effect 'perched' subject to the continued existence of their environment. Extinctions occur to an extent proportional to the speed of regression, degree of stenotopy attained, or a combination of the two.

The ecological model proposed by Hallam (1978) to account for Jurassic ammonite extinctions bears several resemblances to that of Johnson. Evidence of progressively increasing stenotopy is recorded by data on phyletic size increase, a K-selected trend. Times of relatively low sea level signify times of restriction and deterioration of neritic habitat, such as increases in the variability of temperature and salinity of extremely shallow water. In familiar ecological parlance the increased environmental stress favours r-selected organisms. This is recorded by the small size of new ammonite taxa that have evolved rapidly from their larger, more stenotopic ancestors, often as a result of heterochronous changes. Thus speciation, involving the origin of new taxa by repro- ductive isolation at times of restricted connections between epicontinental seas, as well as extinction, is promoted by regression. Comparable models to those of Johnson and Hallam have been put forward for a variety of Palaeozoic and Mesozoic marine invertebrates, for which an empirical relationship between regressions and increase in extinction rate has frequently been noted (see references in Hallam 1989a; see also Brett and Baird 1995).

A further factor that may help to account for the very low extinction rates of shallow-water molluscs in the Quaternary is the high extinction rate, at least regionally, in the Pliocene (Jablonski 1994). The survivors could have been relatively stenotopic and therefore comparatively extinction-resistant. Indeed, Stanley (1990) has argued for spacing in time of major extinctions bound up with this factor.

While a strong association between sea-level change and marine mass extinction seems undeniable, it has become increasingly apparent that the key factor in many mass extinctions may be less the regression than the reduction in habitat area provoked by the development of anoxic waters during the subsequent transgression. Anoxia as an extinction mechanism will be discussed in the next section, but here reference must be made to those Phanerozoic extinction events mentioned in this book for which regression appears to have been significant.

The earliest Phanerozoic event is in the late Early Cambrian (early Royonian) – the so-called Hawke Bay event. End-Ordovician regression, probably associated with the growth of a Gondwana ice sheet, caused considerable extinctions among shallow platform faunas and was before an important oceanic event affecting outer- to off-shelf faunas, including plankton (Fortey 1989). Probably the best link comes from the recently identified late Maokouan (Guadalupian) crisis experienced by the low-latitude faunas of carbonate platforms and ramps. These habitats were eliminated by the widespread regression at this time, as particularly well seen in the southern United States (where the Guadalupian reefs and their diverse fauna disappeared) and South China. The end-Permian mass extinction, in contrast, was formerly related to a major regression but recent studies have demonstrated that this was an early Changxingian event and that the extinctions occurred during the subsequent transgression towards the close of the Changxingian.

The Triassic–Jurassic boundary is marked by a significant global regression–transgression couplet, and the regression component cannot be ruled out as a major cause of marine extinction. It could well have had at least as great a role as subsequent transgression and anoxia, as discussed in Chapter 6. Important regional extinction events in Europe associated with regression occurred in the Carnian and Tithonian. The Cretaceous–Tertiary boundary is also marked by a global regression–transgression couplet of probably even greater magnitude than that embracing the Triassic–Jurassic boundary. At least some of the marine extinctions are likely to be attributable to regression, as discussed in Chapter 9.

Anoxia and transgression

The widespread development of dysoxic to anoxic waters is regarded as the principal cause of most marine extinctions (Table 11.1). Unlike many other extinction mechanisms, the evidence is generally diverse and includes faunal, sedimentological and geochemical data. Invariably such dysoxic/anoxic episodes are associated with sea-level rise, and the anoxia–transgression nexus is in fact one of the principal recurrent themes of the stratigraphic record (see, for example, Hallam 1981b and Wignall 1994).

Indeed, it has been argued that there are many more transgressive black shale events than there are mass extinctions and therefore, as with flood basalts, the coincidence of some examples may be no more than a fluke. However, the crucial point is that only the globally widespread black shales are associated with mass extinctions, with the possible exception of some Cretaceous oceanic anoxic events, although in these cases the precise synchrony of the scattered occurrences of black shales is open to doubt (Chapter 8).

The ultimate cause of transgressive anoxia is unresolved. The Jenkyns (1980) model, in which flooding of vegetated coastal plains triggers elevated productivity in shallow-marine seas, is popular, but it fails to predict the early Palaeozoic and Proterozoic examples of transgressive black shales. Neither does it explain why such facies can develop equally well in carbonate settings, where there is little terrestrial nutrient input, and in clastic settings. Further discussion of this enigmatic phenomenon is beyond the scope of this book (cf. Wignall 1994; Tyson 1995), although the possible connection with global warming has already been alluded to.

Marine anoxia is generally a feature of the deeper levels of the water column where the vertical advection of oxygen declines. However, to cause significant extinction among inner-shelf benthos, the oxygen-restricted conditions must occur in unusually shallow water. Such conditions have been invoked (see, for example, Schlager 1981; Narkiewicz and Hoffman 1989; Vogt 1989) but only rarely has evidence been presented (see, for example, Wignall and Twitchett 1996). Further consequences of global marine anoxia lie in the changes of nutrient fluxes; in particular, the enhanced recycling of phosphorus in anoxic conditions after a critical lag period is doubtless an important factor (cf. Chapter 5 and Fig. 5.17). Productivity changes and particularly productivity collapse are therefore probably an additional factor in the lethality of the anoxic kill mechanism.

Conclusions

This chapter has reviewed the principal extinction mechanisms in isolation, although it should be apparent that many are inextricably linked, not least the anoxia–transgression connection. The contribution of global warming to the development of marine anoxia/dysoxia was probably also important for the latest Hirnantian, Permian and Palaeocene events and may have been equally important at other times. However, one of the most important signals to emerge from this overview is the occurrence of major sea-level fluctuations during biotic crises. Sea-level changes have occurred throughout the Phanerozoic (see, for example, Haq *et al.* 1988; Hallam 1992) but the fluctuations associated with the big five mass extinctions and many of the more minor events are extraordinary for their rapidity, magnitude and clearly demonstrable global extent. The fluctuations at the Devonian–Carboniferous boundary are particularly well constrained and illustrate eustatic fluctuations of at least several hundred metres within the duration of a single conodont zone (Chapter 4). Major sea-level changes similarly occurred within the interval of the last planktonic foraminiferal zone of the Maastrichtian. The D–C events are a transgressive–regressive–transgressive triplet, but for the majority of extinction episodes regressive–transgressive couplets are the norm (Hallam 1989a, 1992). The extinctions are intimately tied to these fluctuations, although the relative importance of each component varies from event to event. Thus, end-Frasnian and end-Permian extinctions principally occurred during transgression, while the end-Triassic and end-Cretaceous extinctions are probably more closely linked with the regressive component of regressive–transgressive cycles.

The rapidity and magnitude of these spectacular sea-level oscillations must have caused rapid habitat tracking by both the marine and terrestrial biota and extinction for those that failed to keep up. Only the deeper-marine benthos are likely to have been immune to these effects. However, there is uncertainty concerning the ultimate cause or causes of this eustatic variation. Its rate is too fast to be ascribed to normal processes of mid-ocean ridge formation and subsidence and only the Hirnantian regressive–transgressive couplet is directly linked with a brief glacial episode. The end-Triassic couplet, in contrast, has been tentatively ascribed to doming and subsequent rifting during the initial break-up of Pangaea (Chapter 6). The end-Permian sea-level changes could possibly be related to Pangaea-wide uplift caused by doming of a Siberian hotspot prior to deflation and eruption of flood basalts (Erwin 1993). However, this model does not explain why the best evidence for Late Permian regression occurs in the western United States, far away from Siberia, nor why the same regressive–transgressive couplet is also seen in the isolated continent of South China. Stress-induced changes in lithosphere density constitute another potential mechanism for the generation of geologically-rapid regressive–transgressive cycles (Cathles and Hallam 1991), although such effects would be confined to single continents.

Finally, it should be noted that many of the extinction mechanisms reviewed herein focus on marine extinctions, although many of them are equally capable of causing terrestrial extinctions. There are two reasons for this: first, much of the literature has focussed on the better-recorded marine events, and second, the widely perceived notion that terrestrial plants in particular are not prone to mass extinctions (Knoll 1984; Traverse 1990). Recent studies have convincingly demonstrated that the second assumption is erroneous: plants have suffered serious losses in all the major extinction episodes since their invasion of the land in the Middle Devonian (see, for example, Algeo *et al.* 1995; Retallack 1995a). Clearly, one of the many outstanding tasks still remaining in mass extinction studies is to integrate both marine and terrestrial data in order to better understand these calamitous events.

References

Aberhan, M. and Fürsich, F.T. (1996). Diversity analysis of Lower Jurassic bivalves of the Andean Basin and the Pliensbachian–Toarcian mass extinction. *Lethaia* **29**, 181–95.

Ager, D.V. (1956–67). *The British Liassic Rhynchonellidae.* Monographs of the Palaeontological Society, London.

Ager, D.V. (1973). *The nature of the stratigraphic record.* Macmillan, London.

Alberti, H. (1979). Devonian trilobites. *Special Papers in Palaeontology*, **23**, 313–24.

Aldiss, D.T., Benson, J.M., and Rundle, C.C. (1984). Early Jurassic pillow lavas and palynomorphs in the Karoo of eastern Botswana. *Nature*, **310**, 302–4.

Aldridge, R.J. and Smith, M.P. (1993). Conodonta. In *The Fossil Record 2* (ed. M.J. Benton), pp. 563–72. Chapman & Hall, London.

Algeo, T.J., Berner, R.A., Maynard, J.B., and Scheckler, S.E. (1995). Late Devonian oceanic anoxic events and biotic crises: 'rooted' in the evolution of vascular land plants? *GSA Today*, **5**, 63–6.

Allasinaz, A. (1992). The Late Triassic–Hettangian bivalve turnover in Lombardy (Southern Alps). *Rivista Italiana di Paleontologia e Stratigrafia*, **97**, 431–54.

Allison, P.A. and Briggs, D.E.G. (1993). Paleolatitudinal sampling bias, Phanerozoic species diversity, and the end-Permian extinction. *Geology*, **21**, 65–8.

Allmon, W.D. (1992). Role of temperature and nutrients in extinction of turritelline gastropods: Cenozoic of the northwestern Atlantic and northeastern Pacific. *Palaeogeography, Palaeoclimatology, Palaeoecology*, **92**, 41–54.

Alt, D., Sears, J.M., and Hyndman, D.W. (1988). Terrestrial maria: the origins of large basalt plateaus, hotspot tracks and spreading ridges. *Journal of Geology*, **96**, 647–62.

Alvarez, L.W., Alvarez, W., Asaro, F., and Michel, H.V. (1980). Extraterrestrial cause for the Cretaceous–Tertiary extinction: experimental results and theoretical interpretation. *Science*, **208**, 1095–1108.

Anderson, J.M. and Cruickshank, A.R.I. (1978). The biostratigraphy of the Permian and the Triassic Part 5. A review of the classification and distribution of Permo-Triassic tetrapods. *Palaeontographica Africana*, **21**, 15–44.

Archibald, J.D. (1992a). Dinosaur diversity and extinction. *Science*, **256**, 160.

Archibald, J.D. (1992b). Dinosaur extinction: how much and how fast? *Journal of Vertebrate Paleontology*, **12**, 263–64.

Archibald, J.D. (1993). The importance of phylogenetic analysis for the assessment of species turnover: a case history of Paleocene mammals in North America. *Paleobiology*, **19**, 1–27.

Archibald, J.D. (1994). Testing K/T extinction hypothesis using the vertebrate fossil record. *Lunar and Planetary Institute Contribution*, **825**, 6.

Archibald. J.D. (1996). *Dinosaur extinction and the end of an era*. Columbia University Press, New York.

Archibald, J.D. and Bryant, L.J. (1990). Differential Cretaceous/Tertiary extinctions of nonmarine vertebrates; evidence from northeastern Montana. *Geological Society of America Special Paper*, **247**, 549–62.

Armstrong, H.A. (1995). High-resolution biostratigraphy (conodonts and graptolites) of the Upper Ordovician and Lower Silurian – evaluation of the Late Ordovician mass extinction. *Modern Geology*, **20**, 41–68.

Armstrong, H.A. (1996). Biotic recovery after mass extinction: the role of climate and ocean-state in the post-glacial (Late Ordovician–Early Silurian) recovery of the conodonts. In *Biotic recovery from mass extinction* (ed. M.B. Hart), pp. 105–17. Geological Society Special Publication, **102**.

Arthur, M.A., Schlanger, S.O. (1979). Cretaceous 'oceanic anoxic events' as causal factors in development of reef-reservoired giant oil fields. *Bulletin of the American Association of Petroleum Geologists*, **63**, 870–85.

Arthur, M.A., Schlanger, S.O., and Jenkyns, H.C. (1987). The Cenomanian–Turonian oceanic anoxic event II. Palaeoceanographic controls on organic-matter production and preservation. *Marine petroleum source rocks* (ed. J. Brooks and A.J. Fleet), pp. 401–20. Geological Society Special Publication, **26**.

Ascoli, P., Poag, C.W., and Remane, J. (1984). Microfossil zonation across the Jurassic–Cretaceous boundary on the Atlantic margins of North America. *Geological Association of Canada Special Paper*, **21**, 31–48.

Ash, S. (1986). Fossils plants and the Triassic–Jurassic boundary. In *The beginning of the age of dinosaurs* (ed. K. Padian), pp. 21–9. Cambridge University Press, Cambridge.

Askin, R.A. (1988). The palynological record across the Cretaceous/Tertiary transition on Seymour Island, Antarctica. *Geological Society of America Memoir*, **169**, 155–62.

Askin, R.A. (1992). Preliminary palynology and stratigraphic implications from a new Cretaceous–Tertiary boundary section from Seymour Island. *Antarctic Journal of the United States*, **25**, 42–4.

Attrep, M., Orth, C.J., and Qintana, L.R. (1991). The Permian–Triassic of the Gartnerkofel-1 Core (Carnic Alps, Austria): geochemistry of common and trace elements. *Abhandlungen der Geologischen Bundesanstalt*, **45**, 123–37.

Aubry, M.-P. (1992). Late Paleogene calcareous nannoplankton evolution: a tale of climatic deterioration. In *Eocene–Oligocene climatic and biotic evolution* (ed. D.R. Prothero and W.A. Berggren), pp. 272–309. Princeton University Press, Princeton.

Ausich, W.I., Kammer, T.W., and Baumiller, T.K. (1994). Demise of the middle Paleozoic crinoid fauna: a single extinction event or rapid faunal turnover? *Paleobiology*, **20**, 345–61.

Badjukov, D.D., Barsukova, L.D., Kolesov, G.M., Nizhegoroda, I.V., Nazarov, M.A., and Lobitzer, H. (1988). Element concentrations at the Triassic–Jurassic boundary in the Kendelbachgraben, Austria. In *Rare events in geology, abstracts of lectures and excusion guide*, pp. 1–2. IGCP Project 199. Geologische Bundesanstalt, Wien.

Bai S. and Ning Z. (1988). Faunal changes and events across the Devonian–Carboniferous boundary of Huangmao Section, Guangxi, South China. *Canadian Society of Petroleum Geologists, Memoir*, **14**, 147–58.

Bakker, R.T. (1993). Plesiosaur extinction cycles–events that mark the beginning, middle and end of the Cretaceous. *Geological Association of Canada Special Paper*, **21**, 641–64.

Balme, B.E. (1970). Palynology of Permian and Triassic strata in the Salt Range and Surghar Range, West Pakistan. In *Stratigraphic boundary problems: Permian and Triassic of West Pakistan* (ed. B. Kummel and C. Teichert), pp. 305–453. University of Kansas, Special Publication, **4**.

Balme, B.E. (1979). Palynology of Permian–Triassic boundary beds of Kap Stosch, east Greenland. *Meddelelser om Grønland*, **200**, 1–37.

Balme, B.E. and Helby, R.J. (1973). Floral modifications at the Permian-Triassic boundary in Australia. *Canadian Society of Petroleum Geologists, Memoir*, **2**, 433–44.

Banerjee, A. and Boyajian, G. (1996). Changing biologic selectivity of extinction in the Foraminifera over the past 150 m.y. *Geology*, **24**, 607–10.

Barnes, C.R. (1988). Stratigraphy and palaeontology of the Ordovician–Silurian boundary interval, Anticosti Island, Quebec, Canada. *Bulletin of the British Museum of Natural History (Geology)*, **43**, 195–219.

Barnes, C.R. and Bergström, S.M. (1988). Conodont biostratigraphy of the uppermost Ordovician and lowermost Silurian. *Bulletin of the British Museum of Natural History (Geology)*, **43**, 325–43.

Barnes, C.R., Fortey, R.A., and Williams, S.H. (1995). The pattern of global bio-events during the Ordovician period. In *Global events and events stratigraphy* (ed. O.H. Walliser), pp. 139–72. Springer-Verlag, Berlin.

Barrera, E. (1994). Global environmental changes preceding the Cretaceous–Tertiary boundary: early–late Maastrichtian transition. *Geology*, **22**, 877–80.

Barrera, E. and Huber, B.T. (1990). Evolution of Antarctic waters during the Maastrichtian: foraminifer oxygen and carbon isotope ratios, ODP Leg 113. *Proceedings of the Ocean Drilling Programme, Initial Report*, **113**, 813–28.

Barrera, E. and Keller, G. (1990). Foraminiferal stable isotope evidence for gradual decrease of marine productivity and Cretaceous species survivorship in the earliest Danian. *Paleoceanography*, **5**, 867–90.

Basu, A.R., Poreda, R.J., Renne, P.R., Teichmann, F., Vosiliev, Y.R., Sobolev, N.V., and Turrin, B.D. (1995). High He-3 plume origin and temporal-spatial evolution of the Siberian flood basalts. *Science*, **269**, 822–5.

Bate, R.H. and Robinson, E. (1978). *A stratigraphical index of British Ostracoda*. Seel House Press, Liverpool.

Batt, R.J. (1989). Ammonite shell morphotype distributions in the Western Interior Greenhorn Sea and some paleoecological implications. *Palaios*, **4**, 32–42.

Batten, R.L. (1973). The vicissitudes of the gastropods during the interval of Guadalupian–Ladinian time. *Canadian Society of Petroleum Geologists, Memoir*, **2**, 596–607.

Batten, R.L. and Stokes, W.M.L. (1987). Early Triassic gastropods from the Sinbad Member of the Moenkopi Formation, San Rafael Swell, Utah. *American Museum Novitates*, **1864**, 1–33.

Baud, A., Magaritz, M. and Holser, W.T. (1989) Permian–Triassic of the Tethys: carbon isotopes studies. *Geologische Rundschau*, **78**, 649–77.

Beauchamp, B. (1994). Permian climatic cooling in the Canadian Arctic. *Geological Society of America Special Paper*, **288**, 229–46.

Beauvais, L. (1984). Evolution and diversification of Jurassic Scleractinia. *Palaeontographica Americana*, **54**, 219–24.

Beck, M.W. (1996). On discerning the cause of late Pleistocene megafaunal extinctions. *Paleobiology*, **22**, 91–103.

Becker, R.T. (1992). Analysis of ammonoid palaeobiogeography in relation to the global Hangenberg (terminal Devonian) and Lower Alum Shale (middle Tournasian) events. *Annales de la Société Géologique de Belgique*, **115**, 459–73.

Becker, R.T. and House, M.R. (1994). Kellwasser events and goniatite successions in the Devonian of the Montagne Noire with comments on possible causations. *Courier Forschungsinstitut Senckenberg*, **16**, 45–77.

Becker, R.T., Feist, R., Flajs, G., House, M.R., and Klapper, G. (1989). Frasnian–Famennian extinction events in the Devonian at Coumiac, southern France. *Comptes Rendus de l'Academie de Sciences, Paris, Series II*, **309**, 259–66.

Becker, R.T., House, M.R., Kirchgasser, W.T., and Playford, P.E. (1991). Sedimentary and faunal changes across the Frasnian/Famennian boundary at the Canning Basin of Western Australia. *Historical Biology*, **5**, 183–96.

Belka, Z. and Wendt, J. (1992). Conodont biofacies patterns in the Kellwasser facies (upper Frasnian/lower Famennian) of the eastern Anti-Atlas, Morocco. *Palaeogeography, Palaeoclimatology, Palaeoecology*, **91**, 143–73.

Bendix-Almgreen, S.E. (1976). Palaeovertebrate faunas of Greenland. In *Geology of Greenland* (ed. A. Escher and W.S. Watt), pp. 536–73. Grønlands Undersøgelse. Odense. Denmark.

Bengtson, S. (1992). Proterozoic and earliest Cambrian skeletal metazoans. In *The Proterozoic biosphere* (ed. J.W. Schopf and C. Klein), pp. 397–411. Cambridge University Press, Cambridge.

Bengtson, S. (ed.) (1994). *Early life on Earth*. Nobel Symposium No. 84. Columbia University Press, New York.

Benson, R.H. (1984). The Phanerozoic 'crisis' as viewed from the Miocene. In *Catastrophes and Earth history* (ed. W.A. Berggren and J.A. van Couvering), pp. 437–46. Princeton University Press, Princeton.

Benton, M.J. (1986). More than one event in the late Triassic mass extinction. *Nature*, **321**, 857–61.

Benton, M.J. (1987). Progress and competition in microevolution. *Biological Reviews*, **62**, 305–38.

Benton, M.J. (1988). Mass extinctions in the fossil record of reptiles: paraphyly, patchiness, and periodicity. In *Extinction and survival in the fossil record* (ed. G.P. Larwood), pp. 269–94. Systematics Association Special Volume, **34**.

Benton, M.J. (1989). Mass extinctions among tetrapods and the quality of the fossil record. *Philosophical Transactions of the Royal Society of London*, **B325**, 369–86.

Benton, M.J. (1991). What really happened in the Late Triassic? *Historical Biology*, **5**, 263–78.

Benton, M.J. (ed.) (1993). *The Fossil Record 2*. Chapman & Hall, London.

Benton, M.J. (1994). Late Triassic to Middle Jurassic extinctions among continental tetrapods: testing the pattern. In *In the shadow of dinosaurs* (ed. N.C. Fraser and H.-D. Sues), pp. 366–97. Cambridge University Press, Cambridge.

Benton, M.J. (1995). Diversification and extinction in the history of life. *Science*, **268**, 52–8.

Benton, M.J. and Simms, M.J. (1995). Testing the marine and continental fossil records. *Geology*, **23**, 601–4.

Benton, M.J. and Storrs, G.W. (1994). Testing the quality of the fossil record: paleontological knowledge is improving. *Geology*, **22**, 111–14.

Berggren, A. and Prothero, D.R. (1992). Eocene–Oligocene climatic and biotic evolution: an overview. In *Eocene-Oligocene climatic and biotic evolution* (ed. D.R. Prothero and W.A. Berggren), pp. 1–28. Princeton University Press, Princeton.

Berner, R.A. (1989). Biogeochemical cycles of carbon and sulfur and their effect on atmospheric oxygen over the Phanerozoic times. *Palaeogeography, Palaeoclimatology, Palaeoecology*, **75**, 97–122.

Berner, R.A. and Canfield, D.E. (1989). A new model for atmospheric oxygen over Phanerozoic time. *American Journal of Science*, **298**, 333–61.

Bernoulli, D. and Jenkyns, H.C. (1974). Alpine, Mediterranean and central Atlantic Mesozoic facies in relation to the early evolution of Tethys. *Special Publication of the Society of Economic Paleontologists and Mineralogists*, **19**, 129–60.

Berry, W.B.N. (1996). Recovery of post-Late Ordovician extinction graptolites: a western North American perspective. In *Biotic recovery from mass extinction* (ed. M.B. Hart), pp. 119–26. Geological Society Special Publication, **102**.

Berry, W.B.N. and Boucot, A.J. (1973). Glacioeustatic control of Late Ordovician–Early Silurian platform sedimentation and faunal change. *Bulletin of the Geological Society of America*, **84**, 275–84.

Berry, W.B.N., Wilde, P., and Quinby-Hunt, M.S. (1990). Late Ordovician graptolite mass mortality and subsequent early Silurian re-radiation. In *Extinction events in Earth history* (ed. E.G. Kauffman and O.H. Walliser), pp. 115–23. Springer-Verlag, Berlin.

Beuf, S., Biju-Duval, B., Stevaux, J., and Kulbicki, G. (1966). Ampleur des glaciations 'Siluriennes' au Sahara: leurs influences et leurs consequences sur la sédimentation. *Institute Française Pétrole Revue*, **21**, 363–81.

Beurlen, K. (1956). Der Faunenschnitt an der Perm-Trias Grenze. *Zeitschrift der Deutschen Geologischen Gesellschaft*, **108**, 88–99.

Bhandari, N., Shukla, P.N., Ghevariya, Z.G., and Sundaram, S.M. (1995). Impact did not trigger Deccan volcanism: evidence from Anjar K/T boundary intertrappen sediments. *Geophysical Research Letters*, **22**, 433–6.

Bice, D.M., Newton, C.R., McCaulay, S., Reiners, P.W., and McRoberts, C.A. (1992). Shocked quartz at the Triassic–Jurassic boundary in Italy. *Science*, **255**, 443–6.

Birkelund, T. and Hakansson, E. (1982). The terminal Cretaceous extinction in Boreal shelf seas – a multicausal event. *Geological Society of America Special Paper*, **190**, 373–84.

Bless, M.J.M., Becker, R.T., Higgs. K.T., Paproth, E., and Streel, M. (1992). Eustatic cycles around the Devonian–Carboniferous boundary and the sedimentary and fossil record in Saurland (Federal Republic of Germany). *Annales de la Société Géologique de Belgique*, **115**, 689–702.

Blum, J.D. and Chamberlain, C.P. (1992). Oxygen isotope constraints on the origin of impact glasses from the Cretaceous–Tertiary boundary. *Science,* **257**, 1104–7.

Blumenstengel, H. (1992). Ostracodes from the Devonian–Carboniferous boundary beds in Thuringia (Germany). *Annales de la Société Géologique de Belgique,* **115**, 483–9.

Boersma, A., Premoli-Silva, I., and Shackleton, N.J. (1987). Atlantic Eocene planktonic foraminiferal paleohydrographic indicators and stable isotope paleoceanography. *Paleoceanography*, **2**, 287–331.

Bogoyavlenskaya, O.V. (1982). Late Devonian to Early Carboniferous stromatoporoids, *Paleontological Journal*, **16**, 29–36.

Bottjer, D.J., Campbell, K.A., Schubert, J.K., and Droser, M.L. (1995). Palaeoecological models, non-uniformitarianism and tracking the changing ecology of the past. In *Marine palaeoenvironmental analysis from fossils* (ed. D.J.W. Bosence and P.A. Allison), pp. 7–26. Geological Society Special Publication, **83**.

Boucot, A.J. (1990). Phanerozoic extinctions: how similar are they to each other? In *Extinction events in Earth history* (ed. E.G. Kauffman and O.H. Walliser), pp. 5–30. Springer-Verlag, Berlin.

Boucot, A.J. and Gray, J. (1978). Comment on: 'Catastrophe theory: application to the Permian mass extinction.' *Geology*, **6**, 646–7.

Boulter, M.C., Spicer, R.A., and Thomas, B.A. (1988). Patterns of plant extinction from some palaeobotanical evidence. In *Extinction and survival in the fossil record* (ed. G.P. Larwood), pp. 1–36. Systematics Association Special Volume, **34**.

Bourgeois, J., Hansen, T.A., Wiberg, P.L., and Kauffman, E.G. (1988). A tsunami deposit at the Cretaceous–Tertiary boundary. *Science*, **241**, 567–70.

Bown, P.R. and Lord, A.R. (1990). The occurrence of calcareous nanofossils in the Triassic/Jurassic boundary interval. *Cahiers de l'Université de Lyon*, Série Scientifique, **3**, 127–36.

Brack, P., Rieber, H., and Mundil, R. (1995). The Anisian/Ladinian boundary interval at Bagolino (Southern Alps, Italy): I. Summary and new results on ammonoid horizons and radiometric age dating. *Albertiana*, **15**, 45–56.

Bralower, T.J. (1988). Calcareous nanofossil biostratigraphy and assemblages of the Cenomanian–Turonian boundary interval: implications for the origin and timing of oceanographic anoxia. *Paleoceanography*, **3**, 275–316.

Bralower, T.J., Arthur, M.A., Leckie, R.M., Sliter, W.V., Allard, D.J., and Schlanger, S.O. (1994). Timing and paleoceanography of oceanic dysoxia/anoxia in the late Barremian and early Aptian (Early Cretaceous). *Palaios*, **9**, 335–69.

Bramlette, M.N. and Martini, E. (1964). The great change in calcareous nannoplankton fossils between the Maastrichtian and Danian. *Micropaleontology*, **10**, 291–322.

Brand, U. (1989). Global climatic change in the Devonian–Mississippian: stable isotope biogeochemistry of brachiopods. *Palaeogeography, Palaeoclimatology, Palaeoecology*, **75**, 311–29.

Brand, U. (1992). Global perspectives of Frasnian–Tournasian oceanography: geochemical analysis of brachiopods. *Annales de la Société Géologique de Belgique*, **115**, 491–6.

Brandner, R. (1984). Meeresspiegelschwankungen und Tektonik in der Trias der NW-Tethys. *Jahrbuch der Geologischen Bundesanstalt*, **126**, 435–75.

Brandner, R. (1988). Plate tectonics and fluctuations of sea level and climate at the Permian–Triassic boundary. *Berichte der Geologischen Bundesanstalt, Wien*, **15**, 3.

Brasier, M.D. (1979). The Cambrian radiation event. In *The origin of the invertebrate groups* (ed. M.R. House), pp. 103–59. Academic Press. London and New York.

Brasier, M.D. (1982). Sea-level changes, facies changes and the late Precambrian–early Cambrian evolutionary explosion. *Precambrian Research*, **17**, 105–23.

Brasier, M.D. (1988). Foraminiferal extinction and ecological collapse during global biological events. In *Extinction and survival in the fossil record* (ed. G.P. Larwood), pp. 37–64. Systematics Association Special Volume, **34**.

Brasier, M.D. (1989). On mass extinction and faunal turnover near the end of the Precambrian. In *Mass extinctions: processes and evidence* (ed. S.K. Donovan), pp. 73–88. Belhaven Press, London.

Brasier, M.D. (1990). Phosphogenic events and skeletal preservation across the Precambrian–Cambrian boundary interval. In *Phosphorite research and development* (ed. A.J.G. Notholt and I. Jarvis), pp. 289–303. Geological Society of London Special Publication, **52**.

Brasier, M.D. (1995). The basal Cambrian transition and Cambrian bio-events. In *Global events and event stratigraphy* (ed. O.H. Walliser), pp. 113–38. Springer-Verlag, Berlin.

Brasier, M.D., Magaritz, M., Corfield, R.M., Luo H., Wu X., Ouyang L., Jiang Z., Hamdi, B., He, T., and Fraser, A.G. (1990). The carbon- and oxygen-isotope record of the Precambrian–Cambrian boundary interval in China and Iran and their correlation. *Geological Magazine*, **127**, 319–32.

Brasier, M.D., Corfield, R.M., Derry, L.A., Rozanov, A.Y., and Zhuravlev, A.Y. (1994). Multiple $\delta^{13}C$ excursions spanning the Cambrian explosion to the Botomian crisis in Siberia. *Geology*, **22**, 455–8.

Brass, G.W., Southam, J.R., and Peterson, W.H. (1982). Warm saline bottom water in the ancient ocean. *Nature*, **296**, 620–3.

Bratton, J.F. (1996). Brachiopods and oxygen levels during the survival interval of the Late Devonian mass extinction recovery in the Great Basin, western U.S.A. 6th North American Paleontological Convention, Abstracts Volume, *Paleontological Society Special Publication*, **8**, 44.

Brauckmann, C., Chlupac, I. and Feist, R. (1992). Trilobites at the Devonian–Carboniferous boundary. *Annales de la Société Géologique de Belgique*, **115**, 507–18.

Brenchley, P.J. (1984). Late Ordovician extinctions and their relationship to the Gondwana glaciation. In *Fossils and climate* (ed. P.J. Brenchley), pp. 291–315. John Wiley and Sons, London.

Brenchley, P.J. (1989). The Late Ordovician extinction. In *Mass extinctions: processes and evidence* (ed. S.K. Donovan), pp. 104–32. Belhaven Press, London.

Brenchley, P.J. and Newall, G. (1984). Late Ordovician environmental changes and their effect on faunas. In *Aspects of the Ordovician system* (ed. D.L. Bruton), pp. 65–79. Geological Contributions of the University of Oslo, **295**.

Brenchley, P.J. and Storch, P. (1989). Environmental changes in the Hirnantian (Upper Ordovician) of the Prague Basin, Czechoslovakia. *Geological Journal*, **24**, 165–82.

Brenchley, P.J., Romano, M., Young, T.P., and Storch, P. (1991). Hirnantian glaciomarine diamictites – evidence for the spread of glaciation and its effect on Upper Ordovician faunas. *Geological Survey of Canada*, Paper **90-9**, 325–36.

Brenchley, P.J., Marshall, J.D., Carden, G.A.F., Robertson, D.B.R., Long, D.G.F., Meidla, T., Hints, L., and Anderson, T.F. (1994). Bathymetric and isotopic evidence for a short-lived Late Ordovician glaciation in a greenhouse period. *Geology*, **22**, 295–8.

Brenchley, P.J., Carden, G.A.F., and Marshall, J.D. (1995). Environmental changes associated with the 'first strike' of the Late Ordovician mass extinction. *Modern Geology*, **20**, 83–100.

Bretsky, P.W. (1968). Evolution of Paleozoic marine communities. *Science*, **159**, 1231–3.

Bretsky, P.W. (1973). Evolutionary patterns in the Paleozoic Bivalvia: documentation and some theoretical considerations. *Bulletin of the Geological Society of America*, **84**, 2079–96.

Brett, C.E. and Baird, G.C. (1995). Coordinated stasis and evolutionary ecology of Silurian to Middle Devonian faunas in the Appalchian Basin. In *New approaches to speciation in the fossil record* (ed. D.H. Erwin and R.L. Anstey), pp. 285–315. Columbia University Press, New York.

Bridges, P.H., Gutteridge, P., and Pickard, N.A.H. (1995). The environmental setting of Early Carboniferous mud-mounds. *Special Publication of the International Association of Sedimentologists*, **23**, 171–90.

Briggs, D.E.G., Fortey, R.A. and Clarkson, E.N.K. (1988). Extinction and the fossil record of the arthropods. In *Extinction and survival in the fossil record* (ed. G.P. Larwood), pp. 171–209. Systematic Association Special Volume, **34**.

Brinkhuis, H. and Zachariasse, W.J. (1988). Dinoflagellate cysts, sea level changes and planktonic foraminifers across the Cretaceous–Tertiary boundary at El Haria, northwest Tunisia. *Marine Micropaleontology*, **13**, 153–91.

Brochwicz-Lewinski, W., Gasiewicz, A., Suffczynski, S., Szatkowski, K., and Zbik, M. (1984). Lacunes et condensations à la limite Jurassique moyen-supérieur dans le Sud de la Pologne: manifestation d'un phénomène mondial? *Comptes Rendus de l'Academie de Sciences, Paris, Series II*, **299**, 1359–62.

Broglio-Loriga, C., Neri, C., Pasini, M., and Posenato, R. (1988) Marine fossil assemblages from the Upper Permian to lowermost Triassic in the western Dolomites (Italy). *Memorie della Società Geologica Italiana*, **34** (for 1986), 5–44.

Brunton, F.R. and Copper, P. (1994). Paleoecologic, temporal, and spatial analysis of Early Silurian reefs of the Chicotte Formation, Anticosti Island, Quebec, Canada. *Facies*, **31**, 57–80.

Bryan, J.R. and Jones, D.S. (1989). Fabric of the Cretaceous–Tertiary marine macrofaunal transition at Braggs, Alabama. *Palaeogeography, Palaeoclimatology, Palaeoecology*, **69**, 279–301.

Buffetaut, E. (1990). Vertebrate extinctions and survival across the Cretaceous–Tertiary boundary. *Tectonophysics*, **171**, 337–45.

Buggisch, W. (1972). Zur Geologie und Geochemie der Kellwasserkalke und ihrer begleitenden Sedimente (Unteres Oberdevon). *Abhandlungen des Hessischen Landesamts für Bodenforschung*, **62**, 1–68.

Buggisch, W. (1991). The global Frasnian–Famennian 'Kellwasser event'. *Geologische Rundschau*, **80**, 49–72.

Burke, W.H., Denison, R.E., Hetherington, F.A., Koeprick, R.B., Nelson, H.F., and Otto, J.B. (1982). Variations in seawater $^{87}Sr/^{86}Sr$ throughout Phanerozoic time. *Geology*, **10**, 516–19.

Calef, C.E. and Bambach, R.K. (1973). Low nutrient levels in lower Palaeozoic (Cambrian–Silurian) oceans. *Geological Society of America Abstracts with Programs*, **5**, 565.

Campbell, I.H., Czamanske, G.K., Fedorenko, V.A., Hill, R.I., and Stepanov, V. (1992). Synchronism of the Siberian Traps and the Permian–Triassic boundary. *Science*, **258**, 1760–3.

Canfield, D.E. and Thamdrup, B. (1994). The production of ^{34}S-depleted sulfide during bacterial disproportionation of elemental sulfur. *Science*, **266**, 1973–5.

Cappetta, H. (1987). Extinctions et renouvellements fauniques chez les selaciens post-jurassiques. *Mémoires de la Société Géologique de France*, **150**, 113–31.

Caputo, M.V. (1985). Late Devonian glaciation in South America. *Palaeogeography, Palaeoclimatology, Palaeoecology*, **51**, 291–317.

Caputo, M.V. and Crowell, J.C. (1985). Migration of glacial centres across Gondwana during the Paleozoic era. *Bulletin of the Geological Society of America*, **96**, 1020–36.

Carlisle, D.B. (1992). Diamonds at the K/T boundary. *Nature*, **357**, 119–20.

Carlson, S.J. (1991). A phylogenetic perspective on articulate brachiopod diversity and the Permo-Triassic extinction. In *The unity of evolutionary biology* (ed. E. Dudley), pp. 119–42. Dioscorides Press, Portland, Oregon.

Caron, M. and Homewood, P. (1983). Evolution of early planktonic foraminifers. *Marine Micropaleontology*, **7**, 453–62.

Carpenter, K. and Breithaupt, B. (1986). Latest Cretaceous occurrence of nodosaurid ankylosaurs (Dinosauria, Ornithischia) in western North America and the gradual extinction of dinosaurs. *Journal of Vertebrate Paleontology*, **6**, 251–7.

Casey, R.E., Wigley, C.R., and Perez-Guzman, A.M. (1983). Biogeographic and ecologic perspective on polycystine radiolarian evolution. *Paleobiology*, **9**, 363–76.

Cathles, L.M. and Hallam, A. (1991). Stress-induced changes in plate density, Vail sequences, epeirogeny and short-lived global sea-level fluctuations. *Tectonics*, **10**, 659–71.

Chai, C., Mao, X., Ma, S., Bai, S., Zhou, Y., Ma, J., and Ning, Z. (1989). Geochemical anomaly at the Devonian/Carboniferous boundary, Hangmao, Guangxi, China. *Historical Biology*, **2**, 89–100.

Chai, C., Zhou, Y., Mao, X., Ma, S., Ma, J., Kong, P., and He, J. (1992). Geochemical constraints on the Permo-Triassic boundary event in South China. In *Permo-Triassic events in the eastern Tethys* (ed. W.C. Sweet, Z. Yang, J.M. Dickins and H. Yin), pp. 158–68. Cambridge University Press, Cambridge.

Chamberlin, T.C. (1909). Diastrophism as the ultimate basis of correlation. *Journal of Geology*, **17**, 689–93.

Chatterton, B.D.E. and Speyer, S.E. (1989). Larval ecology, life history strategies, and patterns of extinction and survivorship among Ordovician trilobites. *Paleobiology*, **15**, 118–32.

Chen, G., Tyburczy, J.A., and Ahrens, T.J. (1994). Shock-induced devolatilisation of calcium sulfate and implications for K–T extinctions. *Earth and Planetary Science Letters*, **128**, 615–28.

Chen, X. and Rong, J. (1991). Concepts and analysis of mass extinction with the Late Ordovician event as an example. *Historical Biology*, **5**, 107–21.

Chen, X. and Zhang, Y. (1995). The Late Ordovician graptolite extinction in China. *Modern Geology*, **20**, 1–10.

Chen, Y. and Teichert, C. (1983). Cambrian Cephalopoda of China. *Palaeontographica-A*, **191**, 1–102.

Chiappe, L.M. (1995). The first 85 million years of avian evolution. *Nature*, **378**, 349–55.

Christensen, W.K. (1976). Palaeobiogeography of Late Cretaceous belemnites of Europe. *Paläontologische Zeitschrift*, **50**, 113–29.

Chuvashov, B. and Riding, R. (1984). Principal floras of Palaeozoic marine calcareous algae. *Palaeontology*, **27**, 487–500.

Claeys, P. and Casier, J.-G. (1994). Microtektite-like impact glass associated with the Frasnian–Fammenian boundary mass extinction. *Earth and Planetary Science Letters*, **122**, 303–15.

Claeys, P., Casier, J.-G., and Margolis, S.V. (1992). Microtektites and mass extinctions from the Late Devonian of Belgium: evidence for a 367 Ma asteroid impact. *Science*, **257**, 1102–4.

Claeys, P., Kyte, F.T., and Casier, J.-G. (1994). Frasnian–Famennian boundary: mass extinction, anoxic oceans, microtektite layers, but not much iridium? *Lunar and Planetary Institute Contributions*, **825**, 22–4.

Claoué-Long, J.C., Zhang, Z., Ma, G., and Du, S. (1991). The age of the Permian–Triassic boundary. *Earth and Planetary Science Letters*, **105**, 182–90.

Clark, D.L. (1987). Conodonts: the final fifty million years. In *Palaeobiology of conodonts* (ed. R.J. Aldridge), pp. 165–74. Ellis Horwood, Chichester.

Clark, D.L., Cheng-Yuan, W., Orth, C.S., and Gilmore, J.S. (1986). Conodont survival and low iridium abundances across the Permian–Triassic boundary in South China. *Science*, **233**, 984–6.

Claypool, G.E., Holser, W.T., Kaplan, I.R., Sakai, H. and Zak, I. (1980). The age curves of sulfur and oxygen isotopes in marine sulfate, and their mutual interpretation. *Chemical Geology*, **28**, 199–259.

Clemens, W.A. (1986). Evolution of the terrestrial vertebrate fauna during the Cretaceous–Tertiary transition. In *Dynamics of extinction* (ed. D.K. Elliott), pp. 63–85. Wiley, New York.

Clemens, W.A. and Nelms, L.G. (1993). Paleoecological implications of Alaskan terrestrial vertebrate fauna in latest Cretaceous time at high paleolatitudes. *Geology*, **21**, 503–6.

Coccioni, R. and Galeotti, S. (1994). K–T boundary extinction: geologically instantaneous or gradual event? Evidence from deep-sea benthic foraminifera. *Geology*, **22**, 779–82.

Cocks, L.R.M. (1985). The Ordovician–Silurian boundary. *Episodes*, **8**, 98–100.

Cocks, L.R.M. (1988). Brachiopods across the Ordovician–Silurian boundary. *Bulletin of the British Museum (Natural History) (Geology)*, **43**, 311–16.

Cocks, L.R.M. and Copper, P. (1981). The Ordovician–Silurian boundary at the eastern end of Anticosti Island. *Canadian Journal of Earth Sciences*, **18**, 1029–34.

Cocks, L.R.M. and Fortey, R.A. (1988). Lower Palaeozoic facies and faunas around Gondwana. In *Gondwana and Tethys* (ed. M.G. Audley-Charles and A. Hallam), pp. 183–200. Geological Society Special Publication, **37**.

Coffin, M.F. and Eldholm, O. (1993). Large igneous provinces. *Scientific American*, **269**, 4, 26–33.

Colbert, E.H. (1973). Tetrapods and the Permian–Triassic transition. *Canadian Society of Petroleum Geologists, Memoir*, **2**, 481–92.

Collinson, M.E. (1986). Catastrophic vegetation changes. *Nature*, **324**, 112.

Collinson, M.E. (1992). Vegetational and floristic changes around the Eocene/Oligocene boundary in western and central Europe. In *Eocene–Oligocene climatic and biotic evolution* (ed. D.R. Prothero and W.A. Berggren), pp. 437–50. Princeton University Press, Princeton.

Colodner, D.C., Boyle, E.A., Edmond, J.M., and Thomson, J. (1992). Post-depositional mobility of platinum, iridium and rhenium in marine sediments. *Nature*, **358**, 402–4.

Conaghan, P.J., Shaw, S.E. and Veevers, J.J. (1993). Sedimentary evidence of the Permian/Triassic global crisis induced by the Siberian hotspot. *Canadian Society of Petroleum Geologists, Memoir*, **17**, 785–95.

Connor, E.F. and McCoy, E.D. (1979). The statistics and biology of the species–area relationship. *American Naturalist*, **113**, 791–833.

Conway Morris, S. (1993). Ediacara-like fossils in Cambrian Burgess Shale-type faunas of North America. *Palaeontology*, **36**, 593–635.

Conze, E. (1959). *Buddhism: its essence and development*. Oxford University Press, Oxford.

Cook, P.J. and Shergold, J.H. (1984). Phosphorus, phosphorites and skeletal evolution at the Precambrian–Cambrian boundary. *Nature*, **308**, 231–6.

Copeland, M.J. (1973). Ostracoda from the Ellis Bay Formation (Ordovician), Anticosti Island, Quebec. *Geological Survey of Canada, Paper* **72-43**, 1–49.

Copeland, M.J. (1989). Silicified Upper Ordovician-Lower Silurian ostracodes from the Avalanche Lake area, southwestern District of Mackenzie. *Bulletin of the Geological Survey of Canada*, **341**, 1–100.

Copper, P. (1977). Paleolatitudes in the Devonian of Brazil and the Frasnian–Famennian mass extinction. *Palaeogeography, Palaeoclimatology, Palaeoecology*, **21**, 165–207.

Copper, P. (1986). Frasnian/Famennian mass extinction and cold-water oceans. *Geology*, **14**, 835–9.

Copper, P. (1995). Five new genera of Late Ordovician–Early Silurian brachiopods from Anticosti Island, eastern Canada. *Journal of Paleontology*, **69**, 846–61.

Cornet, B. and Olsen, P.E. (1990). Early to Middle Carnian (Triassic) flora and fauna of the Richmond and Taylorsville basins, Virginia and Maryland, USA. *Virginia Museum of Natural History Guidebook*, **1**, 1–83.

Courtillot, V.E. (1990). What caused the mass extinction? A volcanic eruption. *Scientific American*, **263**, October, 85–92.

Courtillot, V.E. and Gaudemer, Y. (1996). Effects of mass extinctions on biodiversity. *Nature*, **381**, 146–8.

Crimes, T.P. (1989). Trace fossils. In *The Precambrian–Cambrian boundary* (eds. J.W. Cowie and M.D. Brasier), pp. 117–65. Clarendon Press, Oxford.

Crowley, T.J. and North, G.R. (1991). *Paleoclimatology*. Oxford University Press, Oxford.

Cuffey, R.J. and McKinney, F.K. (1979). Devonian Bryozoa. *Special Papers in Palaeontology*, **23**, 307–11.

Dagys, A.S. (1993). Geographic differentiation of Triassic brachiopods. *Palaeogeography, Palaeoclimatology, Palaeoecology*, **100**, 79–87.

Dagys, A.S. and Dagys, A.A. (1988). Biostratigraphy of the lowermost Triassic and the boundary between Palaeozoic and Mesozoic. *Memorie della Società Geologica Italiana*, **34** (for 1986), 313–20.

Dagys, A.S. and Dagys, A.A. (1994). Global correlation of the terminal Triassic. *Mémoires de Géologie* (Lausanne), **22**, 25–34.

Dean, W.T., Donovan, D.T. and Howarth, M.K. (1961). The Liassic ammonite zones and subzones of the north-west European Province. *Bulletin of the British Museum (Natural History), Geology*, **4**, 438–505.

Debrenne, F. (1991). Extinction of Archaeocyatha. *Historical Biology*, **5**, 95–106.

Denison, R.E., Koepnick, R.B., Burke, W.H., Hetherington, E.A., and Fletcher, A. (1994). Construction of the Mississippian, Pennsylvanian and Permian seawater $^{87}Sr/^{86}Sr$ curve. *Chemical Geology*, **112**, 145–67.

Derry, L.A., Brasier, M.D., Corfield, R.M., Rozanov, A.Y., and Zhuravlev, A.Y. (1994). Sr and C isotopes in Lower Cambrian carbonates from the Siberian craton: a paleoenvironmental record during the 'Cambrian explosion'. *Earth and Planetary Science Letters*, **128**, 671–81.

D'Hondt, S. and Lindinger, A. (1994). A stable isotopic record of the Maastrichtian ocean–climate system: South Atlantic DSDP Site 528. *Palaeogeography, Palaeoclimatology, Palaeoecology*, **112**, 363–78.

D'Hondt, S., Pilson, M.E.Q., Sigurdsson, H., Hanson, A.K., and Carey, S. (1994). Surface-water acidification and extinction at the Cretaceous–Tertiary boundary. *Geology*, **22**, 983–6.

Diamond, J.M. (1984a). 'Normal' extinctions of isolated populations. In *Extinctions* (ed. M.H. Nitecki), pp. 191–246. University of Chicago Press, Chicago.

Diamond, J.M. (1984b). Historic extinctions: a Rosetta Stone for understanding prehistoric extinctions. In *Quaternary extinctions: a prehistoric revolution* (ed. P.S. Martin and R.G. Klein), pp. 824–62. University of Arizona Press, Tueson.

Diamond, J.M. (1989). The present, past and future of human-caused extinctions. *Philosophical Transactions of the Royal Society of London*, **B325**, 469–77.

Dimichele, W.A. and Aronson, R.B. (1992). The Pennsylvanian–Permian vegetational transition: a terrestrial analogue to the onshore–offshore hypothesis. *Evolution*, **46**, 807–24.

Ding M. (1992). Conodont sequences in the Upper Permian and Lower Triassic of South China and the nature of conodont faunal changes at the systemic boundary. In *Permo-Triassic boundary events in the eastern Tethys* (ed. W.C. Sweet, Z. Yang, J.M. Dickins and H. Yin), pp. 109–19. Cambridge University Press, Cambridge

Dingus, L. (1984). Effects of stratigraphic completeness on interpretation of extinction rates across the Cretaceous–Tertiary boundary. *Paleobiology*, **10**, 420–38.

Dobruskina, I.A. (1987). Phytogeography of Eurasia during the Early Triassic. *Palaeogeography, Palaeoclimatology, Palaeoecology*, **58**, 75–86.

Dockery, D.T. (1986). Punctuated succession of Paleogene molluscs in the northern Gulf Coastal Plain. *Palaios*, **1**, 582–9.

Dockery, D.T. and Hansen, T.A. (1987). Eocene–Oligocene molluscan extinctions: comment and reply. *Palaios*, **1**, 620–2.

Donnelly, T.H., Shergold, J.H., Southgate, P.N., and Barnes, C.J. (1990). Events leading to global phosphogenesis around the Proterzoic–Cambrian boundary. In *Phosphorite research and developments* (ed. A.J.G. Notholt and I. Jarvis), pp. 273–87. Geological Society of London Special Publication, **52**.

Donovan, A.D., Baum, G.R., Blechschmidt, G.L., Lowtit, T.S., Pflum, C.E. and Vail, P.R. (1988). Sequence stratigraphic setting of the Cretaceous–Tertiary boundary in central Alabama. *Society of Economic Paleontologists and Mineralogists Special Publication*, **42**, 299–308.

Donovan, S.K. (1989). Palaeontological criteria for the recognition of mass extinction. In *Mass extinctions: processes and evidence* (ed. S.K. Donovan), pp. 19–36. Belhaven Press, London.

Donovan, S.K. (1996). Testing the marine and continental fossil records: comment. *Geology*, **24**, 381.

Douglas, R.G. and Savin, S.M. (1973). Oxygen and carbon isotope analyses of Cretaceous and Tertiary foraminfera from the central north Pacific. *Initial Reports of the Deep Sea Drilling Project*, **17**, 591–605.

Douglas, R.G. and Woodruff, F. (1981). Deep-sea benthic foraminifera. In *The Sea* (ed. C. Emiliani), pp. 1233–327. Wiley, New York.

Doyle, P. (1990). The biogeography of the Aulacocerida (Coleoida). *Atti II Conv. Int. F.E.A. Pergola*, **87**, 263–71.

Doyle, P. (1990–92). The British Toarcian (Lower Jurassic) belemnites. *Palaeontographical Society Monograph*.

Dunning, G.R. and Hodych, J.P. (1990). U/Pb zircon and baddeleyite ages for the Palisades and Gettysburg sills of the northeastern United States: implications for the age of the Triassic/Jurassic boundary. *Geology*, **18**, 795–8.

Eckert, J.D. (1988). Late Ordovician extinction of North American and British crinoids. *Lethaia*, **21**, 147–67.

Eder, W. and Franke, W. (1982). Death of Devonian reefs. *Neues Jahrbuch für Geologie und Paläontologie Abhandlungen*, **163**, 241–3.

Edgecombe, G.D. (1992). Trilobite phylogeny and the Cambrian–Ordovician 'event': cladistic reappraisal. In *Extinction and phylogeny* (ed. M.J. Novacek and Q.D. Wheeler), pp. 144–77. Columbia University Press, New York.

Ehrmann, W.U. and Mackensen, A. (1992). Sedientological evidence for the formation of an east Antarctic ice sheet in Eocene/Oligocene time. *Palaeogeography, Palaeoclimatology, Palaeoecology*, **93**, 85–112.

Eicher, D.L. and Diner, R. (1991). Environmental factors controlling Cretaceous limestone-marlstone rhythms. In *Cycles and events in stratigraphy* (ed. G. Einsele, W. Ricken, and A. Seilacher), pp. 79–93. Springer-Verlag, Berlin.

Elder, W.P. (1987). The paleoecology of the Cenomanian–Turonian (Cretaceous) Stage boundary extinctions at Black Mesa, Arizona. *Palaios*, **2**, 24–40.

Elder, W.P. (1989). Molluscan extinction patterns across the Cenomanian–Turonian Stage boundary in the western interior of the United States. *Paleobiology*, **15**, 299–320.

Elder, W.P. (1991). Molluscan paleoecology and sedimentation patterns of the Cenomanian–Turonian extinction intervals in the southern Colorado Plateau region. *Geological Society of America Special Paper*, **260**, 113–37.

Elderfield, H. (1986). Strontium isotope stratigraphy. *Palaeogeography, Palaeoclimatology, Palaeoecology*, **57**, 71–90.

Eldholm, O. and Thomas, E. (1993). Environmental impact of volcanic margin formation. *Earth and Planetary Science Letters*, **117**, 319–29.

Eldredge, N. (1991). *The miner's canary*. Prentice Hall Press, New York.

Elias, R.J. (1989). Extinctions and origins of solitary rugose corals, latest Ordovician to earliest Silurian in North America. In *Fossil Cnidarians 5* (ed. P.A. Jell and J.W. Pickett), pp. 319–26. Association of Australasian Palaeontologists, Memoir, **8**.

Elliot, D.H., Askin, R.A., Kyte, F.T., and Zinsmeister, W.J. (1994). Iridium and dinocysts at the Cretaceous–Tertiary boundary at Seymour Island, Antarctica: implications for the K–T event. *Geology*, **22**, 675–8.

Elliot, W.C. (1993). Origin of the Mg-smectite at the Cretaceous/Tertiary (K/T) boundary at Stevns Klint, Denmark. *Clays and Clay Minerals*, **41**, 442–52.

Ellis, J. and Schramm, D.N. (1995). Could a nearby supernova explosion have caused a mass extinction? *Proceedings of the National Academy of Sciences of the USA*, **92**, 235–8.

El Shaarawy, Z. (1981). Foraminifera and ostracods of the topmost Triassic and basal Jurassic of England, Wales and Austria. PhD thesis, University of Birmingham.

Embry, A.F. (1988). Triassic sea-level changes: evidence from the Canadian Arctic Archipelago. *Society of Economic Paleontologists and Mineralogists, Special Publication*, **42**, 249–59.

Embry, A.F. and Suneby, L.B. (1994). The Triassic–Jurassic boundary in the Sverdrup Basin, Arctic Canada. *Canadian Society of Petroleum Geologists, Memoir*, **17**, 857–68.

Emiliani, C., Krans, B., and Shoemaker, E.M. (1981). Sudden death at the end of the Mesozoic. *Earth and Planetary Science Letters*, **55**, 317–34.

Erwin, D.H. (1989). Regional paleocology of Permian gastropod genera, southwestern United States and the end-Permian mass extinction. *Palaios*, **4**, 424–38.

Erwin, D.H. (1990). Carboniferous–Triassic gastropod diversity patterns and the Permo-Triassic mass extinction. *Paleobiology*, **16**, 187–203.

Erwin, D.H. (1993). *The great Paleozoic crisis: life and death in the Permian*. Columbia University Press, New York.

Erwin, D.H. (1994). The Permo-Triassic extinction. *Nature*, **367**, 231–6.

Erwin, D.H. and Droser, M.L. (1993). Elvis taxa. *Palaios*, **8**, 623–4.

Erwin, D.H. and Pan H. (1996). Recoveries and radiations: gastropods after the Permo-Triassic mass extinction. In *Biotic recovery from mass extinction events* (ed. M.B. Hart), pp. 223–9. Geological Society Special Publication, **102**.

Eshet, Y. (1992). The palynofloral succession and palynological events in the Permo-Triassic boundary interval in Israel. In *Permo-Triassic boundary events in the eastern Tethys* (eds. W.C. Sweet, Z. Yang, J.M. Dickins and H. Yin), pp. 134–45. Cambridge University Press, Cambridge.

Eshet, Y., Rampino, M.R., and Visscher, H. (1995). Fungal event and palynological record of ecological crisis and recovery across the Permian–Triassic boundary. *Geology*, **23**, 967–70.

Ezaki, Y. (1993a). The last representatives of Rugosa in Abadeh and Julfa, Iran: survival and extinction. *Courier Forschungsinstitut Senckenberg*, **164**, 75–80.

Ezaki, Y. (1993b). Sequential disappearance of Permian Rugosa in Iran and Transcaucasus, west Tethys. *Bulleting of the Geological Survey of Japan*, **44**, 447–53.

Ezaki, Y. (1994). Patterns and palaeoenvironmental implications of end-Permian extinction of Rugosa in South China. *Palaeogeography, Palaeoclimatology, Palaeoecology*, **107**, 165–77.

Ezaki, Y., Kawamura, T., and Nakamura, K. (1994). Kapp Starostin Formation in Spitsbergen: a sedimentary and faunal record of Late Permian paleoenvironments in an Arctic region. *Canadian Society of Petroleum Geologists, Memoir*, **17**, 647–55.

Fabricius, F., Friedrichsen, V., and Jacobshagen, V. (1970). Paläotemperaturen und Paläoklima in Obertrias und Lias der Alpen. *Geologische Rundschau*, **59**, 805–26.

Fagerstrom. J.A. (1987). *The evolution of reef communities*. John Wiley and Sons, New York.

Fagerstrom, J.A. (1994). The history of Devonian–Carboniferous reef communities: extinctions, effects, recovery. *Facies*, **30**, 177–92.

Fan, J., Ma, X., Zhang, Y., and Zhang W. (1982). The Upper Permian reefs in west Hubei, China. *Facies*, **6**, 1–14.

Fan, J., Rigby, J.K. and Qi, J. (1990). The Permian reefs of South China and comparisons with the Permian reef complex of the Guadalupe Mountains, west Texas and New Mexico. *Brigham Young University, Geology Studies*, **36**, 15–56.

Farrimond, P., Eglinton, G., Brassell, S.C., and Jenkyns, H.C. (1990). The Cenomanian/Turonian anoxic event in Europe: an organic geochemical study. *Marine and Petroleum Geology*, **7**, 75–89.

Farson, N.M. (1986). Faunenwandel oder Faunenkrise? Faunistische Untersuchung der Grenze Frasnium-Famennium im mittleren Südasien. *Newsletters in Stratigraphy*, **16**, 113–31.

Fastovsky, D.E. (1987). Paleoenvironments of vertebrate-bearing strata during the Cretaceous–Paleogene transition, eastern Montana and western North Dakota. *Palaios*, **2**, 282–95.

Faure, K., de Wit, M.J. and Willis, J.P. (1995). Late Permian global coal hiatus linked to ^{13}C-depleted CO_2 flux into the atmosphere during the final consolidation of Pangaea. *Geology*, **23**, 507–10.

Fedorowski, J. (1989). Extinction of Rugosa and Tabulata near the Permian/Triassic boundary. *Acta Palaeontologica Polonica*, **34**, 47–70.

Feist, R. (1991). The Late Devonian trilobite crises. *Historical Biology*, **5**, 197–214.

Feist, R. and Clarkson, E.N.K. (1989). Environmentally controlled phyletic evolution, blindness and extinction in Late Devonian tropidocoryphine trilobites. *Lethaia*, **22**, 359–73.

Feldman, R.M. and Woodburne, M.O. (ed.) (1988). The geology and paleontology of Seymour Island, Antarctica. *Geological Society of America, Memoir*, **169**.

Fischer, A.G. (1964). Brackish oceans as the cause of the Permo-Triassic faunal crisis. In *Problems in palaeoclimatology* (ed. A.E.M. Nairn), pp. 566–74. Interscience, London.

Fischer, A.G. and Arthur, M.A. (1977). Secular variations in the pelagic realm. *Society of Economic Paleontologists and Mineralogists, Special Publication*, **25**, 19–50.

Fitch, F.J. and Miller, J.A. (1984). Dating Karoo igneous rocks by the conventional K-Ar and ^{40}Ar/^{39}Ar age spectrum methods. *Geological Society of South Africa Special Publication*, **13**, 247–66.

Flajs, G. and Feist, R. (1988). Index conodonts, trilobites and environment of the Devonian-Carboniferous boundary beds at La Serre (Montagne Noire, France). *Courier Forschungsinstitut Senckenberg*, **100**, 53–107.

Flessa, K.W, and Jablonski, D. (1995). Biogeography of Recent marine bivalve molluscs and its implications for paleobiogeography and the geography of extinction: a progress report. *Historical Biology*, **10**, 25–47.

Flügel, E. (1994). Pangean shelf carbonates: controls and paleoclimatic significance of Permian and Triassic reefs. *Geological Society of America Special Papers*, **288**, 247–266.

Flügel, E. and Reinhardt, J. (1989). Uppermost Permian reefs in Skyros (Greece) and Sichuan (China): implications for the Late Permian extinction event. *Palaios*, **4**, 502–18.

Flügel, E. and Stanley, G.D. (1984). Re-organization, development and evolution of post-Permian reefs and reef-organisms. *Palaeontographica Americana*, **54**, 177–86.

Fois, E. and Gaetani, M. (1984). The recovery of reef-building communities and the role of cnidarians in carbonate sequences of the Middle Triassic (Anisian) in the Italian Dolomites. *Paleontographica Americana*, **54**, 191–200.

Fordyce, R.E. (1992). Cetacean evolution and Eocene/Oligocene environments. In *Eocene–Oligocene climatic and biotic evolution* (ed. D.R. Prothero and W.A. Berggren), pp. 368–81. Princeton University Press, Princeton.

Forney, G.G. (1975). Permo-Triassic sea-level change. *Journal of Geology*, **83**, 773–9.

Fortey, R.A. (1989). There are extinctions and extinctions: examples from the lower Palaeozoic. *Philosophical Transactions of the Royal Society of London*. B325. 327–55.

Fowell, S.J. and Olsen, P.E. (1993). Time calibration of Triassic–Jurassic microfloral turnover, eastern North America. *Tectonophysics*, **222**, 361–9.

Frakes, L.A., Francis, J.E., and Sytkus, J.J. (1992). *Climate modes of the Phanerozoic*. Cambridge University Press, Cambridge.

Frederiksen, N.O. (1989). Changes in floral diversities, floral turnover rates, and climates in Campanian and Maastrichtian time, North Slope of Alaska. *Cretaceous Research*, **10**, 249–66.

Frey, R.C. (1986). The paleoecology of a Late Ordovician shale unit from southwest Ohio and southeastern Indiana. *Journal of Paleontology*, **61**, 242–67.

Gale, A.S., Jenkyns, H.C., Kennedy, W.J., and Corfield, R.M. (1993). Chemostratigraphy versus biostratigraphy: data from around the Cenomanian: Turonian boundary. *Journal of the Geological Society*, **150**, 29–32.

Gallagher, W.B. (1991). Selective extinction and survival across the Cretaceous/Tertiary boundary in the northern Atlantic Coastal Plain. *Geology*, **19**, 967–70.

Gardiner, B.G. (1990). Placoderm fishes: diversity through time. In *Major evolutionary radiations* (ed. P.D. Taylor and G.P. Larwood), pp. 305–19. Systematics Association Special Volume, **42**.

Gardiner, B.G. (1993). Osteichthyes: basal Actinopterygians. In *The fossil record 2* (ed. M.J. Benton), pp. 611–19. Chapman and Hall, London.

Geldsetzer, H.H.J., Goodfellow, W.D., McLaren, D.J., and Orchard, M.J. (1987). Sulfur-isotope anomaly associated with the Frasnian-Famennian extinction, Medicine Lake, Alberta, Canada. *Geology*, **15**, 393–96.

Geldsetzer, H.H.J., Goodfellow, W.D. and McLaren, D.J. (1993). The Frasnian–Famennian extinction event in a stable cratonic shelf setting: Trout River, Northwest Territories, Canada. *Palaeogeography, Palaeoclimatology, Palaeoecology*, **104**, 81–95.

Gilinsky, N.L. and Good, I.J. (1991). Problems of origination, persistence, and extinction of families of marine invertebrate life. *Paleobiology*, **17**, 145–66.

Gingerich, P.D. (1984). Pleistocene extinctions in the context of origination–extinction equilibria in Cenozoic mammals. In *Quaternary extinctions: a prehistoric revolution* (ed. P.S. Martin and R.G. Klein), pp. 211–22. University of Arizona Press, Tucson.

Glaessner, M.F. (1984). *The dawn of animal life*. Cambridge University Press, Cambridge.

Glen, W. (1994). What the impact/volcanism/mass extinction debates are about. In *The mass extinction debates: how science works in a crisis* (ed. W. Glen), pp. 7–38. Stanford University Press, Stanford.

González-Leon, C.M., Taylor, D.G., and Stanley, G.D. (1996). The Antimonio Formation in Sonora, Mexico, and the Triassic–Jurassic boundary. *Canadian Journal of Earth Sciences*, **33**, 418–28.

Goodfellow, W.D., Geldsetzer, H.H.J., McLaren, D.J., Orchard, M.J., and Klapper, G. (1989). Geochemical and isotopic anomalies associated with the Frasnian–Famennian extinction. *Historical Biology*, **2**, 51–72.

Goodfellow, W.D., Nowlan, G.S., McCracken, A.D., Lenz, A.C., and Grégoire, D.C. (1992). Geochemical anomalies near the Ordovician–Silurian boundary, Northern Yukon Territory, Canada. *Historical Biology*, **6**, 1–23.

Gould, S.J. (1989). *Wonderful life: the Burgess Shale and the nature of history*. Norton, New York.

Gradstein, F.M., Agterberg, F.P., Ogg, J.G., Hardenbol, J., van Veen, P., Thierry, J., and Huang, Z. (1994). A Mesozoic time-scale. *Journal of Geophysical Research*, **99**, 24051–74.

Graham, J.B., Dudley, R., Aguiler, N.M. and Gan, C. (1995). Implications of the late Palaeozoic oxygen pulse for physiology and evolution. *Nature*, **375**, 117–20.

Graham, R.W. and Lundelius, E.L. (1984). Coevolutionary disequilibrium and Pleistocene extinctions. In *Quaternary extinctions: a prehistoric revolution* (ed. P.S. Martin and R.G. Klein), pp. 223–49. University of Arizona Press, Tucson.

Grahn, Y. (1988). Chitinozoan stratigraphy in the Ashgill and Llandovery. *Bulletin of the British Museum of Natural History (Geology)*, **43**, 317–23.

Grasmück, K. and Trümpy, R. (1969). Triassic stratigraphy and general geology of the country around Fleming Fjord (east Greenland). *Meddelelser om Grønland*, **168(2)**, 5–71.

Grayson, D.K. (1984). Explaining Pleistocene extinctions. In *Quaternary extinctions: a prehistoric revolution* (ed. P.S. Martin and R.G. Klein), pp. 807–23. University of Arizona Press, Tucson.

Gregor, H.-J. (1992). The Ries and Steinheim meteorite impacts and their effects on environmental conditions in time and space. In *Phanerozoic global bio-events* (ed. O.H. Walliser). Abstract volume, IGCP Project 216, Göttingen.

Grieve, R.A.F. (1987). Terrestrial impact structures. *Annual Reviews of Earth and Planetary Sciences*, **15**, 245–70.

Groos-Uffenorde, H. and Schindler, E. (1990). The effect of global events on entomozoacean Ostracoda. In *Ostracoda and global events* (ed. R. Whatley and C. Maybury), pp. 101–12. Chapman & Hall, London.

Grossman, E.L. (1994). The carbon and oxygen isotope record during the evolution of Pangaea: Carboniferous to Trassic. *Geological Society of America Special Paper*, **288**, 207–28.

Grotzinger, J.P., Bowring, S.A., Saylor, B.Z., and Kaufman, A.J. (1995). Biostratigraphic and geochronologic constraints on early animal evolution. *Science*, **270**, 598–604.

Gruszczynski, M., Halas, S., Hoffman, A., and Malkowski, K. (1989). A brachiopod calcite record of the oceanic carbon and oxygen isotope shifts at the Permian/Triassic transition. *Nature*, **337**, 64–8.

Gruszczynski, M., Hoffman, A., Malkowski, K. and Veizer. J. (1992). Seawater strontium isotopic perturbation at the Permian–Triassic boundary, west Spitsbergen, and its implications for the interpretation of strontium isotope data. *Geology*, **20**, 779–82.

Gullo, M. and Kozur, H. (1993). First evidence of Scythian conodonts in Sicily. *Neues Jahrbuch für Geologie und Paläontologie, Monatsheft*, **1993**, 477–88.

Habib, D., Moshkovitz, S., and Kramer, C. (1992), Dinoflagellate and calcareous nanofossil response to sea-level change in the Cretaceous–Tertiary boundary sections. *Geology*, **20**, 165–8.

Hakansson, E. and Thomsen, E. (1979). Distribution and types of bryozoan communities at the boundary in Denmark. In *Boundary events* (ed. T. Birkelund and R.G. Bromley), pp. 78–91. University of Copenhagen.

Hallam, A. (1961). Cyclothems, transgressions and faunal change in the Lias of north west Europe. *Transactions of the Edinburgh Geological Society*, **18**, 132–74.

Hallam, A. (1976). Stratigraphic distribution and ecology of European Jurassic bivalves. *Lethaia*, **9**, 245–59.

Hallam, A. (1978). How rare is phyletic gradualism? Evidence from Jurassic bivalves. *Paleobiology*, **4**, 16–25.

Hallam, A. (1981a). The end-Triassic bivalve extinction event. *Palaeogeography, Palaeoclimatology, Palaeoecology*, **35**, 1–44.

Hallam, A. (1981b). *Facies interpretation and the stratigraphic record*. W.H. Freeman, Oxford.

Hallam, A. (1984a). Pre-Quaternary sea-level changes. *Annual Review of the Earth and Planetary Sciences*, **12**, 205–43.

Hallam, A. (1984b). The causes of mass extinction. *Nature*, **308**, 686–7.

Hallam, A. (1986). The Pliensbachian and Tithonian extinction events. *Nature*, **319**, 765–8.

Hallam, A. (1987a). Radiations and extinctions in relation to environmental change in the marine Jurassic of north west Europe. *Paleobiology*, **13**, 152–68.

Hallam, A. (1987b). End-Cretaceous mass extinction event: argument for terrestrial causation. *Science*, **238**, 1237–42.

Hallam, A. (1988). A compound scenario for the end-Cretaceous mass extinctions. *Revista Española de Paleontologia*, no. Extraordinario, Palaeontology and evolution: extinction events, 7–20.

Hallam, A. (1989a). The case for sea-level change as a dominant causal factor in mass extinction of marine invertebrates. *Philosophical Transactions of the Royal Society of London*, **B325**, 437–55.

Hallam, A. (1989b). *Great geological controversies* (2nd edition). Oxford University Press, Oxford.

Hallam, A. (1990a). The end-Triassic mass extinction event. *Geological Society of America Special Paper*, **247**, 577–83.

Hallam, A. (1990b). Correlation of the Triassic–Jurassic boundary of England and Austria. *Journal of the Geological Society*, **148**, 420–2.

Hallam, A. (1990c). Biotic and abiotic factors in the evolution of early Mesozoic marine molluscs. In *Causes of evolution: a paleontological perspective* (ed. R.M. Moss and W.D. Allmon), pp. 249–60. University of Chicago Press, Chicago.

Hallam, A. (1991). Why was there a delayed radiation after the end-Palaeozoic extinction? *Historical Biology*, **5**, 257–62.

Hallam, A. (1992). *Phanerozoic sea-level changes*. Columbia University Press, New York.

Hallam, A. (1994a). Strontium isotope profiles of Triassic–Jurassic boundary sections in England and Austria. *Geology*, **22**, 1079–82.

Hallam, A. (1994b). *An outline of Phanerozoic biogeography*. Oxford Biogeography Series No. 10. Oxford University Press, Oxford.

Hallam, A. (1995a). Major bio-events in the Triassic and Jurassic. In *Global events and event stratigraphy* (ed. O.H. Walliser), pp. 265–83. Springer-Verlag, Berlin.

Hallam, A. (1995b). Oxygen-restricted facies of the basal Jurassic of north west Europe. *Historical Biology*, **10**, 247–57.

Hallam, A. (1996). Recovery of the marine fauna in Europe after the end-Triassic and early Toarcian mass extinctions. In *Biotic recovery from mass extinction events* (ed. M.B. Hart), pp. 231–6. Geological Society of London Special Publication, **102**.

Hallam, A. and Goodfellow, W.D. (1990). Facies and geochemical evidence bearing on the end-Triassic disappearance of the Alpine reef ecosystem. *Historical Biology*, **4**, 131–8.

Hallam, A. and Miller, A.I. (1988). Extinction and survival in the Bivalvia. In *Extinction and survival in the fossil record* (ed. G.P. Larwood), pp. 121–38. Systematics Association Special Volume, **34**.

Hallam, A. and Perch-Nielsen, K. (1990). The biotic record of events in the marine realm at the end of the Cretaceous: calcaerous, siliceous and organic-walled microfossils and macroinvertebrates. *Tectonophysics*, **171**, 347–57.

Hallock, P., Premoli-Silva, I. and Shackleton, N.J. (1991). Similarities between planktonic and larger foraminiferal evolutionary trends through Paleogene paleoceanographic changes. *Palaeogeography, Palaeoclimatology, Palaeoecology*, **83**, 49–64.

Halstead, L.B. (1993). Agnatha. In *The Fossil Record 2* (ed. M.J. Benton), pp. 573–82. Chapman & Hall, London.

Hamilton, G.B. (1982). Triassic and Jurassic calcareous nanofossils. In *A stratigraphic index of calcareous nanofossils* (ed. A.R. Lord), pp. 17–39. Ellis Horwood, Chichester.

Hance, L., Muchez, P., Coen, M., Fang, X., Groessens, E., Hou, H., Poty, E., Steemans, P., Streel, M., Tan, Z., Tourneur, F., van Steenwinkel, M., and Xu, S. (1994). Biostratigraphy and sequence stratigraphy at the Devonian–Carboniferous transition in southern China (Hunan Province). Comparison with southern Belgium. *Annales de la Société Géologique de Belgique*, **115**, 359–78.

Hancock, J.M. and Kauffman, E.G. (1979). The great transgressions of the Late Cretaceous. *Journal of the Geological Society*, **136**, 175–86.

Hansen, T.A. (1987). Extinction of late Eocene to Oligocene molluscs: relationship to shelf area, temperature changes and impact events. *Palaios*, **2**, 69–75.

Hansen, T.A. (1988). Early Tertiary radiation of marine molluscs and the long-term effects of the Cretaceous–Tertiary extinction. *Paleobiology*, **14**, 37–51.

Hansen, T.A. (1992). The patterns and causes of molluscan extinction across the Eocene/ Oligocene boundary. In *Eocene–Oligocene climatic and biotic evolution* (ed. D.R. Prothero and W.A. Berggren), pp. 341–8. Princeton University Press, Princeton.

Hansen, T.A., Farrand, R.B., Montgomery, H.A., Billman, H.G., and Blechschmidt, G. (1987). Sedimentology and extinction patterns across the Cretaceous–Tertiary boundary interval in east Texas. *Cretaceous Research*, **8**, 229–52.

Hansen, T.A., Farrell, B.R., and Upshaw, B. (1993). The first 2 million years after the Cretaceous–Tertiary boundary in east Texas and paleoecology of the molluscan recovery. *Paleobiology*, **19**, 251–65.

Haq, B.U., Premoli-Silva, I. and Lohman, G.P. (1977). Calcareous plankton paleobiogeographic evidence for major climatic fluctuations in the early Cenozoic Atlantic Ocean. *Journal of Geophysical Research*, **82**, 3861–76.

Haq, B.U., Hardenbol, J. and Vail, P.R. (1987). Chronology of fluctuating sea levels since the Triassic (250 million years ago to present). *Science*, **235**, 1156–67.

Haq, B.U., Hardenbol, J. and Vail, P.R. (1988). Mesozoic and Cenozoic chronostratigraphy and cycles of sea-level change. In *Sea-level changes: an integrated approach* (ed. C.K. Wilgus *et al.*), pp. 71–108. Society of Economic Paleontologists and Mineralogists, Special Publication, **42**.

Harland, W.B., Armstrong, R.L., Cox, A.V., Craig, L.E., Smith, A.G., and Smith, D.G. (1989). *A geological time-scale*. Cambridge University Press, Cambridge.

Harper, C.W. Jr. (1987). Might Occam's canon explode the Death Star? A moving-average model of biotic extinctions. *Palaios*, **2**, 600–4.

Harper, D.A.T. and Rong J. (1995). Patterns of change in the brachiopod faunas through the Ordovician–Silurian interface. *Modern Geology*, **20**, 83–100.

Harries, P.J. (1993). Dynamics of survival following the Cenomanian–Turonian (Upper Cretaceous) mass extinction event. *Cretaceous Research*, **14**, 563–83.

Harries, P.J. and Kauffman, E.G. (1990). Patterns of survival and recovery following the Cenomanian–Turonian (Late Cretaceous) mass extinction in the Western Interior Basin, United States. In *Extinction events in Earth history* (eds. E.G. Kauffman and O.H. Walliser), pp. 277–98. Springer-Verlag, Berlin.

Hart, M.B. (1980). A water depth model for the evolution of planktonic Foraminifera. *Nature*, **286**, 252–4.

Hart, M.B. (1991). The Late Cenomanian calcisphere global bioevent. *Proceedings of the Ussher Society*, **7**, 413–17.

Hartenberger, J.-L. (1986). Crises biologiques en milieu continental au cours du Paléogène: exemple des mammifères d'Europe. *Bulletin du Centre de Recherche, Exploration et Production Elf-Aquitaine*, **10**, 489–500.

Hayami, I. (1989). Outlook on the post-Paleozoic historical biogeography of pectinids in the Western Pacific Region. In *Current aspects of biogeography in West Pacific and East Asian regions* (ed. H. Ohba), pp. 3–25. University Museum, Tokyo, Nature and Culture No. **1**.

He, J. (1989). Restudy of the Permian–Triassic boundary clay in Meishan, Changxing, Zhejiang, China. *Historical Biology*, **2**, 73–87.

Heinberg, C. (1979). Bivalves from the latest Maastrichtian of Stevns Klint and their stratigraphic affinities. In *Cretaceous–Tertiary boundary events* (eds. T. Birkelund and R.G. Bromley) vol. 1, pp. 58–64. University of Copenhagen.

Heissig, K. (1986). No effects of the Ries impact event on the local mammals. *Modern Geology*, **10**, 171–9.

Helby, R., Morgan, R., and Partridge, A.D. (1987). A palynological zonation of the Australian Mesozoic. In *Studies in Australian palynology* (ed. P.A. Jell), pp. 1–94. Memoir of the Association of Australian Palynology, **4**.

Hess, J., Bender, M.L., and Schilling, J.G. (1986). Evolution of the ratio of strontium-87 to strontium-86 in seawater from Cretaceous to present. *Science*, **231**, 979–84.

Hickey, L.J. (1981). Land plant evidence compatible with gradual, not catastrophic, change at the end of the Cretaceous. *Nature*, **292**, 529–31.

Hickey, L.J. (1984). Changes in the angiosperm flora across the Cretaceous–Tertiary boundary. In *Catastrophes in Earth History* (ed. W.A. Berggren and J.A. van Couvering), pp. 279–313. Princeton University Press, Princeton.

Hilbrecht, H. and Hoefs, J. (1986). Geochemical and palaeontological studies of $\delta^{13}C$ anomaly in Boreal and North Tethyan Cenomanian – Turonian sediments in Germany and adjacent areas. *Palaeogeography, Palaeoclimatology, Palaeoecology*, **53**, 169–89.

Hildebrandt, A.R., Penfield, G.T., Kring, D.A., Pilkington, M., Camargo, A., Jacobsen, S.B. and Boynton, W.V. (1991). Chicxulub Crater: a possible Cretaceous/Tertiary boundary impact crater on the Yucutan Peninsula, Mexico. *Geology*, **19**, 867–71.

Hildebrandt, A.R., Pilkington, M., Cannors, M., Ortiz-Aleman, C. and Chavez, R.E. (1995). Size and structure of the Chicxulub crater revealed by horizontal gravity gradients and cenotes. *Nature*, **376**, 415–17.

Hillebrandt, A. von (1990). The Triassic/Jurassic boundary in northern Chile. *Cahiers de l'Université Catholique de Lyon*, Série Scientifique, **3**, 27–53.

Hillebrandt, A. von (1994). The Triassic/Jurassic boundary and Hettangian biostratigraphy in the area of the Utcubamba Valley (northern Peru). *Geobios*, M.S., **17**, 297–307.

Hodych, J.P. and Dunning, G.R. (1992). Did the Manicougan impact trigger end-of-Triassic mass extinction? *Geology*, **20**, 51–4.

Hoffman, A. (1985). Patterns of family extinction depend on definition and geological time-scale. *Nature*, **315**, 659–62.

Hoffman, A. (1986). Neutral model of Phanerozoic diversification: implications for macroevolution. *Neues Jahrbuch für Geologie und Paläontologie, Abhandlungen*, **172**, 219–44.

Hoffman, A. (1989a). *Arguments on evolution: a paleontologist's perspective*. Oxford University Press, Oxford.

Hoffman, A. (1989b). Mass extinctions: the view of a sceptic. *Journal of the Geological Society*, **146**, 21–35.

Hoffman, A. (1989c). What, if anything, are mass extinctions? *Philosophical Transactions of the Royal Society*, **B325**, 253–61.

Hoffman, A. and Ghiold, J. (1985). Randomness in the pattern of 'mass extinctions' and 'waves of origination'. *Geological Magazine*, **122**, 1–4.

Hoffman, A., Gruszczynski, M., and Malkowski, K. (1991). On the interrelationship between temporal trends in $\delta^{13}C$, $\delta^{18}O$, and $\delta^{34}S$ in the world ocean. *Journal of Geology*, **99**, 355–70.

Holland, S.M. (1995). The stratigraphic distribution of fossils. *Paleobiology*, **21**, 92–109.

Hollis, C.J. (1996). Radiolarian faunal change through the Cretaceous–Tertiary transition of eastern Marlborough, New Zealand. In *Cretaceous-Tertiary mass extinctions: biotic change and environmental changes* (ed. N. MacLeod and G. Keller), pp. 173–204. Norton, London.

Holser, W.T. (1977). Catastrophic chemical events in the history of the ocean. *Nature*, **267**, 403–8.

Holser, W.T. (1984). Gradual and abrupt shifts in ocean chemistry during Phanerozoic time. In *Patterns of change in Earth evolution* (ed. H.D. Holland and A.F. Trendall), pp. 123–43. Springer-Verlag, Berlin.

Holser, W.T. and Magaritz, M. (1985). The Late Permian carbon isotope anomaly in the Bellerophon Basin, Carnic and Dolomite Alps. *Jahrbuch der Geologischen Bundesanstalt, Wien*, **128**, 75–82.

Holser, W.T. and Magaritz, M. (1987). Events near the Permian – Triassic boundary. *Modern Geology*, **11**, 155–80.

Holser, W.T. and Magaritz, M. (1992). Cretaceous/Tertiary and Permian/Triassic boundary events compared. *Geochimica et Cosmochimica Acta*, **56**, 3297–309.

Holser, W.T., Magaritz, M., and Clark, D.L. (1986). Carbon-isotope stratigraphic correlations in the Late Permian. *American Journal of Science*, **286**, 390–402.

Holser, W.T., Schidlowski, M., Mackenzie, F.T., and Maynard, J.B. (1988). Geochemical cycles of carbon and sulfur. In *Chemical cycles in the evolution of the Earth* (ed. C.B. Gregor, R.M. Garrels, F.T. Mackenzie and J.B. Maynard), pp. 105–173. John Wiley and Son, New York.

Holser, W.T., Schönlaub, H.-P., Attrep, M. Jr, Boeckelmann, K., Klein, P., Magaritz, M., Pak, E., Schramm, J.-M., Stattgegger, K. and Schmöller, R. (1989). A unique geochemical record at the Permian/Triassic boundary. *Nature*, **337**, 39–44.

Holser, W.T., Schönlaub, H.P., Boeckelmann, K., Magaritz, M., and Orth, C.J. (1991). The Permian–Triassic of the Gartnerkofel-1 core (Carnic Alps, Austria): synthesis and conclusions. *Abhandlungen der Geologischen Bundesanstalt*, **45**, 213–32.

Hooker, J.J. (1992). British mammalian paleocommunities across the Eocene–Oligocene transition and their environmental implications. In *Eocene–Oligocene climatic and biotic evolution* (ed. D.R. Prothero and W.A. Berggren), pp. 494–511. Princeton University Press, Princeton.

Hori, S.R. (1993). Toarcian oceanic event in deep-sea sediments. *Bulletin of the Geological Survey of Japan*, **44**, 555–70.

Horowitz, A.S. and Pachut, J.F. (1993). Specific, generic, and familial diversity of Devonian bryozoans. *Journal of Paleontology*, **67**, 42–52.

House, M.R. (1971). Devonian faunal distributions. In *Faunal provinces in space and time* (ed. F.A. Middlemiss, P.F. Rawson, and G. Newall), pp. 77–94. Geological Journal Special Issue, **4**.

House, M.R. (1975). Faunas and time in the marine Devonian. *Proceedings of the Yorkshire Geological Society*, **40**, 459–90.

House, M.R. (1985). Correlation of mid Paleozoic ammonoid evolutionary events with global sedimentary perturbations. *Nature*, **313**, 17–22.

House, M.R. (1989). Ammonoid extinction events. *Philosophical Transactions of the Royal Society of London*, **B325**, 307–26.

House, M.R. (1992). Earliest Carboniferous goniatite recovery after the Hangenberg Event. *Annales de la Société Géologique de Belgique*, **115**, 559–79.

Hsü, K.J. and McKenzie, J.A. (1985). A 'Strangelove' ocean in the earliest Tertiary. *American Geophysical Union Geophysical Monograph*, **32**, 487–92.

Hsü, K.J. and McKenzie, J.A. (1990). Carbon-isotope anomalies at era boundaries; global catastrophes and their ultimate cause. *Geoogical Society of America Special Paper*, **247**, 61–70.

Hsü, K.J., Oberhänsli, H., Gao, J., Chen, H., and Krähebühl, U. (1985). Strangelove ocean before the Cambrian explosion. *Nature*, **316**, 809–11.

Hubbard, A.E. and Gilinsky, N.L. (1992). Mass extinctions as statistical phenomena: an examination of the evidence using χ^2 tests and bootstrapping. *Paleobiology*, **18**, 148–60.

Huber, B.T., Hodell, D.A., and Hamilton, C.P. (1995). Middle–Late Cretaceous climate of the southern high latitudes: stable isotope evidence for minimal equator-to-pole thermal gradients. *Bulletin of the Geological Society of America*, **107**, 1164–91.

Hunt, A.P. and Lucas, S.G. (1992). No tetrapod extinction event at the Carnian–Norian boundary (Late Triassic): evidence from the western United States and India. *Abstracts of the 29th International Geological Congress, Kyoto*, 66.

Hurlbert, S.H. and Archibald, J.D. (1995). No statistical support for sudden (or gradual) extinction of dinosaurs. *Geology*, **23**, 881–4.

Hut, P., Alvarez, W., Elder, W.P., Hansen, T.A., Kauffman, E.G., Keller, G., Shoemaker, E.M., and Weisman, P.R. (1987). Comet showers as causes of mass extinctions. *Nature*, **329**, 118–26.

Hutchison, J.H. (1992). Western North American reptile and amphibian record across the Eocene/Oligocene boundary and its implications. In *Eocene–Oligocene climatic and biotic evolution* (ed. D.R. Prothero and W.A. Berggren), pp. 451–63. Princeton University Press, Princeton.

Ishiga, H. and Yamakita, S. (1993). Permian/Triassic boundary in pelagic sediments, Southwest Japan – an introduction. *Bulletin of the Geological Survey of Japan*, **44**, 419–23.

Isozaki, Y. (1994). Superanoxia across the Permo-Triassic boundary: recorded in accreted deep-sea pelagic chert in Japan. *Canadian Society of Petroleum Geologists, Memoir*, **17**, 805–12.

Isozaki, Y. (1997). Permo–Triassic boundary superanoxia and stratified superocean: records from lost deep sea, *Science*, **276**, 235–8.

Isozaki, Y., Maruyama. S., and Furuoka. F. (1990). Accreted oceanic materials in Japan. *Tectonophysics*, **181**, 179–205.

Izett, G.A. (1990). The Cretaceous/Tertiary boundary interval, Raton Basin, Colorado and New Mexico, and its content of shock-metamorphosed minerals – evidence relevant to the K–T impact–extinction theory. *Geological Society of America Special Paper*, **249**, 1–93.

Jablonski, D. (1985). Marine regressions and mass extinctions: a test using the modern biota. In *Phanerozoic diversity patterns* (ed. J.W. Valentine), pp. 335–54. Princeton University Press, Princeton.

Jablonski, D. (1986a). Background and mass extinctions: the alternation of macroevolutionary regimes. *Science*, **231**, 129–33.

Jablonski, D. (1986b). Causes and consequences of mass extinctions: a comparative approach. In *Dynamics of extinction* (ed. D.K. Elliot), pp. 183–230. Wiley, New York.

Jablonski, D. (1989). The biology of mass extinction: a palaeontological view. *Philosophical Transactions of the Royal Society*, **B325**, 357–68.

Jablonski, D. (1991). Extinctions: a paleontological perspective. *Science*, **253**, 754–7.

Jablonski, D. (1994). Extinctions in the fossil record. *Philosophical Transactions of the Royal Society*, **B344**, 11–17.

Jablonski, D. and Raup, D.M. (1995). Selectivity of end-Cretaceous marine bivalve extinctions. *Science*, **268**, 389–391.

Jablonski, D., Sepkoski, J.J. Jr., Bottjer, D.J., and Sheehan, P.M. (1983). Onshore–offshore patterns in the evolution of Phanerozoic shelf communities. *Science*, **222**, 1123–5.

Jackson, J.B.C., Jung, P., Coates, A.G., and Collins, L.S. (1993). Diversity and extinction of tropical & American mollusks and emergence of the Isthmus of Panama. *Science*, **260**, 1624–6.

Jansa, L.F. (1993). Cometary impacts into ocean: their recognition and the threshold constraint for biological extinction. *Palaeogeography, Palaeoclimatology, Palaeoecology*, **104**, 271–86.

Jansa, L.F., Aubry, M.-P., and Gradstein, F.M. (1990). Comets and extinctions, cause and effect? *Geological Society of America Special Paper*, **247**, 223–32.

Jarvis, I., Carson, G.A., Cooper, K., Hart, M.B., Leary, P., Tocher, B.A., Horne, D., and Rosenfeld, A. (1988). Microfossil assemblages and the Cenomanian–Turonian (Late Cretaceous) oceanic anoxic event. *Cretaceous Research*, **9**, 3–103.

Jeans, C.V., Long, D., Hall, M.A., Bland, D.J., and Cornford, C. (1991). The geochemistry of the Plenus Marls at Dover, England: evidence of fluctuating oceanographic conditions and of glacial control during the development of the Cenomanian–Turonian $\delta^{13}C$ anomaly. *Geological Magazine*, **128**, 603–32.

Jefferies, R.P.S. (1962). The palaeoecology of the *Actinocamax plenus* subzone (lowest Turonian) in the Anglo-Paris Basin. *Palaeontology*, **4**, 609–47.

Jehanno, C., Boclet, D., Froget, L., Lambert, B., Robin, E., Rocchia, R., and Turpin, L. (1992). The Cretaceous–Tertiary boundary at Beloc, Haiti: no evidence for an impact in the Caribbean area. *Earth and Planetary Science Letters*, **109**, 229–41.

Jenkins, D.G. and Murray, J.W. (ed.) (1989). *Stratigraphic atlas of fossil foraminifera*. Ellis Horwood, Chichester.

Jenkyns, H.C. (1980). Cretaceous anoxic events: from continents to oceans. *Journal of the Geological Society of London*, **137**, 171–88.

Jenkyns, H.C. (1988). The early Toarcian (Jurassic) anoxic event: stratigraphic, sedimentary and geochemical evidence. *American Journal of Science*, **288**, 101–51.

Jenkyns, H.C., Gale, A.S., and Corfield, R.M. (1994). Carbon- and oxygen-isotope stratigraphy of the English Chalk and Italian Scaglia and its palaeoclimatic significance. *Geological Magazine*, **131**, 1–34.

Ji, Q. (1989). On the Frasnian–Famennian mass extinction event in South China. *Courier Forschungsinstitut Senckenberg*, **117**, 275–301.

Ji, Z. and Barnes, C.R. (1993). A major conodont extinction event during the Early Ordovician within the midcontinent realm. *Palaeogeography, Palaeoclimatology, Palaeoecology*, **104**, 37–47.

Jiménez, A.P., Jiménez de Cisneros, C., Rivas, P., and Vera, J.A. (1996). The early Toarcian anoxic event in the westernmost Tethys (Subbetic): paleogeographic and paleobiogeographic significance. *Journal of Geology*, **104**, 399–416.

Jin, Y., Glenister, B.F., Kotlyar, G.V. and Sheng J. (1994a). An operational scheme of Permian chronostratigraphy. *Palaeoworld*, **4**, 1–13.

Jin, Y., Zhu Z. and Mei S. (1994b). The Maokouan–Lopingian sequences in South China. *Palaeoworld*, **4**, 138–52.

Jin, Y., Zhang, J. and Shang, Q. (1994c). Two phases of the end-Permian mass extinction. *Canadian Society of Petroleum Geologists, Memoir*, **17**, 813–22.

Joachimski, M.M. and Buggisch, W. (1993). Anoxic events in the late Frasnian – causes of the Frasnian–Famennian faunal crisis? *Geology*, **21**, 675–8.

Joachimski, M.M. and Buggisch, W. (1994). Comparison of inorganic and organic carbon isotope patterns across the Frasnian/Famennian boundary. *Erlanger Geologische Abhandlungen*, **122**, 35.

Johnson, A.L.A. and Simms, M.J. (1989). The timing and cause of Late Triassic marine invertebrate extinctions: evidence from scallops and crinoids. In *Mass extinctions: processes and evidence* (ed. S.K. Donovan), pp. 174–94. Belhaven, London.

Johnson, C.C. and Kauffman, E.G. (1990). Originations, radiations and extinctions of Cretaceous rudistid bivalve species in the Caribbean Province. In *Extinction events in Earth history* (ed. E.G. Kauffman and O.H. Walliser), pp. 305–24. Springer-Verlag, Berlin.

Johnson, C.C., Barron, E.J., Kauffman, E.G., Arthur, M.A., Fawcett, P.J., and Yasuda, M.K. (1996). Middle Cretaceous reef collapse linked to ocean heat transport. *Geology*, **24**, 376–80.

Johnson, J.G. (1974). Extinction of perched faunas. *Geology*, **2**, 479–82.

Johnson, J.G. (1979). Devonian brachiopod biostratigraphy. *Special Papers in Palaeontology*, **23**, 291–306.

Johnson, J.G., Klapper, G., and Sandberg, C.A. (1985). Devonian eustatic fluctuations in Euramerica. *Bulletin of the Geological Society of America*, **96**, 567–87.

Johnson, K.R. (1993). Extinction at antipodes. *Nature*, **366**, 511–12.

Johnson, K.R. and Hickey, L.J. (1990). Megafloral change across the Cretaceous/Tertiary boundary in the northern Great Plains and Rocky Mountains, U.S.A. *Geological Society of America Special Paper*, **247**, 433–44.

Jones, D.S. and Nichol, D. (1986). Origination, survivorship, and extinction of rudist taxa. *Journal of Paleontology*, **60**, 107–15.

Kaiho, K. (1988). Uppermost Cretaceous to Paleogene bathyal benthic foraminiferal biostratigraphy of Japan and New Zealand: latest Paleocene–middle Eocene benthic foraminiferal species turnover. *Revue de Paléobiologie*, volume spéciale, **2**, 553–9.

Kaiho, K. (1991). Global changes of Paleogene aerobic/anaerobic benthic foraminifera and deep-sea circulation. *Palaeogeography, Palaeoclimatology, Palaeoecology*, **83**, 65–85.

Kaiho, K. (1992). A low extinction rate of intermediate-water benthic foraminifera at the Cretaceous/Tertiary boundary. *Marine Micropaleontology*, **18**, 229–59.

Kaiho, K. (1994). Planktonic and benthic foraminiferal extinction events during the last 100 m.y. *Palaeogeography, Palaeoclimatology, Palaeoecology*, **111**, 45–71.

Kaiho, K. and Hasegawa, T. (1994). End-Cenomanian benthic foraminiferal extinctions and oceanic dysoxic events in the northwestern Pacific Ocean. *Palaeogeography, Palaeoclimatology, Palaeoecology*, **111**, 29–43.

Kajiwara, Y. and Kaiho, K. (1992). Oceanic anoxia at the Cretaceous/Tertiary boundary supported by the sulfur isotopic record. *Palaeogeography, Palaeoclimatology, Palaeoecology*, **99**, 151–62.

Kajiwara, Y., Yamakita, S., Ishida, K., Ishiga, H., and Imai, A. (1994). Development of a largely anoxic stratified ocean and its temporary massive mixing at the Permian/Triassic boundary supported by the sulfur isotopic record. *Palaeogeography, Palaeoclimatology, Palaeoecology*, **111**, 367–79.

Kakuwa, Y. (1996). Permian–Triassic mass extinction event recorded in bedded shert sequence in southwest Japan. *Palaeogeography, Palaeoclimatology, Palaeoecology*, **121**, 35–51.

Kaljo, D. (1996). Diachronous recovery patterns in Early Silurian corals, graptolites and acritarchs. In *Biotic recovery from mass extinction events* (ed. M.B. Hart), pp. 127–33. Geological Society Special Publication, **102**.

Kaljo, D. and Klaaman, E. (1973). Ordovician and Silurian corals. In *Atlas of palaeobiogeography* (ed. A. Hallam), pp. 37–47. Elsevier, Amsterdam.

Kaljo, D., Boucot, A.J., Corfield, R.M., Le Herisse, A., Koren, T.N., Kriz, J., Männik, P., Märss, T., Nestor, V., Shaver, R.H., Siveter, D.J., and Viira, V. (1996). Silurian bio-events. In *Global events and event stratigraphy* (ed. O.H. Walliser), pp. 173–224. Springer-Verlag, Berlin.

Kalvoda, J. (1990). Late Devonian–Early Carboniferous paleobiogeography of benthic foraminifera and climatic oscillations. In *Extinction events in Earth history* (ed. E.G. Kauffman and O.H. Walliser), pp. 183–7. Springer-Verlag, Berlin.

Kammer, T.W., Brett, C.E., Boardman, D.R. and Mapes, R.H. (1986). Ecologic stability of the dysaerobic biofacies during the late Paleozoic. *Lethaia*, **19**, 109–21.

Kamo, S.L. and Krogh, T.E. (1995). Chicxulub crater source for shocked zircon crystals from the Cretaceous–Tertiary boundary layer, Saskatchewan: evidence from new U-Pb data. *Geology*, **23**, 281–4.

Kauffman, E.G. (1984). The fabric of Cretaceous marine extinctions. In *Catastrophes and Earth history* (ed. W.A. Berggren and J.A. van Couvering), pp. 151–246. Princeton University Press, Princeton.

Kauffman, E.G. (1986). High resolution event stratigraphy: regional and global Cretaceous bio-events. In *Global bio-events* (ed. O.H. Walliser), pp. 279–336. Springer-Verlag, Berlin.

Kauffman, E.G. and Erwin, D.H. (1995). Surviving mass extinctions. *Geotimes*, **40**, 14–17.

Kauffman, E.G. and Hart, M.B. (1995). Cretaceous bio-events. In *Global events and event stratigraphy* (ed. O.H. Walliser), pp. 285–312. Springer-Verlag, Berlin.

Kauffman, S. (1995). *At home in the universe: the search for laws of complexity*. Oxford University Press, New York.

Kaufman, A.J., Jacobsen, S.B., and Knoll, A.H. (1993). The Vendian record of Sr- and C-isotope variations in seawater: implications for tectonics and climate. *Earth and Planetary Science Letters*, **120**, 409–30.

Keith, M.L. (1982). Violent volcanism, stagnant oceans and some inferences regarding petroleum, strata-bound ores and mass extinctions. *Geochimica et Cosmochimica Acta*, **46**, 2621–37.

Keller, G. (1988). Extinction, survivorship and evolution of planktonic foraminifers across the Cretaceous/Tertiary boundary at El Kef, Tunisia. *Marine Micropaleontology*, **13**, 239–63.

Keller, G. (1989). Extended period of extinctions across the Cretaceous/Tertiary boundary in planktonic foraminifera at continental shelf sections: implications for impact and volcanism theories. *Bulletin of the Geological Society of America*, **101**, 1408–19.

Keller, G., MacLeod, N., and Barrera, E. (1992). Eocene–Oligocene faunal turnover in planktonic foraminifera, and Antarctic glaciation. In *Eocene–Oligocene climatic and biotic evolution* (ed. D.R. Prothero and W.A. Berggren), pp. 218–44. Princeton University Press, Princeton.

Keller, G., MacLeod, N., Lyons, J.B. and Officer, C.B. (1993a) Is there evidence for Cretaceous–Tertiary boundary-age deep-water deposits in the Caribbean and Gulf of Mexico? *Geology*, **21**, 776–80.

Keller, G., Barrera, E., Schmitz, B., and Mattson, E. (1993b). Gradual mass extinction, species survivorship, and long-term environmental changes across the Cretaceous–Tertiary boundary in high latitudes. *Bulletin of the Geological Society of America*, **105**, 979–97.

Keller, G., Li, L. and MacLeod, N. (1996). The Cretaceous/Tertiary boundary stratotype section at El Kef, Tunisia: how catastrophic was the mass extinction? *Palaeogeography, Palaeoclimatology, Palaeoecology*, **119**, 221–54.

Kelley, P.H. and Hansen, T.A. (1996). Recovery of the naticid gastropod predator–prey system from the Cretaceous–Tertiary and Eocene–Oligocene extinctions. In *Biotic recovery from mass extinction events* (ed. M.B. Hart), pp. 373–86. Geological Society Special Publication, **102**.

Kelley, P.H. and Raymond, A. (1991). Migration, origination and extinction of Southern Hemisphere brachiopods during the middle Carboniferous. *Palaeogeography, Palaeoclimatology, Palaeoecology*, **86**, 23–39.

Kelly, S.R.A. (1977). The bivalves of the Spilsby Sandstone Formation and contiguous deposits. PhD thesis, University of London.

Kelly, S.R.A. (1984–92). Bivalvia of the Spilsby Sandstone and Sandringham Sands (Late Jurassic–Early Cretaceous) of southern England. *Palaeontographical Society Monograph*.

Kennedy, W.J. (1984). Ammonite faunas and the 'standard zones' of the Cenomanian to Maastrichtian stages in their type areas, with some proposals for their definition of the stage boundaries by ammonites. *Bulletin of the Geological Society of Denmark*, **33**, 147–61.

Kennett, J.P. and Stott, L.D. (1991). Abrupt deep-sea warming, palaeoceanographic changes and benthic extinctions at the end of the Palaeocene. *Nature*, **353**, 225–9.

Kerr, R.A. (1995). Chesapeake Bay impact crater confirmed. *Science*, **269**, 1672.

Kier, P.M. (1984). Echinoids from the Triassic (St Cassian) of Italy, their lantern supports and a revised phylogeny of Triassic echnoids. *Smithsonian Contributions to Paleobiology*, **56**, 1–41.

King, A.H. (1993). Mollusca: Cephalopoda (Nautiloidea). In *The fossil record 2* (ed. M.J. Benton), pp. 169–88. Chapman & Hall, London.

King, G.M. (1991). Terrestrial tetrapods and the end Permian mass event: a comparison of analyses. *Historical Biology*, **5**, 239–55.

Kirkland, J.I. (1991). Lithostratigraphic and biostratigraphic framework for the Mancos Shale (Late Cenomanian to Middle Turonian) at Black Mesa, northeastern Arizona. *Geological Society of America Special Paper*, **260**, 85–112.

Kitchell, J.A., Clark, D.L., and Gambos, A.M. (1986). Biological selectivity and extinction: a link between background and mass extinction. *Palaios*, **1**, 504–11.

Klapper, G., Feist, R., Becker, R.T., and House, M.R. (1993). Definition of the Frasnian–Famennian Stage boundary. *Episodes*, **16**, 433–41.

Knoll, A.H. (1984). Patterns of extinction in the fossil record of vascular plants. In *Extinctions* (ed. M.H. Nitecki), pp. 23–68. University of Chicago Press. Chicago.

Knoll, A.H. (1989). Evolution and extinction in the marine realm: some constraints imposed by phytoplankton. *Philosophical Transactions of the Royal Society of London*, **B325**, 279–90.

Knoll, A.H. and Walter, M.R. (1992). Latest Proterozoic stratigraphy and Earth history. *Nature*, **356**, 673–8.

Knoll, A.H., Fairchild, I.J., and Swett, K. (1993). Calcified microbes in Neoproterozoic carbonates: implications for our understanding of the Proterozoic/Cambrian transition. *Palaios*, **8**, 512–25.

Knoll, A.H., Bambach, R.K., Canfield, D.E., and Grotzinger, J.P. (1996). Comparative Earth history and Late Permian mass extinction. *Science*, **273**, 452–7.

Koch, C.F. (1991). Species extinction across the Cretaceous–Tertiary boundary: observed patterns versus predicted sampling effects, stepwise or otherwise? *Historical Biology*, **5**, 355–61.

Koch, C.F. and Morgan, J.P. (1988). On the expected distribution of species ranges. *Paleobiology*, **14**, 126–38.

Koeberl, C. (1993). Chicxulub Crater, Yucatan: tektites, impact glasses, and the geochemistry of target rocks and breccias. *Geology*, **21**, 211–14.

Koepnick, R.B., Denison, R.E., Burke, W.H., Herrington, E.A., Nelson, H.F., Otto, J.B., and Waite, L.E. (1985). Construction of the seawater $^{87}Sr/^{86}Sr$ curve for the Cenozoic and Cretaceous: supporting data. *Chemical Geology (Isotope Geosciences Section)*, **58**, 55–81.

Korn, D. (1992). The ammonoid faunal changes near the Devonian–Carboniferous boundary. *Annales de la Société géologique de Belgique*, **115**, 581–93.

Koutsoukos, E.A.M., Leary, P.N., and Hart, M.B. (1990). Latest Cenomanian–earliest Turonian low-oxygen tolerant benthic foraminifera: A case study from the Sergipe basin (N.E. Brazil) and the western Anglo-Paris Basin (southern England). *Palaeogeography, Palaeoclimatology, Palaeoecology*, **77**, 145–77.

Kozur, H. (1979). The main events in the Upper Permian and Triassic conodont evolution and its bearing to the Upper Permian and Triassic stratigraphy. *Rivista Italiana Paleontologia e Stratigraphia*, **85**, 741–66.

Kozur, H. (1985). Biostratigraphic evaluation of the upper Paleozoic conodonts, ostracods and holothurian sclerites of the Bükk Mts. Part II: Upper Paleozoic ostracods. *Acta Geologica Hungarica*, **28**, 225–56.

Kozur, H. (1991). Permian deep-water ostracods from Sicily (Italy) Part 2: Biofacial evaluation and remarks to the Silurian to Triassic paleopsychrospheric ostracods. *Geologisch-Paläontologische Mitteilungen Innsbruck*, **3**, 25–38.

Kozur, H. (1993). Report of P–T boundary working group. *Albertiana*, **12**, 33.

Kozur, H. (1994a). The correlation of the Zechstein with the marine standard. *Jahrbuch der Geologischen Bundesanstalt*, **137**, 85–103.

Kozur, H. (1994b). The Permian/Triassic boundary and possible causes of the faunal change near the P/T boundary. *Permophiles*, **24**, 51–4.

Kramm, U. and Wedepohl, K.H. (1991). The isotopic evidence of strontium and sulfur in seawater of Late Permian (Zechstein) age. *Chemical Geology*, **90**, 253–62.

Kring, D.A., Hildebrandt, A.R., and Boynton, W.V. (1994). Provenance of mineral phases in the Cretaceous–Tertiary boundary sediments exposed on the southern peninsula of Haiti. *Earth and Planetary Science Letters*, **128**, 629–41.

Kriz, J. (1979). Devonian Bivalvia. *Special Papers in Palaeontology*, **23**, 255–8.

Krogh, T.E., Kamo, S.L., Sharpton, V.L., Martin, L.E. and Hildebrandt, A.R. (1993). U-Pb ages of single shocked zircons linking distal K/T ejecta to the Chicxulub crater. *Nature*, **366**, 731–4.

Kuhnt, W. (1992). Abyssal recolonization by benthic foraminifera after the Cenomanian/Turonian boundary anoxic event in the North Atlantic. *Marine Micropaleontology*, **19**, 257–274.

Kummel, B. (1957). Paleoecology of Lower Triassic formations of southeastern Idaho and adjacent areas. *Geological Society of America, Memoir*, **67**(3), 437–68.

Kummel, B. and Teichert, C. (1970). Stratigraphy and paleontology of the Permian–Triassic boundary beds, Salt Range and Trans-Indus Ranges, West Pakistan. *University of Kansas, Department of Geology, Special Publication*, **4**, 1–110.

Kürschner, W., Becker, R.T., Buhl, D., and Veizer, J. (1992). Strontium isotopes in conodonts: Devonian–Carboniferous transition, the northern Rhenish Slate mountains, Germany. *Annales de la Société Géologique de Belgique*, **115**, 595–621.

Labandeira, C. and Sepkoski, J.J. Jr (1993). Insect diversity in the fossil record: myth and reality. *Science*, **261**, 310–15.

Landing, E. (1994). Precambrian–Cambrian boundary global stratotype ratified and a new perspective of Cambrian time. *Geology*, **22**, 179–82.

Leary, P.N. (1989). An assessment of the morphology of the northwestern Anglo-Paris Basin, eustatic movements and position of the oxygen minimum zone during the late Cenomanian (Cretaceous) using the ontogeny of *Rotalipora cushmani* (Morrow). *Revue de Micropaléontologie*, **32**, 134–39.

Leckie, R.M. (1989). An oceanographic model for the early evolutionary history of planktonic Foraminifera. *Palaeogeography, Palaeoclimatology, Palaeoecology*, **73**, 107–38.

Leckie, R.M., Schmidt, M.G., Finkelstein, D. and Yuretich, R. (1991). Paleoceanographic and paleoclimatic interpretations of the Mancos Shale (Upper Cretaceous), Black Mesa Basin, Arizona. *Geological Society of America Special Paper*, **260**, 139–52.

Legendre, S. and Hartenberger, J.-L. (1992). Evolution of mammalian faunas in Europe during the Eocene and Oligocene. In *Eocene–Oligocene climatic and biotic evolution* (ed. D.R. Prothero and W.A. Berggren), pp. 516–28. Princeton University Press, Princeton.

Le Loeuff, J., Buffetaut, E., and Martin, M. (1994). The last stages of dinosaur faunal history in Europe: a succession of Maastrichtian dinosaur assemblages from the Corbières (southern France). *Geological Magazine*, **131**, 625–30.

Leroux, H., Rocchia, R., Froget, L., Orue-Etzebarria, X., Doukham, J.-C., and Robin, E. (1995a). The K/T boundary at Beloc (Haiti): compared stratigraphic distributions of the boundary markers. *Earth and Planetary Science Letters*, **131**, 255–68.

Leroux, H., Warme, J.E., and Doukham, J.-C. (1995b). Shocked quartz in the Alamo breccia, southern Nevada: evidence for a Devonian impact event. *Geology*, **23**, 1003–6.

Lespérance, P.J. (1985). Faunal distributions across the Ordovician–Silurian boundary, Anticosti Island and Perce, Quebec, Canada. *Canadian Journal of Earth Sciences*, **22**, 838–49.

Lespérance, P.J. (1988). Trilobites. *Bulletin of the British Museum of Natural History (Geology)*, **43**, 359–76.

Lespérance, P.J., Barnes, C.R., Berry, W.B.N., Boucot, J., and Mu, E. (1987). The Ordovician–Silurian boundary stratotype: consequences of its approval by the IUGS. *Lethaia*, **20**, 217–22.

Lethiers, F. and Feist, R. (1991). La crise des ostracodes benthiques au passage Frasnien–Famennien de Coumiac (Montagne Noire, France mériodionale). *Comptes Rendus de l'Academies de Sciences, Paris*. **312**, 1057–63.

Lethiers, F. and Whatley, R. (1994). The use of Ostracoda to reconstruct the oxygen levels of late Palaeozoic oceans. *Marine Micropaleontology*, **24**, 57–69.

Lewan, M.D. (1986). Stable carbon isotopes of amorphous kerogens from Phanerozoic sedimentary rocks. *Geochimica et Cosmochimica Acta*, **50**, 1583–91.

Lingham-Soliar, T. (1994). Going out with a bang: the Cretaceous–Tertiary extinction. *Biologist*, **41** (5), 215–18.

Lipps, J.H., Bengtson, S. and Farmer, J.D. (1992). The Precambrian–Cambrian evolutionary transition. In *The Proterozoic biosphere* (ed. J.W. Schopf and C. Klein), pp. 453–7. Cambridge University Press, Cambridge.

Little, C.T.S. (1996). The Pliensbachian–Toarcian (Lower Jurassic) extinction event. *Geological Society of America Special Paper*, **307**, 505–12.

Little, C.T.S. and Benton, M.J. (1995). Early Jurassic mass extinction: a global long-term event. *Geology*, **23**, 495–8.

Litwin, R.J., Traverse, A. and Ash, S.R. (1991). Preliminary palynological zonation of the Chinle Formation, south western U.S.A., and its correlation to the Newark Supergroup (eastern U.S.A.). *Reviews of Palaeobotany and Palynology*, **68**, 269–87.

Liu Y. and Schmitt, R.A. (1992). Permian/Triassic boundary, Carnic Alps Austria, revisited; correlations with Ce anomalies, $\delta^{13}C$, and Siberian Trap flood basalts. *Lunar and Planetary Science*, **23**, 789–92.

Long, D.G.F. (1993). Oxygen and carbon isotopes and event stratigraphy near the Ordovician/Silurian boundary, Anticosti Island, Quebec. *Palaeogeography, Palaeoclimatology, Palaeoecology*, **104**, 49–59.

Long, D.G.F. and Copper, P. (1987a). Stratigraphy of the Upper Ordovician upper Vaureal and Ellis Bay formations, eastern Anticosti Island, Quebec. *Canadian Journal of Earth Sciences*, **24**, 1807–20.

Long, D.G.F and Copper, P. (1987b). Late Ordovician sandwave complexes on Anticosti Island, Quebec: a marine tidal embayment? *Canadian Journal of Earth Sciences*, **24**, 1821–32.

Longoria, J.F. and Gamper, M.A. (1995). Planktonic foraminiferal faunas across the Cretaceous–Tertiary succession of Mexico; implications for the Cretaceous–Tertiary boundary problem. *Geology*, **23**, 329–32.

Loper, D.E. and McCartney, K. (1986). Mantle plumes and the periodicity of magnetic field reversals. *Geophysical Research Letters*, **13**, 1525–8.

Loper, D.E., McCartney, K., and Buzyna, G. (1988). A model of correlated episodicity in magnetic-field reversals, climate, and mass extinctions. *Journal of Geology*, **96**, 1–15.

Lottman, J., Sandberg, C.A., Schindler, E., Walliser, O.H. and Ziegler, W. (1986). Devonian events at the Euse area (excursion to the Rheinisches Schiefergebirge). In *Global bio-events* (ed. O.H. Walliser), pp. 17–21. Springer-Verlag, Berlin.

Lucas, S.G. (1994). Triassic tetrapod extinctions and the compiled correlation effect. *Canadian Society of Petroleum Geologists, Memoir*, **17**, 869–75.

Ludvigsen, R. and Westrop, S.R. (1983). Trilobite biofacies of the Cambro-Ordovician boundary interval in northern North America. *Alcheringa*, **7**, 301–19.

Lutz, T.M. (1987). Limitations to the statistical analysis of episodic and periodic models of geologic time series. *Geology*, **15**, 1115–17.

Lyons, J.B. and Officer, C.B. (1992). Mineralogy and petrology of the Haiti Cretaceous/Tertiary section. *Earth and Planetary Science Letters*, **109**, 205–24.

Macdougall, J.D. (1988). Seawater strontium isotopes, acid rain, and the Cretaceous–Tertiary boundary. *Science*, **239**, 485–88.

MacLeod, K.G. (1994). Extinction of inoceramid bivalves in Maastrichtian strata of the Bay of Biscay region of France and Spain. *Journal of Paleontology*, **68**, 1048–66.

MacLeod, N. (1995). Biogeography of Cretaceous/Tertiary (K/T) planktonic foraminifera. *Historical Biology*, **10**, 49–101.

MacLeod, N. and Keller, G. (1991). How complete are Cretaceous/Tertiary boundary sections? A chronostratigraphic estimate based on graphic correlation. *Bulletin of the Geological Society of America*, **103**, 1439–57.

MacLeod, N. and Keller, G. (1994). Comparative biogeographic analysis of planktonic foraminiferal survivorship across the Cretaceous/Tertiary (K/T) boundary. *Paleobiology*, **20**, 143–77.

MacLeod, N. and Keller. G. (ed.) (1996). *Cretaceous–Tertiary mass extinctions*. Norton, London.

Macleod, N. and 21 other authors (1997). The Cretaceous–Tertiary biotic transition. *Journal of the Geological Society*, **154**, 265–92.

Magaritz, M. (1989). ^{13}C minima following extinction events: a clue to faunal radiation. *Geology*, **17**, 337–40.

Magaritz, M., Holser, W.T., and Kirschvink, J.L. (1986). Carbon-isotope events across the Precambrian/Cambrian boundary on the Siberian Platform. *Nature*, **320**, 258–59.

Magaritz, M., Krishnamurthy, R.V., and Holser, W.T. (1992). Parallel trends in organic and inorganic carbon isotopes across the Permian/Triassic boundary. *American Journal of Science*, **292**, 727–39.

Malkowski, K., Gruszczynski, M., Hoffman, A., and Halas, S. (1989). Oceanic stable isotope composition and a scenario for the Permo-Triassic crisis. *Historical Biology*, **2**, 289–309.

Manspeizer, W. (1988). Triassic–Jurassic rifting and opening of the Atlantic; an overview. In *Triassic–Jurassic rifting; continental breakup and the origin of the Atlantic Ocean and passive margins* (ed. W. Manspeizer), pp. 41–79. Elsevier, Amsterdam.

Marin, L.E., Sharpton, V.L., Urrutia-Fucugauchi, J., Silora, P., and Carney, C. (1994). The 'Upper Cretaceous Unit' in the Chicxulub multiring basin: new age based on planktonic foraminiferal assemblages. *Lunar and Planetary Institute Contribution*, **825**, 77.

Marshall, C.R. (1994). Confidence intervals on stratigraphic ranges: partial relaxation of the assumptions of randomly distributed fossil horizons. *Paleobiology*, **20**, 459–69.

Marshall, C.R. (1995). Distinguishing between sudden and gradual extinctions in the fossil record: predicting the position of the Cretaceous–Tertiary iridium anomaly using the ammonite fossil record on Seymour Island, Antarctica. *Geology*, **23**, 313–16.

Marshall, J.D. and Middleton, P.D. (1990). Changes in marine isotopic composition and the Late Ordovician glaciation. *Journal of the Geological Society*, **147**, 1–4.

Marshall, J.D., Brenchley, P.J., Mason, P., Wolff, G.A., Astini, R.A., Hints, L., and Meidla, T. (in press). Global carbon isotopic events associated with mass extinction and glaciation in the Late Ordovician. *Palaeogeography, Palaeoclimatology, Palaeoecology*.

Marshall, L.G. (1984). Who killed cock robin? In *Quaternary extinctions* (ed. P.S. Martin and R.G. Klein), pp. 785–806. University of Arizona Press, Tucson.

Martin, E.E. and Macdougall, J.D. (1991). Seawater Sr isotopes at the Cretaceous–Tertiary boundary. *Earth and Planetary Science Letters*, **104**, 166–80.

Martin, E.E. and Macdougall, J.D. (1995). Sr and Nd isotopes at the Permian/Triassic boundary: a record of climate change. *Chemical Geology*, **125**, 73–100.

Martin, P.S. (1984a). Prehistoric overkill: the global model. In *Quaternary extinctions: a prehistoric revolution* (ed. P.S. Martin and R.G. Klein), pp. 354–403. University of Arizona Press, Tucson.

Martin, P.S. (1984b). Catastrophic extinctions and late Pleistocene blitzkreig: two radiocarbon tests. In *Extinctions* (ed. M. Nitecki), pp. 153–89. University of Chicago Press, Chicago.

Martin, T.E. (1981). Species–area slopes and their coefficients: a caution on their interpretation. *American Naturalist*, **118**, 823–27.

Masse, J.P. (1989). Relations entre modifications biologiques et phénomènes géologiques sur les plates-formes carbonatées du domaine périméditerranéen au passage Bédoulien-Gargasien. *Géobios Mémoire Spéciale*, **11**, 279–94.

Mathey, B., Alzouma, K., Lang, J., Meister, C., Néraudeau, D., and Pascal, A. (1995). Unusual faunal associations during the upper Cenomanian–lower Turonian floodings on the Niger ramp (central west Africa). *Palaeogeography, Palaeoclimatology, Palaeoecology*, **119**, 63–76.

Maxwell, W.D. (1992). Permian and Early Triassic extinction of non-marine tetrapods. *Palaeontology*, **35**, 571–84.

McCracken, A.D. and Barnes, C.R. (1981). Conodont biostratigraphy and palaeoecology of the Ellis Bay Formation, Anticosti Island, Quebec with special reference to the Late Ordovician–Early Silurian chronostratigraphy and the systemic boundary. *Bulletin of the Geological Survey of Canada*, **329–2**, 51–134.

McDonald, J.N. (1984). The recorded North American selection regime and late Quaternary megafaunal extinctions. In *Quaternary extinctions: a prehistoric revolution* (ed. P.S. Martin and R.G. Klein), pp. 404–39. University of Arizona Press, Tucson.

McGhee, G.R. Jr (1982). The Frasnian–Famennian extinction event: a preliminary analysis of Appalachian marine ecosystems. *Geological Society of America, Special Paper*, **190**, 491–500.

McGhee, G.R. Jr (1988). The Late Devonian extinction event: evidence for abrupt ecosystem collapse. *Paleobiology*, **14**, 250–7.

McGhee, G.R. Jr (1996). *The Late Devonian mass extinction*. Columbia University Press, New York.

McGhee, G.R. Jr, Orth, C.J., Qunitana, L.R., Gilmore, J.S., and Olsen, E.J. (1986). Late Devonian 'Kellwasser event' mass extinction horizon in Germany: no geochemical evidence for a large-body impact. *Geology*, **14**, 776–9.

McGowran, B. (1990). Fifty million years ago. *American Scientist*, January–February, 30–39.

McGowran, B., Moss, G., and Beecroft, A. (1992). Late Eocene and early Oligocene in southern Australia. In *Eocene–Oligocene climatic and biotic evolution* (ed. D.R. Prothero and W.A. Berggren), pp. 178–201. Princeton University Press, Princeton.

McHone, J.F., Nieman, R.A., Lewis, C.F., and Yates, A.M. (1989). Stishovite at the Cretaceous–Tertiary boundary. *Science*, **239**, 485–8.

McIntosh, G.C. and Macurda, D.B. (1979). Devonian echinoderm biostratigraphy. *Special Papers in Palaeontology*, **23**, 331–4.

McKerrow, W.S. (1979). Ordovician and Silurian changes in sea level. *Journal of the Geological Society*, **136**, 137–46.

McKinney, M.L. (1985). Mass extinction patterns of marine invertebrate groups and some implications for a causal phenomenon. *Paleobiology*, **11**, 227–33.

McKinney, M.L. (1987). Taxonomic selectivity and continuous variation in mass and background extinctions of marine taxa. *Nature*, **325**, 143–5.

McKinney, M.L. (1995). Extinction selectivity among lower taxa: gradational patterns and rarefaction in extinction estimates. *Paleobiology*, **21**, 300–13.

McKinney, M.L., McNamara, K.J., Carter, B.D., and Donovan, S.K. (1992). Evolution of Paleogene echinoids: a global and regional review. In *Eocene–Oligocene climatic and biotic evolution* (ed. D.R. Prothero and W.A. Berggren), pp. 341–8. Princeton University Press, Princeton.

McLaren, D.J. (1970). Time, life, and boundaries. *Journal of Paleontology*, **44**, 801–15.

McLaren, D.J. (1982). Frasnian–Famennian extinctions. *Geological Society of America, Special Paper*, **190**, 477–84.

McLaren, D.J. (1983). Bolides and biostratigraphy. *Bulletin of the Geological Society of America*, **94**, 313–24.

McLaren, D.J. and Goodfellow, W.D. (1990). Geological and biological consequences of giant impacts. *Annual Reviews of Earth and Planetary Sciences*, **18**, 123–71.

McMenamin, M.A.S. and McMenamin, D.L.S. (1990). *The emergence of animals: the Cambrian breakthrough.* Columbia University Press, New York.

McRoberts, C.A. (1994). The Triassic-Jurassic ecostratigraphic transition in the Lombardian Alps, Italy. *Palaeogeography, Palaeoclimatology, Palaeoecology*, **110**, 145–66.

McRoberts, C.A. and Newton, C.R. (1995). Selective extinction among end-Triassic European bivalves. *Geology*, **23**, 102–4.

McRoberts, C.A., Newton, C.R., and Allasinaz, A. (1995). End-Triassic bivalve extinction: Lombardian Alps, Italy. *Historical Biology*, **9**, 297–317.

Mead, J.I. and Meltzer, D.J. (1984). North American Late Quaternary extinctions and the radiocarbon record. In *Quaternary extinctions: a prehistoric revolution* (ed. P.S. Martin and R.G. Klein), pp. 440–50. University of Arizona Press, Tucson.

Meisel, T., Krähenbühl, U., and Nazarov, M.A. (1995). Combined osmium and strontium isotopic study of the Cretaceous–Tertiary boundary at Sumbar, Turkmenistan: a test for impact vs. a volcanic hypothesis. *Geology*, **23**, 313–16.

Melchin, M.J., McCracken, A.D., and Oliff, F.J. (1991). The Ordovician–Silurian boundary on Cornwallis and Truro islands, Arctic Canada: preliminary data. *Canadian Journal of Earth Sciences*, **28**, 1854–62.

Mendelson, C. and Schopf, J.W. (1992). Proterozoic and Early Cambrian acritarchs. In *The Proterozoic biosphere* (ed. J.W. Schopf and C. Klein), pp. 219–31. Cambridge University Press, Cambridge.

Menning, M. (1995). A numerical time-scale for the Permian and Triassic periods: an integrative time analysis. In *The Permian of northern Pangaea 1* (ed. P.A. Scholle, T.M. Peryt, and D.S. Ulmer-Scholle), pp. 77–97. Springer-Verlag, Berlin.

Meyen, S.V. (1973). The Permian–Triassic boundary and its relation to the Paleophyte– Mesophyte floral boundary. In *The Permian and Triassic systems and their mutual boundary* (ed. A. Logan and L.V. Hills), pp. 662–7. *Canadian Society of Petroleum Geologists, Memoir*, **2**.

Meyerhoff, A.A., Lyons, J.B., and Officer, C.B. (1994). Chicxulub structure: a volcanic sequence of Late Cretaceous age. *Geology*, **22**, 3–4.

Miller, K.G. (1992). Middle Eocene to Oligocene stable isotopes, climate and deep-water history. In *Eocene–Oligocene climatic and biotic evolution* (ed. D.R. Prothero and W.A. Berggren), pp. 160–77. Princeton University Press, Princeton.

Miller, K.G., Janecek, T.R., Katz, M.E., and Keil, D.J. (1987). Abyssal circulation and benthic foraminiferal changes near the Paleocene/Eocene boundary. *Paleoceanography*, **2**, 741–61.

Miller, K.G., Berggren, W.H., Zhang, J., and Palmer-Julson, A. (1991). Biostratigraphy and isotope stratigraphy of upper Eocene microtektites at Site 612: how many impacts? *Palaios*, **6**, 17–38.

Milner, A.R. (1990). The radiation of temnospondyl amphibians. In *Major evolutionary radiations* (ed. P.D. Taylor and G.P. Larwood), pp. 321–49. Systematics Association Special Volume, **42**.

Montgomery, H., Pessagno, E., Soegaard, K., Smith, C., Muñuz, I., and Pessagno, J. (1992). Misconceptions concerning the Cretaceous/Tertiary boundary at the Brazos River, Falls County, Texas. *Earth and Planetary Science Letters*, **109**, 593–600.

Moore, R.C. (1954). Evolution of late Paleozoic invertebrates in response to major oscillations of shallow seas. *Bulletin of the Museum of Comparative Zoology, Harvard*, **122**, 259–86.

Moore, T.C., van Andel, T.H., Sancetta, C., and Pisias, N. (1978). Cenozoic hiatuses in pelagic sediments. *Micropaleontology*, **24**, 113–38.

Morante, R. (1996). Permian and Early Triassic isotopic records of carbon and strontium in Australia and a scenario of events about the Permian–Triassic boundary. *Historical Biology*, **11**, 289–310.

Morante, R. and Hallam, A. (1996). The organic carbon isotope record across the Triassic–Jurassic boundary in Austria and its bearing on the cause of the mass extinction. *Geology*, **24**, 391–4.

Morante, R., Veevers, J.J., Andrew, A.S., and Hamilton, P.J. (1994). Determination of the Permian–Triassic boundary in Australia from carbon isotope stratigraphy. *Australian Petroleum Exploration Association Journal*, **34**, 330–6.

Morbey, S.J. (1975). The palynostratigraphy of the Rhaetian stage, Upper Triassic, in the Kendelbachgraben, Austria. *Palaeontographica, Abteilung B*, **152**, 1–75.

Morrow, J.R. and Sandberg, C.A. (1996). Conodont faunal turnover and diversity changes through the Frasnian–Famennian (F/F) mass extinction and recovery episodes. 6th North American Paleontological Convention Abstracts Volume, *Special Publication of the Paleontological Society*, **8**, 284.

Mostler, H., Schauring, B., and Urlichs. M. (1978). Zur Mega-, Mikrofauna and Mikroflora der Kössener Schichten (Alpine, Obertrias) van Weissloferbach in Tirol unter besonderer Berücksichtigung der in der *suessi* und *marshi* Zone auftretenden Conodonten. *Österreichisch Akademie des Wissenschaften, Schriftenreihe des Erdwissenschaflichen Kommission*, **4**, 141–71.

Muchez, P., Boulvain, F., Dreesen, R., and Hou, H. (1996). Sequence stratigraphy of the Frasnian–Famennian transitional strata: a comparison between South China and southern Belgium. *Palaeogeography, Palaeoclimatology, Palaeoecology*, **123**, 289–96.

Murray, P. (1984). Extinctions downunder: a bestiary of extinct Late Pleistocene monotremes and marsupials. In *Quaternary extinctions: a prehistoric revolution* (ed. P.S. Martin and R.G. Klein), pp. 600–28. University of Arizona Press, Tucson.

Musashino, M. (1990). The Panthalassa – a cerium-rich Atlantic type Ocean: sedimentary environments of the Tamba Group, southwest Japan. *Tectonophysics*, **181**, 165–77.

Musashino, M. (1993). Chemical composition of the 'Toishi-type' siliceous shale – Part 1. *Bulletin of the Geological Survey of Japan*, **44**, 699–705 [English abstract].

Nakamura, K., Kimura, G., and Winsnes, T.S. (1987). Brachiopod zonation and age of the Permian Kapp Starostin Formation (Central Spitsbergen). *Polar Research*, new series, **5**, 207–19.

Nakazawa, K. (1985). The Permian and Triassic systems in the Tethys – their paleogeography. In *The Tethys: her paleogeography and paleobiogeography from Paleozoic to Mesozoic* (eds. K. Nakazawa and J.M. Dickins). pp. 93–111. Tokai University Press, Tokyo.

Nakazawa, K., Bando, Y., and Matsuda, T. (1980). The *Otoceras woodwardi* Zone and the time-gap at the Permian–Triassic boundary in east Asia. *Geology and Palaeontology of Southeast Asia*, **21**, 75–90.

Nakazawa, K. and Runnegar, B. (1973). The Permian–Triassic boundary: a crisis for bivalves? In *The Permian and Triassic systems and their mutual boundary* (ed. A. Logan and L.V. Hills), pp. 608–21. Canadian Society of Petroleum Geologists, Memoir **2**.

Nakrem, H.A. (1994). Environmental distribution of bryozoans in the Permian of Spitsbergen. In *Biology and palaeobiology of Bryozoans* (ed. P.J. Howard *et al.*), pp. 133–137. Proceedings of the 9th International Bryozoology Conference. Olsen and Olsen, Fredensburg.

Nakrem, H.A., Nilsson, I., and Mangerud, G. (1992). Permian biostratigraphy of Svalbard (Arctic Norway). *International Geological Review*, **34**, 933–55.

Narbonne, G.M. and Hofmann, H.J. (1987). Ediacaran biota of the Wernecke Mountains, Yukon, Canada. *Palaeontology*, **30**, 647–76.

Narkiewicz, M. and Hoffman, A. (1989). The Frasnian/Famennian transition: the sequence of events in southern Poland and its implications. *Acta Geologica Polonica*, **39**, 13–28.

Nemirovskaya, T. and Nigmadganov, I. (1994). The mid-Carboniferous conodont event. *Courier Forschungsinstitut Senckenberg*, **168**, 319–33.

Neri, C., Pasini, M., and Posenato, R. (1986). The Permian–Triassic boundary and the early Scythian sequence – Tesero Section, Dolomites. In *Permian and Permian–Triassic boundary in the south-Alpine segment of western Tethys*, pp. 111–16. Excursion Guidebook, IGCP Project 203.

Newell, N.D. (1952). Periodicity in invertebrate evolution. *Journal of Paleontology*, **26**, 371–85.

Newell, N.D. (1962). Paleontological gaps and geochronology. *Journal of Paleontology*, **36**, 592–610.

Newell, N.D. (1967). Revolutions in the history of life. *Geological Society of America Special Paper*, **89**, 63–91.

Newell, N.D. and Boyd, D.W. (1995). Pectinoid bivalves of the Permian–Triassic crisis. *Bulletin of the American Musem of Natural History*, **227**, 5–95.

Newton, C.R., Whalen, M.T., Thompson, J.B., Prins, N., and Dellalla, D. (1987). Systematics and paleoecology of Norian (Late Triassic) bivalves from a tropical island arc: Wallowa Terrane, Oregon. *Paleontological Society Memoir*, **22**, 1–83.

Nicholas, C.J. (1994). New stratigraphic constraints on the Durness Group of NW Scotland. *Scottish Journal of Geology*, **30**, 73–85.

Nichols, D.J. and Fleming, R.F. (1990). Plant microfossil record of the terminal Cretaceous event in the western United States and Canada. *Geological Survey of America Special Paper*, **89**, 63–91.

Nicol, D. (1979). A survey of suspension-feeding animals. *Florida Scientist*, **42**, 177–86.

Nicoll, R.S. and Playford, P.E. (1993). Upper Devonian iridium anomalies, conodont zonation and the Frasnian–Famennian boundary in the Canning Basin, Western Australia. *Palaeogeography, Palaeoclimatology, Palaeoecology*, **104**, 105–13.

Niklas, K.J., Tiffrey, B.H. and Knoll, A.H. (1985). Patterns of vascular plant diversification: an analysis of the species level. In *Phanerozoic diversity patterns* (ed. J.W. Valentine), pp. 97–128. Princeton University Press, Princeton.

Noé, S.U. (1987). Facies and paleogeography of the marine Upper Permian and of the Permian–Triassic boundary in the Southern Alps (Bellerophon Formation, Tesero Horizon). *Facies*, **16**, 89–142.

Noé, S.U. (1988). The Permian–Triassic boundary in the Southern Alps – a study of foraminiferal evolution. *Berichte der Geologischen Bundesanstalt, Wien*, **15**, 19.

Norris, M.S. and Hallam, A. (1995). Facies variation across the Middle–Upper Jurassic boundary in western Europe and the relationship to sea-level changes. *Palaeogeography, Palaeoclimatology, Palaeoecology*, **116**, 189–245.

Officer, C.B., Hallam, A., Drake, C.L. and Devine, J.D. (1987). Late Cretaceous and paroxysmal Cretaceous/Tertiary extinctions. *Nature*, **326**, 143–9.

O'Keefe, J.D. and Ahrens, T.J. (1989). Impact production of CO_2 by the Cretaceous/Tertiary extinction bolide and the resultant heating of the Earth. *Nature*, **338**, 247–9.

Okimura. Y., Ishii, K. and Ross, C.A. (1985). Biostratigraphical significance and faunal provinces of Tethyan Late Permian small Foraminifera. In *The Tethys, her paleogeography and paleobiogeography from Paleozoic to Mesozoic* (ed. K. Nakazawa and J.M. Dickins), pp. 115–58. Tokai University Press, Tokyo.

Oliver, W.A. Jr (1990). Extinctions and migrations of Devonian rugose corals in the eastern Americas realm. *Lethaia*, **23**, 167–78.

Oliver, W.A. Jr and Pedder, A.E.H. (1994). Crises in the Devonian history of Rugosa corals. *Paleobiology*, **20**, 178–90.

Olsen, P.E. and Sues, H.-D. (1986). Correlation of continental Late Triassic and Early Jurassic sediments and patterns of the Triassic–Jurassic tetrapod transition. In *The beginning of the age of dinosaurs* (ed. K. Padian), pp. 321–51. Cambridge University Press, Cambridge.

Olsen, P.E., Shubin, N.H. and Anders, M.H. (1987). New Early Jurassic tetrapod assemblages constrain Triassic–Jurassic tetrapod extinction event. *Science*, **237**, 1025–9.

Olsen, P.E., Fowell, S.J., and Cornet, B. (1990). The Triassic/Jurassic boundary in continental rocks of eastern North America; a progress report. *Geological Society of America Special Paper*, **247**, 585–94.

Olson, E.C. (1982). Extinctions of Permian and Triassic nonmarine vertebrates. *Geological Society of America, Special Paper*, **190**, 501–11.

Olson, E.C. (1989). Problems of Permo-Triassic terrestrial vertebrate extinctions. *Historical Biology*, **2**, 17–35.

Olsson, R.K. and Liu C. (1993). Controversies on the placement of the Cretaceous–Paleogene boundary and the K/P mass extinction of planktonic foraminifera. *Palaios*, **8**, 127–39.

Opik, A.A. (1966). The early Upper Cambrian crisis and its correlation. *Royal Society of New South Wales Journal and Proceedings*, **100**, 9–14.

Orchard, M.J. (1994). Conodonts from *Otoceras* beds: are they Permian? *Permophiles*, **24**, 49–51.

Ormiston, A.R. and Oglesby, R.J. (1995). Effect of Late Devonian paleoclimate on source rock quality and location. *American Association of Petroleum Geologists Studies in Geology*, **40**, 105–32.

Orth, C.J., Gilmore, J.S., Quintana, L.R., and Sheehan, P.M. (1986). Terminal Ordovician extinction: geochemical analysis of the Ordovician/Silurian boundary, Anticosti Island, Quebec. *Geology*, **14**, 433–6.

Orth, C.J., Attrep, M., Mao, X., Kauffman, E.G., Diner, R., and Elder. W.P. (1988). Iridium abundance maxima in the Upper Cenomanian extinction interval. *Geophysical Research Letters*, **15**, 346–9.

Orth, C.J., Attrep, M., and Quintana, L.R. (1990). Iridium abundance patterns across bio-event horizons in the fossil record. *Geological Society of America Special Paper*, **247**, 45–59.

Orth, C.J., Attrep, M., Quintana, L.R., Elder, W.P., Kauffman, E.G., Diner, R., and Villamil, T. (1993). Elemental abundance anomalies in the late Cenomanian extinction interval – a search for the source(s). *Earth and Planetary Science Letters*, **117**, 189–204.

Orth, C.J., Knight, J.D., Quintana, L.R., Gilmore, J.S., and Palmer, A.R. (1984). A search for iridium abundance anomalies at two Late Cambrian biomere boundaries in western Utah. *Science*, **223**, 163–5.

Ortiz, N. (1994). Mass extinction of benthic foraminifera at the Paleocene/Eocene boundary. In *Extinction and the fossil record* (ed. E. Molina), pp. 201–18. Seminario Interdisciplinario Universidad Zaragoza 5, Zaragoza.

Ouyang, S. and Utting, J. (1990). Palynology of Upper Permian and Lower Triassic rocks, Meishan, Changxing County, Zhejiang Province, China. *Reviews of Palaeobotany and Palynology*, **66**, 65–103.

Owen, A.W. and Robertson, D.B.R. (1995). Ecological changes during the end-Ordovician extinction. *Modern Geology*, **20**, 21–40.

Owen-Smith, N. (1987). Pleistocene extinctions: the pivotal role of megaherbivores. *Paleobiology*, **13**, 351–62.

Padian, K. and Clemens, W.A. (1985). Terrestrial vertebrate diversity: episodes and insights. In *Phanerozoic diversity patterns* (ed. J.W. Valentine), pp. 41–96. Princeton University Press, Princeton.

Palmer, A.R. (1965). Biomere, a new kind of biostratigraphic unit. *Journal of Paleontology*, **39**, 149–53.

Palmer, A.R. (1982). Biomere boundaries: a possible test for extraterrestrial perturbation of the biosphere. *Geological Society of America Special Paper*, **190**, 469–76.

Palmer, A.R. (1984). The biomere problem: evolution of an idea. *Journal of Paleontology*, **58**, 599–611.

Palmer, A.R. and James, N.P. (1980). The Hawke Bay event: a circum-Iapetus regression near the Lower–Middle Cambrian boundary. In *The Caledonides in the USA* (ed. D.R. Wones), pp. 15–18. Department of Geological Sciences, Virginia Polytechnic Institute and State University Memoirs, **2**.

Pan, H. and Erwin, D.H. (1994). Gastropod diversity patterns in South China during the Chihsian–Ladinian and their mass extinction. *Palaeoworld*, **4**, 249–62.

Pande, P.K. and Kalia, P. (1994). Upper Permian and Lower Triassic nodosariid foraminifera from the Kashmir Himalya, India. *Neues Jahrbuch für Geologie und Paläontologie, Abhandlungen*, **191**, 313–29.

Paproth, E., Feist, R. and Flajs, G. (1991). Decision on the Devonian–Carboniferous boundary stratotype. *Episodes*, **14**, 331–5.

Paris, F., Giraud, C., Feist, R., and Winchester-Seeto, T. (1996). Chitinozoan bio-event in the Frasnian–Famennian boundary beds at La Serre (Montagne Noire, southern France). *Palaeogeography, Palaeoclimatology, Palaeoecology*, **121**, 131–45.

Parrish, J.M., Parrish, J.T., Hutchison, J.H., and Spicer, R.A. (1987). Late Cretaceous vertebrate fossils from the North Slope of Alaska and implications for dinosaur ecology. *Palaios*, **2**, 377–89.

Pasini, M. (1985). Biostratigrafia con i Foraminiferi del limite Fm. a Bellerophon – Fm. di Werfen fra Reccaro e la Val Bardia (Alpi Meridionali). *Rivista Italiana Paleontologia e Stratigrafia*, **90**, 463–80.

Patterson, C. and Smith, A.B. (1987). Is the periodicity of extinctions a taxonomic artefact? *Nature*, **330**, 248–51.

Paul, C.R.C. (1982). The adequacy of the fossil record. In *Problems of phylogenetic reconstruction* (ed. K.A. Joysey and A.E. Friday), pp. 75–117. Academic Press, London.

Paul, C.R.C. (1988). Extinction and survival in echinoderms. In *Extinction and survival in the fossil record* (ed. G.P. Larwood), pp. 155–70. Systematics Association Special Volume, **34**.

Paul, C.R.C. and Mitchell, S.F. (1994). Is famine a common factor in marine mass extinctions? *Geology*, **22**, 679–82.

Paul, C.R.C., Mitchell, S.F., Lamolda, M., and Gorostidi, A. (1994). The Cenomanian–Turonian boundary event in northern Spain. *Geological Magazine*, **131**, 801–7.

Paull, R.K. and Paull, R.A. (1994a). *Hindeodus parvus* – proposed index fossil for the Permian–Triassic boundary. *Lethaia*, **27**, 271–2.

Paull, R.K. and Paull, R.A. (1994b). Lower Triassic transgressive–regressive sequences in the Rocky Mountains, eastern Great Basin, and Colorado Plateau, USA. In *Mesozoic systems of the Rocky Mountains Region, USA* (ed. M.V. Caputo, J.A. Peterson and K.J. Franczyk), pp. 169–80. Society of Economic Paleontologists and Mineralogists, Denver, Colorado.

Pedder, A.E.H. (1982). The rugose coral record across the Frasnian–Famennian boundary. *Geological Society of America, Special Paper*, **190**, 485–90.

Pedersen, K.R. and Lund, J.J. (1980). Palynology of the plant-bearing Rhaetian to Hettangian Kap Stewart Formation, Scoresby Sund, east Greenland. *Review of Palaeobotany and Palynology*, **32**, 1–69.

Perch-Nielsen, K. (1986). Geologic events and the distribution of calcareous nanofossils – some speculations. *Bulletin de la Centre de Recherches Exploration et Production Elf-Aquitaine*, **10**, 421–30.

Perri, M.C. (1991). Conodont biostratigraphy of the Werfen Formation (Lower Triassic), southern Alps, Italy. *Bolletino della Società Paleontologia Italiana*, **30**, 23–46.

Perry, D.G. and Chatterton, B.D. (1979). Late Early Triassic brachiopod and conodont fauna, Thaynes Formation, southeastern Idaho. *Journal of Paleontology*, **53**, 307–19.

Peryt, D., Wyrwicka, K., Orth, C., Attrep, M. Jr and Quintana, L.R. (1994). Foraminiferal changes and geochemical profiles across the Cenomanian/Turonian boundary in central and south-east Poland. *Terra Nova*, **6**, 158–65.

Philip, J. and Airaud-Crumière, C. (1991). The demise of the rudist-bearing carbonate platforms at the Cenomanian/Turonian boundary: a global control. *Coral Reefs*, **10**, 115–25.

Phillips, J. (1841). *Figures and descriptions of the Palaeozoic fossils of Cornwall, Devon and West Somerset*. Longman, Brown, Green and Longman, London.

Pitrat, C.W. (1970). Phytoplankton and the late Paleozoic wave of extinction. *Palaeogeography, Palaeoclimatology, Palaeoecology*, **8**, 49–55.

Pitrat, C.W. (1973). Vertebrates and the Permo-Triassic extinction. *Palaeogeography, Palaeoclimatology, Palaeoecology*, **14**, 249–64.

Playford, P.E., McLaren, D.J., Orth, C.J., Gilmore, J.S., and Goodfellow, W.D. (1984). Iridium anomaly in the Upper Devonian of the Canning Basin, Western Australia. *Science*, **226**, 437–9.

Poag, C.W. and Aubry, M.-P. (1995). Upper Eocene impactites of the U.S. east coast: depositional origins, biostratigraphic framework, and correlation. *Palaios*, **10**, 16–43.

Pojeta, J. Jr and Palmer, T.J. (1976). The origin of rock boring in mytilacean pelecypods. *Alcheringa*, **1**, 167–79.

Pope, K.O., Baines, K.H., Ocampo, A.C., and Ivanov, B.A. (1994). Impact winter and the Cretaceous/Tertiary extinctions; results of a Chicxulub asteroid impact model. *Earth and Planetary Scence Letters*, **128**, 719–25.

Posenato, R. (1991). Endemic to cosmopolitan brachiopods across the P/Tr boundary in the southern Alps (Italy). In *Proceedings of shallow Tethys 3*, pp. 125–39. Saito Ho-on Kai Special Publication, **3**.

Pospichal, J.J. (1994). Calcareous nanofossils at the K–T boundary, El Kef: no evidence for stepwise, gradual, or sequential extinctions. *Geology*, **22**, 99–102.

Pospichal, J.J. (1996). Calcareous nanofossils and clastic sediments at the Cretaceous–Tertiary boundary, northeastern Mexico. *Geology*, **24**, 255–8.

Poty, E. (1996). The Strunian Rugosa recovery after the late Frasnian crisis in Belgium and surrounding areas. 6th North American Paleontological Convention Abstracts Volume. *Special Publication of the Paleontological Society*, **8**, 310.

Prothero, D.R. (1994). *The Eocene–Oligocene transition*. Columbia University Press, New York.

Prothero, D.R. and Berggren, W.A. (ed.) (1992). *Eocene–Oligocene climatic and biotic evolution*. Princeton University Press, Princeton.

Rampino, M.R. and Stothers, R.B. (1988). Flood basalt volcanism during the past 250 million years. *Science*, **241**, 663–8.

Rampino, M.R., Eshet, Y., and Lapenis, A. (1994). Abrupt global ecosystem disaster and mass extinction at the Permian/Triassic boundary. *GSA Annual Meeting, Abstracts with Programs*, Seattle, A-396.

Ramsbottom, W.H.C. and Saunders, W.B. (1985). Evolution and evolutionary biostratigraphy of Carboniferous ammonoids. *Journal of Paleontology*, **59**, 123–59.

Raup, D.M. (1978). Cohort analysis of generic survivorship. *Paleobiology*, **4**, 1–15.

Raup, D.M. (1979). Size of the Permo-Triassic bottleneck and its evolutionary implications. *Science*, **206**, 217–18.

Raup, D.M. (1986). Biological extinction in Earth history. *Science*, **231**, 1528–33.

Raup, D.M. (1989). The case for extraterrestrial causes of extinction. *Philosophical Transactions of the Royal Society of London*, **B325**, 421–35.

Raup, D.M. (1991). *Extinction: bad luck or bad genes?* W.W. Norton and Co., New York.

Raup, D.M. (1992). Large-body impact and extinction in the Phanerozoic. *Paleobiology*, **18**, 80–8.

Raup, D.M. and Boyajian, G.E. (1988). Patterns of generic extinction in the fossil record. *Paleobiology*, **14**, 19–125.

Raup, D.M. and Jablonski, D. (1993). Geography of end-Cretaceous marine bivalve extinctions. *Science*, **260**, 971–3.

Raup, D.M. and Sepkoski, J.J. Jr (1982). Mass extinction in the marine fossil record. *Science*, **215**, 1501–3.

Raup, D.M. and Sepkoski, J.J. Jr (1984). Periodicity of extinctions in the geologic past. *Proceedings of the National Academy of Sciences USA*, **81**, 801–5.

Raup, D.M. and Sepkoski, J.J. Jr (1986). Periodic extinctions of families and genera. *Science*, **231**, 833–6.

Raup, D.M. and Sepkoski, J.J. Jr (1988). Testing for periodicity of extinction. *Science*, **241**, 94–6.

Raymond, A. and Metz, C. (1995). Laurussian land-plant diversity during the Silurian and Devonian – mass extinction, sampling bias, or both. *Paleobiology*, **21**, 74–91.

Raymond, A., Kelley, P.H. and Blanton, C.K. (1989). Polar glaciers and life at the equator: the history of Dinantian and Namurian (Carboniferous) climate. *Geology*, 17, 408–11.

Raymond, A., Kelley, P.H. and Lutken, C.B. (1990). Dead by degress: articulate brachiopods, paleoclimate and the mid-Carboniferous extinction event. *Palaios*, 5, 111–23.

Reinhardt, J.W. (1988). Uppermost Permian reefs and Permo-Triassic sedimentary facies from the southeastern margin of the Sichuan Basin, China. *Facies*, 18, 231–86.

Reinhardt, J.W. (1989). End-Paleozoic regression and carbonate to siliciclastic transition: a Permian–Triassic boundary on the eastern Tethyan Yangzi Platform (Sichuan, China). *Zentralblatt für Geologie und Paläontologie*, 1988, 861–70.

Renne, P.R. and Basu, A.R. (1991). Rapid eruption of the Siberian Traps flood basalts at the Permo-Triassic boundary. *Science*, 253, 176–79.

Renne, P.R., Zhang, Z., Richardson, M.A., Black, M.T., and Basu, A.R. (1995). Synchrony and causal relations between Permo-Triassic boundary crises and Siberian flood volcanism. *Science*, 269, 1413–16.

Retallack, G.J. (1994a) Were the Ediacaran fossils lichens? *Paleobiology*, 20, 523–44.

Retallack, G.J. (1994b). A pedotype approach to latest Cretaceous and earliest Tertiary paleosols in eastern Montana. *Bulletin of the Geological Society of America*, 106, 1377–97.

Retallack, G.J. (1995a) Permian–Triassic crisis on land. *Science*, 267, 77–80.

Retallack, G.J. (1995b). Earliest Triassic *Isoetes* and quillwort adaptive radiation. *Annual Meeting of the Geological Society of America, New Orleans, Programmes and Abstract Volume.*

Retallack, G.J. (1996). Paleoenvironmental change across the Permian–Triassic boundary on land in southeastern Australia and Antarctica. *International Geological Congress Symposium, Beijing, Abstract Volume*, 109.

Retallack, G.J., Veevers, J.J., and Morante, R. (1996). Global coal gap between Permian–Triassic extinction and Middle Triassic recovery of peat-forming plants. *Bulletin of the Geological Society of America*, 108, 195–207.

Rhodes, F.H.T. (1967). Permo-Triassic extinction. In *The fossil record* (ed. W.B. Harland), pp. 57–76. Geological Society of London.

Rhodes, M.C. and Thayer, C.W. (1991). Mass extinctions: ecological selectivity and primary production. *Geology*, 19, 877–80.

Rich, P.V., Rich, T.H., Wagstaff, B.E., Mason, J.M., Douthill, C.B., Gregory, R.T., and Felton, E.A. (1988). Evidence for low temperatures and biological diversity in Cretaceous high latitudes of Australia. *Science*, 242, 1403–6.

Rigby, J.K. Jr, Newman, K.R., Smit, J., van der Kaars, W.A., Sloan, R.E., and Rigby, J.K. (1987). Dinosaurs from the Paleocene part of the Hell Creek Formation, McCone County, Montana. *Palaios*, 2, 296–302.

Rigby, J.K. Jr and Senowbari-Daryan, B. (1995). Permian sponge biogeography and biostratigraphy. In *The Permian of northern Pangaea 1* (ed. P.A. Scholle, T.M. Peryt, and D.S. Ulmer-Scholle), pp. 153–66. Springer-Verlag, Berlin.

Robaszynski, F., Hardenbol, J., Caron, M., Amédro, F., Dupuis, C., González Donoso, J.M., Linares, D., and Gartner, S. (1993). Sequence stratigraphy in a distal environment: the Cenomanian of the Kalaat Senan Region (central Tunisia). *Bulletin de la Centre Recherches Exploration-Production Elf-Aquitaine*, 17, 395–433.

Robertson, D.B.R., Brenchley, P.J., and Owen, A.W. (1991). Ecological disruption close to the Ordovician–Silurian boundary. *Historical Biology*, 5, 131–44.

Robin, E., Bonté, P., Froget, L., Jéhanno, C. and Rocchia, R. (1992). Formation of spinels in cosmic objects during atmospheric entry: a clue to the Cretaceous–Tertiary boundary event. *Earth and Planetary Science Letters*, **108**, 181–90.

Robison, V.D. (1995). The Exshaw Formation: a Devonian/Mississippian hydrocarbon source in the Western Canada Basin. In *Petroleum source rocks* (ed. B.J. Katz), pp. 9–24. Springer-Verlag, Berlin.

Rocchia, R., Boclet, D., Bonté, P., Castellarin, A. and Jéhanno, C. (1986). An iridium anomaly in the Middle–Lower Jurassic of the Venetian region, northern Italy. *Journal of Geophysical Research*, **91**, B13, E359–362.

Rohling, E.J., Zachariasse, W.J., and Brinkhuis, H. (1991). A terrestrial scenario for the Cretaceous–Tertiary boundary collapse of the marine pelagic ecosystem. *Terra Nova*, **3**, 41–8.

Romer, A.S. (1966). *Vertebrate paleontology* (3rd edition). University of Chicago Press, Chicago.

Rong, J. and Harper, D.A.T. (1988). A global synthesis of the latest Ordovician Hirnantian brachiopod faunas. *Transactions of the Royal Society of Edinburgh*, **79**, 383–402.

Rosenzweig, M.L. and McCord, R.D. (1991). Incumbent replacement: evidence for long-term evolutionary progress. *Paleobiology*, **17**, 202–13.

Ross, C.A. and Ross, J.R.P. (1995a). Foraminiferal zonation of the late Paleozoic depositional sequences. *Marine Micropaleontology*, **26**, 469–78.

Ross, C.A. and Ross, J.R.P. (1995b). Permian sequence stratigraphy. In *The Permian of northern Pangaea 1* (ed. P.A. Scholle, T.M. Peryt, and D.S. Ulmer-Scholle), pp. 98–123. Springer-Verlag, Berlin.

Ross, D.J. and Skelton, P.W. (1993). Rudist formations of the Cretaceous: a palaeoecological, sedimentological and stratigraphical review. *Sedimentology Review*, **1**, 73–91.

Ross, J.R.P. (1995). Permian bryozoa. In *The Permian of northern Pangaea 1* (ed. P.A. Scholle, T.M. Peryt, and D.S. Ulmer-Scholle), pp. 196–209. Springer-Verlag, Berlin.

Rougerie, F. and Fagerstrom, J.A. (1994). Cretaceous history of Pacific Basin guyot reefs: a reappraisal based on geothermal endo-upwelling. *Palaeogeography, Palaeoclimatology, Palaeoecology*, **112**, 239–60.

Rowell, A.J. and Brady, M.J. (1976). Brachiopods and biomeres. *Brigham Young University, Geological Studies*, **23**, 165–80.

Runnegar, B.N. and Fedonkin, M.A. (1992). Proterozoic metazoan body fossils. In *The Proterozoic biosphere* (ed. J.W. Schopf and C. Klein), pp. 369–88. Cambridge University Press, Cambridge.

Russell, D.A. (1979). The enigma of the extinction of the dinosaurs. *Annual Review of Earth and Planetary Sciences*, **7**, 163–82.

Ryder, G., Fastershy, D. and Gartner, S. (1996). The Cretaceous–Tertiary event and other catastrophes in Earth history. *Geological Society of America Special Paper*, **307**.

Saito, T., Yamanoi, T., and Kaiho, K. (1986). End-Cretaceous devastation of terrestrial flora in the boreal Far East. *Nature*, **323**, 253–5.

Sakagami, S. (1985). Paleogeographic distribution of Permian and Triassic Ectoprocta (Bryozoa). In *The Tethys: her paleogeography and paleobiogeography from Paleozoic to Mesozoic* (ed. K. Nakazawa and J.M. Dickins), pp. 171–183. Tokai University Press, Tokyo.

Saltzman, M.R., Davidson, J.P., Holden, P., Runnegar, B., and Lohmann, K.C. (1995). Sea-level driven changes in ocean chemistry at an Upper Cambrian extinction horizon. *Geology*, **23**, 893–986.

Sandberg, C.A., Ziegler, W., Dreesen, R., and Butler, J.L. (1988). Part 3: Late Frasnian mass extinction: Conodont event stratigraphy, global changes, and possible causes. *Courier Forschungsinstitut Senckenberg*, **102**, 263–307.

Sandy, M.R. (1988). Tithonian brachiopods. *Mémoires de la Société Géologique de France*, N.S. **154**, 71–4.

Sanfilippo, A., Westberg-Smith, M.J., and Riedel, W.R. (1985). Cenozoic radiolaria. In *Plankton stratigraphy* (ed. H.M. Boli, J.B. Saunders, and K. Perch-Nielsen), pp. 105–30. Cambridge University Press, Cambridge.

Sano, H. and Kanmera, K. (1996). Microbial controls on Panthalassa Carboniferous–Permian oceanic buildups, Japan. *Facies*, **34**, 239–56.

Sarjeant, W.A.S., Volkheimer, W. and Zhang W. (1992). Jurassic palynomorphs of the circum-Pacific region. In *The Jurassic of the Circum-Pacific* (ed. G.E.G. Westermann), pp. 273–92. Cambridge University Press, Cambridge.

Satterley, A.K., Marshall, J.D., and Farichild, I.J. (1994). Diagenesis of an Upper Triassic reef complex, Wilde Kirche, Northern Calcaerous Alps, Austria. *Sedimentology*, **41**, 935–50.

Saunders, W.B. and Ramsbottom, W.H.C. (1986). The mid-Carboniferous eustatic event. *Geology*, **14**, 208–12.

Savoy, L.E. and Harris, A.G. (1993). Conodont biofacies and taphonomy along a carbonate ramp to black shale basin (latest Devonian and earliest Carboniferous), southern-most Canadian Cordillera and adjacent Montana. *Canadian Journal of Earth Sciences*, **30**, 2404–22.

Savrda, C.E. (1993). Ichnosedimentological evidence for a noncatastrophic origin of Cretaceous–Tertiary boundary sands in Alabama. *Geology*, **21**, 1075–8.

Schaeffer, B. (1973). Fishes and the Permian–Triassic boundary. In *The Permian and Triassic systems and their mutual boundary* (ed. A. Logan and L.V. Hills), pp. 493–7. Canadian Society of Petroleum Geologists, Memoir, **2**.

Schäfer, P. and Fois, E. (1987). Systematics and evolution of Triassic Bryozoa. *Geologica et Palaeontologica*, **27**, 173–225.

Schindewolf, O.H. (1954). Über die möglichen Ursachen der grossen erdgeschichtlichen Faunenschnitte. *Neues Jahrbuch für Geologie und Paläontologie Monatshefte*, **1954**, 457–465.

Schindler, E. (1990). The late Frasnian (Upper Devonian) Kellwasser crisis. In *Extinction events in Earth history* (ed. E.G. Kauffman and O.H. Walliser), pp. 151–9. Springer-Verlag, Berlin.

Schindler, E. (1993). Event-stratigraphic markers within the Kellwasser crisis near the Frasnian/Famennian boundary (Upper Devonian) in Germany. *Palaeogeography, Palaeoclimatology, Palaeoecology*, **104**, 115–25.

Schlager, W. (1981). The paradox of drowned reefs and carbonate platforms. *Bulletin of the Geological Society of America*, **92**, 197–211.

Schlanger, S.O. and Jenkyns, H.C. (1976). Cretaceous oceanic anoxic events: causes and consequences. *Geologie en Mÿnbouw*, **55**, 179–84.

Schlanger, S.O., Arthur, M.A., Jenkyns, H.C. and Scholle, P.A. (1987). The Cenomanian–Turonian anoxic event, I. Stratigraphy and distribution of organic carbon-rich beds and the marine $\delta^{13}C$ excursion. In *Marine petroleum source rocks* (ed. J. Brooks and A.J. Fleet), pp. 371–99. Geological Society Special Publication, **26**.

Schmitz, B. (1988). Origin of microlayering in worldwide distributed Ir-rich marine Cretaceous–Tertiary boundary clays from the western interior of the U.S.A. *Geology*, **16**, 1068–72.

Schmitz, B., Anderson, P., and Dahl, J. (1988). Iridium, sulfur isotopes and rare earth elements in the Cretaceous–Tertiary boundary clays at Stevns Klint, Denmark. *Geochimica et Cosmochimica Acta*, **52**, 229–36.

Schmitz, B. (1992). Chalcophite elements and Ir in continental Cretaceous–Tertiary boundary clays from the Western Interior of the USA. *Geochimica et Cosmochimica Acta*, **56**, 1695–1703.

Scholle, P.A. (1995). Carbon and sulfur isotope stratigraphy of the Permian and adjacent intervals. In *The Permian of northern Pangaea 1* (ed. P.A. Scholle, T.M. Peryt, and D.S. Ulmer-Scholle), pp. 133–49. Springer-Verlag, Berlin.

Scholle, P.A. and Arthur, M.A. (1980). Carbon isotope fluctuations in Cretaceous pelagic limestones; potential stratigraphic and petroleum exploration tool. *Bulletin of the American Association of Petroleum Geologists*, **64**, 67–87.

Schönlaub, H.P., Klein, P., Magaritz, M., Orth, C., and Attrep, M. (1988). The D–C boundary event (360 Ma) in the Carnic Alps (Austria). *Berichte der Geologischen Bundesanstalt*, **15**, 24–5.

Schopf, J.M. (1973). The contrasting plant assemblages from Permian and Triassic deposits in southern continents. In *The Permian and Triassic systems and their mutual boundary* (ed. A. Logan and L.V. Hills), pp. 379–97. Canadian Society of Petroleum Geologists, Memoir, **2**.

Schopf, J.W. (1992). Patterns of Proterozoic microfossil diversity: an initial, tentative analysis. In *The Proterozoic biosphere* (ed. J.W. Schopf and C. Klein), pp. 529–52. Cambridge University Press, Cambridge.

Schopf, T.J.M. (1974). Permo-Triassic extinctions: relation to sea-floor spreading. *Journal of Geology*, **82**, 129–43.

Schubert, J.K. and Bottjer, D.J. (1995). Aftermath of the Permian–Triassic mass extinction event: paleoecology of Lower Triassic carbonates in the western USA. *Palaeogeography, Palaeoclimatology, Palaeoecology*, **116**, 1–40.

Schubert, J.K., Bottjer, D.J., and Simms, M.J. (1992). Paleobiology of the oldest articulate crinoid. *Lethaia*, **25**, 97–110.

Scotese, C.R. and McKerrow, W.S. (1990). Revised world maps and introduction. In *Palaeozoic palaeogeography and biogeography* (ed. W.S. McKerrow and C.R. Scotese), pp. 1–21. Geological Society Memoir, **12**.

Scotese, C.R., Bambach, R.K., Barton, C., van der Voo, R., and Ziegler, A.M. (1979). Paleozoic base maps. *Journal of Geology*, **87**, 217–77.

Scott, R.W. (1995). Global environmental controls on Cretaceous reefal ecosystems. *Palaeogeography, Palaeoclimatology, Palaeoecology*, **119**, 187–99.

Scrutton, C.T. (1988). Patterns of extinction and survival in Palaeozoic corals. In *Extinction and survival in the fossil record* (ed. G.P. Larwood), pp. 65–88. Clarendon Press, Oxford.

Seilacher, A. (1984). Late Precambrian and Early Cambrian Metazoa: preservational or real extinctions? In *Patterns of change in Earth evolution* (ed. H.D. Holland and A.F. Trendall), pp. 159–68. Springer Verlag, Berlin.

Seilacher, A. (1989). Vendozoa: organismic construction in the Proterozoic biosphere. *Lethaia*, **22**, 229–39.

Seilacher, A. (1992). Vendobiota and Psammocorallia: last constructions of Precambrian evolution. *Journal of the Geological Society of London*, **149**, 607–13.

Selden, P.A. (1993). Arthropoda (Aglaspida, Pycnogonida and Chelicerata). In *The Fossil Record 2* (ed. M.J. Benton), pp. 297–320. Chapman & Hall, London.

Senowbari-Daryan, B., Zühlke, R., Bechstädt, T., and Flügel, E. (1993). Anisian (Middle Triassic) buildups of the Northern Dolomites (Italy): the recovery of reef communities after the Permian/Triassic crisis. *Facies*, **28**, 181–256.

Sepkoski, J.J. Jr (1981). A factor analytic description of the Phanerozoic marine fossil record. *Paleobiology*, **7**, 36–53.

Sepkoski, J.J. Jr (1982a). A compendium of fossil marine families. *Milwaukee Public Museum Contributions to Biology and Geology*, **51**, 1–125.

Sepkoski, J.J. Jr (1982b). Mass extinctions in the Phanerozoic oceans: a review. In *Geological implications of large asteroids and comets on the Earth* (ed. L.T. Silver and P.H. Schulz), pp. 283–9. Geological Society of America, Special Paper, **190**.

Sepkoski, J.J. Jr (1984). A kinetic model of Phanerozoic taxonomic diversity. III. Post-Paleozoic families and mass extinctions. *Paleobiology*, **10**, 246–67.

Sepkoski, J.J. Jr (1986). Phanerozoic overview of mass extinctions. In *Patterns and processes in the history of life* (ed. D.M. Raup and D. Jablonski), pp. 277–95. Springer-Verlag, Berlin.

Sepkoski, J.J. Jr (1987). Is the periodicity of extinction a taxonomic artefact? Response. *Nature*, **330**, 251–2.

Sepkoski, J.J. Jr (1989). Periodicity in extinction and the problem of catastrophism in the history of life. *Journal of the Geological Society of London*, **146**, 7–19.

Sepkoski, J.J. Jr (1992a). A compendium of fossil marine families. *Milwaukee Public Museum Contributions to Biology and Geology*, **83**, 1–156.

Sepkoski, J.J. Jr (1992b). Proterozoic–Early Cambrian diversification of metazoans and metaphytes. In *The Proterozoic biosphere* (ed. J.W. Schopf and C. Klein), pp. 553–61. Cambridge University Press, Cambridge.

Sepkoski, J.J. Jr (1993). Ten years in the library: new data confirm paleontological patterns. *Paleobiology*, **19**, 43–51.

Sepkoski, J.J. Jr (1996). Patterns of Phanerozoic extinction: a perspective from global data bases. In *Global events and event stratigraphy* (ed. O.H. Walliser), pp. 35–52. Springer-Verlag, Berlin.

Sepkoski, J.J. Jr and Raup, D.M. (1986). Periodicity in marine extinction events. In *Dynamics of extinction* (ed. D.K. Elliot), pp. 3–36. Wiley, New York.

Shackleton, N.J. (1986). Paleogene stable isotope events. *Palaeogeography, Palaeoclimatology, Palaeoecology*, **57**, 91–102.

Sharpton, V.L. and Ward, P.D. (ed.) (1990). *Global catastrophes in Earth History*. Geological Society of America Special Paper **247**.

Sharpton, V.L., Dalrymple, G.B., Marin, L.E., Ryder, G., Schuraytz, B.C., and Urrutia-Fucugauchi, J. (1992). New links between Chicxulub impact structure and the Cretaceous/Tertiary boundary. *Nature*, **359**, 819–21.

Sheehan, P.M. (1975). Brachiopod synecology in a time of crisis (Late Ordovician–Early Silurian). *Paleobiology*, **1**, 205–12.

Sheehan, P.M. (1979). Swedish Late Ordovician marine benthic assemblages and their bearing on brachiopod zoogeography. In *Historical biogeography, plate tectonics and the changing environment* (ed. J. Gray and A.J. Boucot), pp. 61–73. Oregon State University Press, Corvallis.

Sheehan, P.M. (1982). Brachiopod macroevolution at the Ordovician–Silurian boundary. *Proceedings of the Third North American Paleontological Convention*, **2**, 477–81.

Sheehan, P.M. (1988). Late Ordovician events and the terminal Ordovician extinction. *New Mexico Bureau of Mines and Mineral Resources, Memoir*, **44**, 405–15.

Sheehan, P.M. and Fastovsky, D.E. (1992). Major extinctions of land-dwelling vertebrates at the Cretaceous–Tertiary boundary, eastern Montana. *Geology*, **20**, 556–60.

Sheehan, P.M. and Hansen, T.A. (1986). Detritus feeding as a buffer to extinction at the end of Cretaceous. *Geology*, **14**, 868–70.

Sheehan, P.M., Fastovsky, D.E., Hoffman, R.G., Berghaus, C.B., and Gabriel, D.L. (1991). Sudden extinction of the dinosaurs: latest Cretaceous, upper Great Plains, U.S.A. *Science*, **254**, 835–9.

Shen S. and Shi G. (1996). Diversity and extinction patterns of Permian Brachiopoda of South China. *Historical Biology*, **12**, 93–110.

Signor, P.W. III (1982). Species richness in the Phanerozoic: compensating for sampling bias. *Geology*, **10**, 625–8.

Signor, P.W. III and Lipps, J.H. (1982). Sampling bias, gradual extinction patterns, and catastrophes in the fossil record. In *Geological implications of large asteroids and comets on the Earth* (ed. L.T. Silver and P.H. Schulz), pp. 291–6. Geological Society of America, Special Paper, **190**.

Sigurdsson, H., D'Hondt, S. and Carey, S. (1992). The impact of the Cretaceous/Tertiary bolide on evaporite terrane and generation of major sulfuric acid aerosol. *Earth and Planetary Sciences Letters*, **109**, 543–59.

Simakov, K.V. (1993). Biochronological aspects of the Devonian–Carboniferous crisis in the regions of the former USSR. *Palaeogeography, Palaeoecology, Palaeoecology*, **104**, 127–37.

Simberloff, D. (1974). Permo-Triassic extinctions: effects of an area on biotic equilibrium. *Journal of Geology*, **82**, 267–74.

Simms, M.J. (1986). Contrasting life-style in Lower Jurassic crinoids: a comparison of benthic and pseudopelagic Isocrinida. *Palaeontology*, **29**, 475–93.

Simms, M.J. (1989). *British Lower Jurassic crinoids*. Monographs of the Palaeontological Society, London.

Simms, M.J. (1991). The radiation of post-Palaeozoic echinoderms. In *Major evolutionary radiations* (ed. G.P. Larwood), pp. 288–304. Clarendon Press, Oxford.

Simms, M.J. and Ruffell, A.H. (1990). Climatic and biotic change in the Late Triassic. *Journal of the Geological Society*, **147**, 321–7.

Simms, M.J. and Sevastopulo, G.D. (1993). The origin of articulate crinoids. *Palaeontology*, **36**, 91–110.

Simms, M.J., Gale, A.S., Gilliland, P., Rose, E.P.F., and Sevastopulo, G.D. (1993). Echinodermata. In *The fossil record 2* (ed. M.J. Benton), pp. 491–528. Chapman & Hall, London.

Sinha, A. and Stott, L.D. (1994). New atmospheric pCO_2 estimates from paleosols during the late Paleocene/early Eocene global warming interval. *Global and Planetary Change*, **9**, 297–307.

Sloan, R.E., Rigby, J.K., Van Valen, L., and Gabriel, D.L. (1986). Gradual dinosaur extinction and simultaneous ungulate radiation in the Hell Creek Formation. *Science*, **232**, 629–33.

Smit, J. (1982). Extinction and evolution of planktonic foraminifera after a major impact at the Cretaceous/Tertiary boundary. In *Geological implications of large asteroids and comets on the Earth* (ed. L.T. Silver and P.H. Schulz), pp. 329–52. *Geological Society of America Special Paper* **190**.

Smit, J., Montanari, A., Swinburne, N.H.M., Alvarez, W., Hildebrandt, A.R., Margolis, S.V., Lowrie, W., and Asaro, F. (1992). Tektite-bearing deep-water clastic unit at the Cretaceous–Tertiary boundary in northeastern Mexico. *Geology*, **20**, 99–103.

Smit, J., Roep, T.B., Alvarez, W., Montanari, A., Claeys, P., Gradjales-Nishimura, J.M., and Bermudez, J. (1996). Coarse-grained, clastic sandstone complex at the K/T boundary around the Gulf of Mexico: deposition by tsunami waves induced by the Chicxulub impact? *Geological Society of America Special Paper*, **307**, 151–82.

Smith, A.B. (1990). Echinoid evolution from the Triassic to the Lower Liassic. *Cahiers de l'Université Catholique de Lyon, Série Scientifique*, **3**, 79–117.

Smith, P.L., Beyers, J-M., Carter, E.S., Jakobs, G.K., Paley, J., Pessagno, E., and Tipper, H.W. (1994). Jurassic taxa ranges and correlation charts for the Circum Pacific. *Newsletters in Stratigraphy*, **31**, 33–70.

Smith, R.M.H. (1995). Changing fluvial environments across the Permian–Triassic boundary in the Karoo Basin, South Africa and possible causes of tetrapod extinctions. *Palaeogeography, Palaeoclimatology, Palaeoecology*, **117**, 81–104.

Sondaar, P.Y. (1977). Insularity and its effects on mammal evolution. In *Major patterns in vertebrate evolution* (ed. M.K. Hecht, P.G. Goody, and B.M. Hecht), pp. 671–707. Plenum Press, New York.

Sorauf, J.E. and Pedder, A.E.H. (1986). Late Devonian rugose corals and the Frasnian–Fammenian crisis. *Canadian Journal of Earth Sciences*, **23**, 1265–87.

Southam, J.R., Peterson, W.H. and Brass, G.W. (1982). Dynamics of anoxia. *Palaeogeography, Palaeoclimatology, Palaeoecology*, **40**, 183–98.

Speijer, R.P. (1994). Extinction and recovery patterns in benthic foraminiferal paleocommunities across the Cretaceous/Paleogene and Paleocene/Eocene boundaries. *Mededelingen van de Faculteit Aardwetenschappen Universiteit Utrecht*, **124**, 1–191.

Spicer, R.A. and Parrish, J.T. (1990). Late Cretaceous–Early Tertiary paleoclimates of northern high latitudes: a quantitative view. *Journal of the Geological Society of London*, **147**, 329–41.

Spicer, R.A. and Corfield, R.M. (1992). A review of terrestrial and marine climates in the Cretaceous with implications for modelling the 'Greenhouse Earth'. *Geological Magazine*, **129**, 169–80.

Springer, M.S. (1990). The effect of random range truncations on patterns of evolution in the fossil record. *Paleobiology*, **16**, 512–20.

Stanley, G.D. (1988). The history of early Mesozoic reef communities, a three step process. *Palaios*, **3**, 170–83.

Stanley, G.D. and Beauvais, L. (1994). Corals from an Early Jurassic coral reef in British Columbia. *Lethaia*, **27**, 35–47.

Stanley, G.D. and Swart, P.K. (1995). Evolution of the coral-zooxanthellae symbiosis during the Triassic: a geochemical approach. *Paleobiology*, **21**, 179–199.

Stanley, S.M. (1984). Temperature and biotic crises in the marine realm. *Geology*, **12**, 205–8.

Stanley, S.M. (1986a). *Extinction*. Scientific American Books, New York.

Stanley, S.M. (1986b). Anatomy of a regional mass extinction: Plio-Pleistocene decimation of the western Atlantic bivalve fauna. *Palaios*, **1**, 17–36.

Stanley, S.M. (1988). Paleozoic mass extinctions: shared patterns suggest global cooling as a common cause. *American Journal of Science*, **288**, 334–52.

Stanley, S.M. (1990). Delayed recovery and the spacing of major extinctions. *Paleobiology*, **16**, 401–14.

Stanley, S.M. and Campbell, L.D. (1981). Neogene mass extinction of western Atlantic molluscs. *Nature*, **293**, 457–9.

Stanley, S.M. and Yang X. (1994). A double mass extinction at the end of the Paleozoic era. *Science*, **266**, 1340–4.

Stearn, C.W. (1987). Effect of the Frasnian–Famennian extinction event on the stromatoporoids. *Geology*, **15**, 677–9.

Stearn, C.W., Halim-Dihardja, M.K., and Nishida, D.K. (1987). An oil-producing stromatoporoid patch reef in the Famennian (Devonian) Wabamun Formation, Normandville Field, Alberta. *Palaios*, **2**, 560–70.

Stehlin, H.G. (1909). Remarques sur les faunules de mammifères des couches éocènes et oligocènes du Bassin de Paris. *Bulletin de la Société Géologique de France*, **9**, 488–520.

Stemmerik, L. (1988). Discussion. Brachiopod zonation and age of the Permian Kapp Starostin Formation (Central Spitsbergen). *Polar Research*, new series, **6**, 179–80.

Stenseth, N.C. and Maynard Smith, J. (1984). Coevolution in ecosystems: Red Queen or stasis? *Evolution*, **38**, 870–80.

Stevens, C.H. (1977). Was development of brackish oceans a factor in Permian extinctions? *Geology*, **8**, 133–8.

Stinnesbeck, W., Barbarin, J.M., Keller, G., Lopez-Oliva, J.G., Pivnik, D.A., Lyons, J.B., Officer, C.B., Adatte, T., Graup, G., Rocchia, R., and Robin, E. (1993). Deposition of channel deposits near the Cretaceous–Tertiary boundary in northeastern Mexico: catastrophic or 'normal' sedimentary deposits? *Geology*, **21**, 797–800.

Stitt, J.H. (1971). Repeating evolutionary pattern in Late Cambrian trilobite biomeres. *Journal of Paleontology*, **45**, 178–81.

Stitt, J.H. (1977). Late Cambrian and earliest Ordovician trilobites, Wichita Mountains area, Oklahoma. *Oklahoma Geological Survey Bulletin*, **124**, 1–79.

Stothers, R.B. (1993). Flood basalts and extinction events. *Geophysical Research Letters*, **20**, 1399–1402.

Strauss, D. and Sadler, P.M. (1989). Classical confidence intervals and Bayesian probability estimates for ends of local taxon ranges. *Mathematical Geology*, **21**, 411–27.

Stucky, R.K. (1992). Mammalian faunas in North America of Bridgerian to early Arikareean 'ages' (Eocene and Oligocene). In *Eocene–Oligocene climatic and biotic evolution* (ed. D.R. Prothero and W.A. Berggren), pp. 464–93. Princeton University Press, Princeton.

Sukhov, L.G., Bespalaya, Y.A., and Dolin, D.A. (1966). Biostratigraphy of volcanic formations in the western part of the Tunguska Syneclise. *Doklady Akademia Nauk S.S.S.R., Earth Sciences*, **169**, 107–10.

Sun, Y., Xu, D., Zhang, Q., Yang, Z., Sheng, J., Chen, C., Rui, L., Liang, X., Zhao, J., and He, J. (1984). The discovery of an iridium anomaly in the Permian–Triassic boundary clay in Changxing, Zhejiang, China and its significance. In *Developments in Geoscience, Contributions*, pp. 235–45. 27th International Geological Congress. Science Press, Beijing.

Surlyk, F. and Johansen, M.B. (1984). End-Cretaceous brachiopod extinctions in the Chalk of Denmark. *Science*, **223**, 1174–7.

Sutherland, F.L. (1994). Volcanism around K/T boundary time – its role in an impact scenario for the K/T extinction events. *Earth Science Reviews*, **36**, 1–26.

Suzuki, N., Ishida, K., and Ishiga, H. (1993). Organic geochemical implications of black shales related to the Permian/Triassic boundary, Tanba Belt, southwest Japan. *Bulletin of the Geological Survey of Japan*, **44**, 707–20.

Sweet, A.R., Braman, D.R. and Lerbekmo, J.F. (1990). Palynofloral response to K/T boundary events: a transitory interruption within a dynamic system. *Geological Society of America Special Paper* **247**, 457–69.

Sweet, W.C. (1970). Uppermost Permian and Lower Triassic conodonts of the Salt Range and Trans-Indus Ranges, West Pakistan. *University of Kansas Department of Geology, Special Publication*, **4**, 207–75.

Sweet, W.C. (1977). Genus Hindeodus, genus Isarcicella. In *Catalogue of conodonts 3*. (ed. W. Ziegler), pp. 203–30. Schweizerbart'sche Verlagsbuchhandlung, Stuttgart.

Sweet, W.C. (1979). Graphic correlation of Permo-Triassic rocks in Kashmir, Pakistan and Iran. *Geologica et Palaeontologica Abhandlungen*, **13**, 239–48.

Sweet, W.C. (1992). A conodont-based high-resolution biostratigraphy for the Permo-Triassic boundary interval. In *Permo-Triassic boundary events in the eastern Tethys* (ed. W.C. Sweet, Z. Yang, J.M. Dickins and H. Yin), pp. 120–33. Cambridge University Press, Cambridge.

Sweet, W.C., Yang, Z., Dickins, J.M., and Yin H. (1992). Permo-Triassic events in the eastern Tethys – an overview. In *Permo-Triassic boundary events in the eastern Tethys* (ed. W.C. Sweet, Z. Yang, J.M. Dickins and H. Yin), pp. 1–8. Cambridge University Press, Cambridge.

Swisher, C.C., Grajales-Nishimura, J.M., Montanori, A., Margolis, S.V., Claeys, P., Alvarez, W., Renne, P., Cedillo-Pardo, E., Florentino, J.-M., Maurasse, R., Curtis, G.H., Smit, J., and McWilliams, M.O. (1992). Coeval ^{40}Ar/^{39}Ar ages of 65.0 million years ago from Chicxulub Crater melt rock and Cretaceous–Tertiary boundary tektities. *Science*, **257**, 954–8.

Tappan, H. (1970). Phytoplankton abundance and late Paleozoic extinctions: a reply. *Palaeogeography, Palaeoclimatology, Palaeoecology*, **8**, 56–66.

Tappan, H. and Loeblich, A.R. Jr (1988). Foraminiferal evolution, diversification, and extinction. *Journal of Paleontology*, **62**, 695–714.

Taylor, P.D. and Larwood, G.P. (1988). Mass extinctions and the pattern of bryozoan evolution. In *Extinction and survival in the fossil record* (ed. G.P. Larwood), pp. 99–119. Systematics Association Special Volume, **34**.

Teichert, C. (1986). Times of crisis in the evolution of the Cephalopoda. *Paläontologische Zeitschrift*, **60**, 227–43.

Teichert, C. (1990). The Permian–Triassic boundary revisited. In *Extinction events in Earth history* (ed. E.G. Kauffman and O.H. Walliser), pp. 199–238. Springer-Verlag, Berlin.

Teichert, C. and Kummel. B. (1973). Permian–Triassic boundary in the Kap Stosch area, east Greenland. In *The Permian and Triassic systems and their mutual boundary* (ed. A. Logan and L.V. Hills), pp. 269–85. Canadian Society of Petroleum Geology, Memoir, **2**.

Teichert, C. and Kummel, B. (1976). Permian–Triassic boundary in the Kap Stosch area, east Greenland. *Meddelelser om Grønland*, **197**, 1–49.

Thackeray, J.F., van der Merve, N.J., Lee-Thorpe, J.A., Sillon, A., Lanham, J.L., Smith, R., Keyser, A., and Monteiro, P.M.S. (1990). Changes in the carbon isotope ratios in the Late Permian recorded in therapsid tooth apatite. *Nature*, **347**, 751–3.

Thickpenny, A. and Leggett, J.K. (1987). Stratigraphic distribution and palaeo-oceanographic significance of European early Palaeozoic organic-rich sediments. *Geological Society Special Publication*, **26**, 231–47.

Thomas, E. (1990). Late Cretaeous–early Eocene mass extinctions in the deep sea. *Geological Society of America Special Paper* **247**, 481–95.

Thomas, E. (1992). Middle Eocene–late Oligocene bathyal benthic foraminifera (Weddell Sea): faunal changes and implications for oceanic circulation. In *Eocene–Oligocene climatic and biotic evolution* (ed. D.R. Prothero and W.A. Berggren), pp. 464–93. Princeton University Press, Princeton.

Thompson, J.B. and Newton, C.R. (1989). Late Devonian mass extinction: episodic climatic cooling or warming? In *Devonian of the world* (ed. N.J. McMillan, A.F. Embry and D.J. Glass), pp. 29–34. Canadian Society of Petroleum Geologists, Memoir, **14**.

Tjalsma, R.C. and Lohman, G.P. (1983). Paleocene–Eocene bathyal and abyssal benthic foraminifera from the Atlantic Ocean. *Micropaleontology, Special Publication*, **4**, 1–90.

Tong, J. (1993). Biotic mass extinction and biotic alteration at the Permo-Triassic boundary. Foraminifera. In *Permo-Triassic events of South China* (ed. Z. Yang, S. Wu, H. Yin, G. Xu, K. Zhang and X. Bi), pp. 90–7. Geological Publishing House, Beijing.

Tosk, T.A. and Anderson, K.A. (1988). Late Early Triassic foraminifers from possible dysaerobic to anaerobic paleoenvironments of the Thaynes Formation, southeast Idaho. *Journal of Foraminiferan Research*, **18**, 286–301.

Tozer, E.T. (1965) Lower Triassic stages and ammonoid zones of Arctic Canada. *Geological Survey of Canada Paper*, **65-12**.

Tozer, E.T. (1967) A standard for Triassic time. *Bulletin of the Canadian Geological Survey*, **156**.

Tozer, E.T. (1974). Definitions and limits of Triassic stages and substages: suggestions prompted by comparison between North America and the Alpine-Mediterranean region. *Österreichische Akademie der Wissenschaften, Schriftenreihe der Erdwissenschaftlichen Kommission*, **2**, 195–206.

Tozer, E.T. (1979). Latest Triassic ammonoid faunas and biochronology, western Canada. *Geological Survey of Canada Paper*, **79-1B**, 127–35.

Tozer, E.T. (1981). Triassic Ammonoidea: geographic and stratigraphic distribution. In *The Ammonoidea* (ed. M.R. House and J.R. Senior), pp. 397–431. Academic Press, London.

Tozer, E.T. (1988). Definition of the Permian–Triassic (P-T) boundary: the question of the age of the *Otoceras* beds. *Memorie della Società Geologica Italiana*, **34** (for 1986), 291–301.

Tozer, E.T. (1994). Canadian Triassic ammonoid fauna. *Bulletin of the Geological Survey of Canada*, **467**.

Traverse, A. (1988). Plant evolution dances to a different beat; plant and animal evolutionary mechanisms compared. *Historical Biology*, **1**, 277–302.

Traverse, A. (1990). Plant evolution in relation to world crises and the apparent resilience of the kingdom Plantae. *Palaeogeography, Palaeoclimatology, Palaeoecology*, **82**, 203–211.

Tredoux, M., de Wit, M.J., Hart, R.J., Lindsay, N.M., and Sellschop, J.F.F. (1989). Chemostratigraphy across the Cretaceous/Tertiary boundary and a critical assessment of the iridium anomaly. *Journal of Geology*, **97**, 585–605.

Trotter, M.M. and McCulloch, R. (1984). Moas, men, and middens. In *Quaternary extinctions: a prehistoric revolution* (ed. P.S. Martin and R.G. Klein), pp. 483–516. University of Arizona Press, Tucson.

Truesdale, G.A., Downing, A.L., and Lowden, G.F. (1955). The solubility of oxygen in pore water and in seawater. *Journal of Applied Chemistry*, **5**, 53–62.

Trümpy, R. (1969). Lower Triassic ammonites from Jameson Land (east Greenland). *Meddelelser om Grønland*, **168**, 77–116.

Tschudy, R.H. and Tschudy, B.D. (1986). Extinction and survival of plant life following the Cretaceous/Tertiary boundary event, Western Interior, North America. *Geology*, **14**, 667–70.

Tschudy, R.H., Pillmore, C.L., Orth, C.J., Gilmore, J.S., and Knight, J.D. (1984). Disruption of the terrestrial plant ecosystem at the Cretaceous–Tertiary boundary, western North America. *Science*, **225**, 1030–2.

Tuckey, M.E. and Anstey, R.L. (1992). Late Ordovician extinction of bryozoans. *Lethaia*, **25**, 111–17.

Turekian, K.K. (1982). Potential of $^{187}Os/^{186}Os$ as a cosmic versus terrestrial indicator in high iridium layers of sedimentary strata. In *Geological implications of large asteroids and comets on the Earth* (ed. L.T. Silver and P.H. Schulz), pp. 243–9. *Geological Society of America Special Paper*, **190**.

Tyson, R.V. (1995). *Sedimentary organic matter*. Chapman & Hall, London.

Tyson, R.V. and Funnell, B.M. (1987). European Cretaceous shorelines, stage by stage. *Palaeogeography, Palaeoclimatology, Palaeoecology*, **59**, 69–91.

Upchurch, G.R. and Wolfe, J.A. (1993). Cretaceous vegetation of the Western Interior and adjacent regions of North America. *Geological Association of Canada Special Paper*, **39**, 243–82.

Valentine, J.W. and Jablonski, D. (1991). Biotic effects of sea level change: the Pleistocene test. *Journal of Geophysical Research*, **96**, 6873–8.

Valentine, J.W. and Moores, E.M. (1973). Provinciality and diversity across the Permian–Triassic boundary. In *The Permian and Triassic systems and their mutual boundary* (ed. A. Logan and L.V. Hills), pp. 759–66. Canadian Society of Petroleum Geology, Memoir, **2**.

Van Cappellen, P. and Ingall, E.D. (1994). Benthic phosphorus regeneration, net primary production, and ocean anoxia: a model of the coupled marine biogeochemical cycles of carbon and phosphorus. *Paleoceanography*, **9**, 677–92.

Van Cappellen, P. and Ingall, E.D. (1996). Redox stabilization of the atmosphere and oceans by phosphorus-limited marine productivity. *Science*, **271**, 493–6.

Van Steenwinkel, M. (1992). The Devonian–Carboniferous boundary: comparison between the Dinant Synclinorium and the northern border of the Rhenish Slate Mountains. *Annales de la Société Géologique de Belgique*, **115**, 665–1.

Van Steenwinkel, M. (1993). The Devonian–Carboniferous boundary in southern Belgium: biostratigraphic identification criteria of sequence boundaries. *Special Publication of the International Association of Sedimentologists*, **18**, 237–46.

Van Valen, L.M. (1973). A new evolutionary law. *Evolutionary Theory*, **1**, 1–30.

Van Valen, L.M. (1984). A resetting of Phanerozoic community evolution. *Nature*, **307**, 50–2.

Veevers, J.J., Conaghan, P.J., and Shaw, S.E. (1994). Turning point in Pangean environmental history at the Permian/Triassic (P/Tr) boundary. *Geological Society of America Special Paper*, **288**, 187–96.

Veizer, J. (1989). Strontium isotopes in seawater through time. *Annual Reviews of Earth and Planetary Science*, **17**, 141–67.

Veizer, J. and Compston, W. (1974). $^{87}Sr/^{86}Sr$ composition of seawater during the Phanerozoic. *Geochimica et Cosmochimica Acta*, **38**, 1461–84.

Venkatesen, T.R., Pande, K., and Gopalan, K. (1993). Did Deccan volcanism predate the K/T transition? *Earth and Planetary Science Letters*, **119**, 181–9.

Vereshchagin, N.K. and Baryshnikov, G.F. (1984). Quaternary extinctions in northern Eurasia. In *Quaternary extinctions* (ed. P.S. Martin and R.G. Klein), pp. 483–516. University of Arizona Press, Tucson.

Vermeij, G.J. (1986). Survival during biotic crises: the properties and evolutionary significance of refuges. In *Dynamics of extinction* (ed. D.K. Elliot), pp. 231–46. John Wiley and Sons, New York.

Vermeij, G.J. (1987). *Evolution and escalation*. Princeton University Press, Princeton.

Vidal, G. and Knoll, A.H. (1982). Radiations and extinctions of plankton in the Late Precambrian and Early Cambrian. *Nature*, **297**, 57–60.

Visscher, H., and Brugman, W.A. (1981). Ranges of selected palynomorphs in the Alpine Triassic of Europe. *Review of Palaeobotany and Palynology*, **34**, 115–28.

Visscher, H. and Brugman, W.A. (1988). The Permian–Triassic boundary in the Southern Alps: a palynological approach. *Memorie della Società Geologica Italiana*, **34** (for 1986), 121–8.

Visscher, H., van Houte, M., Brugman, W.A., and Poort, R.J. (1993). Rejection of a Carnian (Late Triassic) 'pluvial event' in Europe. *Review of Palaeobotany and Palynology*, **83**, 217–26.

Visscher, H., Brinkhuis, H., Dilcher, D.L., Elsik, W.C., Eshet, Y., Looy, C.V., Rampino, M.R., and Traverse, A. (1996). The terminal Paleozoic fungal event: Evidence of terrestrial ecosystem destabilization and collapse. *Proceedings of the National Academy of Sciences of the USA*, **93**, 2155–8.

Vogt, P.R. (1972). Evidence for global synchronism in mantle plume convection and possible significance for geology. *Nature*, **240**, 338–42.

Vogt, P.R. (1989). Volcanogenic upwelling of anoxic, nutrient-rich water: a possible factor in carbonate-bank/reef demise and benthic faunal extinctions? *Bulletin of the Geological Society of America*, **101**, 1225–45.

Walker, R.G. and James, N.P. (ed.) (1992). *Facies models: response to sea-level change*. Geological Association of Canada, Ottowa.

Wallace, M.W., Keays, R.R., and Gostin, R.V. (1991). Stromatolitic iron oxides: evidence that sea-level changes can cause sedimentary iridium anomalies. *Geology*, **19**, 551–4.

Walliser, O.H. (1996). Global events in the Devonian and Carboniferous. In *Global events and event stratigraphy* (ed. O.H. Walliser), pp. 225–50. Springer-Verlag, Berlin.

Wang C.-Y. (1995). Conodonts of Permian–Triassic boundary beds and biostratigraphic boundary. *Acta Palaeontologica Sinica*, **34**, 129–51 [in Chinese, English abstract].

Wang, K. (1992). Glassy microspherules (microtektites) from an Upper Devonian limestone. *Science*, **256**, 1546–9.

Wang, K., Orth, C.J., Attrep, M., Chatterton, B.D.E., Hou, H., and Geldsetzer, H.H.J. (1991). Geochemical evidence for a catastrophic biotic event at the Frasnian/Famennian boundary in South China. *Geology*, **19**, 776–9.

Wang, K., Chatterton, B.D.E., Attrep, M., and Orth, C.J. (1992). Iridium abundance maxima at the latest Ordovician mass extinction horizon. Yangtze Basin, China – terrestrial or extraterrestrial? *Geology*, **20**, 39–42.

Wang, K., Chatterton, B.D.E., Attrep, M., and Orth, C.J. (1993a). Late Ordovician mass extinction in the Selwyn Basin, northwestern Canada – geochemical, sedimentological and palaeontological evidence. *Canadian Journal of Earth Science*, **30**, 1870–80.

Wang, K., Orth, C.J., Attrep, M., Chatterton, B.D.E., Wang, X., and Li, J. (1993b). The great latest Ordovician extinction on the South China Plate: chemostratigraphic studies of the Ordovician–Silurian boundary interval on the Yangtze Platform. *Palaeogeography, Palaeoclimatology, Palaeoecology*, **104**, 61–79.

Wang, K., Attrep, M., and Orth, C.J. (1993c). Global iridium anomaly, mass extinction, and redox change at the Devonian–Carboniferous boundary. *Geology*, **21**, 1071–4.

Wang, K., Geldsetzer, H.H.J., and Krouse, H.R. (1994). Permian–Triassic extinction: organic $\delta^{13}C$ evidence from British Columbia, Canada. *Geology*, **22**, 580–4.

Wang, K., Chatterton, B.D.E., Attrep, M., and Orth, C.J. (1995). Geochemical analysis through the 'transitional zone' of conodont faunal turnover in the Ordovician–Silurian boundary interval, Anticosti Island, Quebec. *Canadian Journal of Earth Sciences*, **32**, 359–67.

Wang, K., Geldsetzer, H.H.J., Goodfellow, W.D., and Krouse, H.R. (1996). Carbon and sulfur isotope anomalies across the Frasnian–Famennian extinction boundary, Alberta, Canada. *Geology*, **24**, 187–91.

Wang, Z. (1996). Recovery of vegetation from the terminal Permian mass extinction in North China. *Review of Palaeobotany and Palynology*, **91**, 121–42.

Ward, P.D. (1995). The K/T trial. *Paleobiology*, **21**, 245–7.

Ward, P.D., Kennedy, W.J., MacLeod, K.G., and Mount, J.F. (1991). Ammonite and inoceramid bivalve extinction patterns in the Cretaceous/Tertiary boundary sections of the Biscay region (southwestern France, northern Spain). *Geology*, **19**, 1181–4.

Ward, W.C., Keller, G., Stinnesbeck, W., and Adatte, T. (1995). Yucatan subsurface stratigraphy: implications and constraints for the Chicxulub impact. *Geology*, **23**, 873–6.

Waterhouse, J.B. (1972). A Permian overtoniid brachiopod in Early Triassic sediments of Axel Heiberg Island and its implications on the Permo–Triassic boundary. *Canadian Journal of Earth Sciences*, **9**, 486–99.

Waterhouse, J.B. and Bonham-Carter, G.F. (1976). Range, proportionate representation, and demise of brachiopod families through the Permian Period. *Geological Magazine*, **113**, 401–28.

Watkins, R. (1994). Evolution of Silurian Pentamerid communities in Wisconsin. *Palaios*, **9**, 488–99.

Webb, S.D. (1984). Ten million years of mammal extinctions in North America. In *Quaternary extinctions* (ed. P.S. Martin and R.G. Klein), pp. 189–210. University of Arizona Press, Tucson.

Webb, S.D. and Barnosky, A.D. (1989). Faunal dynamics of Pleistocene mammals. *Annual Review of Earth and Planetary Science*, **17**, 413–38.

Weems, R.E. (1992). The 'terminal Triassic catastrophic event' in perspective: a review of Carboniferous through Early Jurassic vertebrate extinction patterns. *Palaeogeography, Palaeoclimatology, Palaeoecology*, **94**, 1–29.

Wei, W. (1995). How many impact-generated microspherule layers in the upper Eocene? *Palaeogeography, Palaeoclimatology, Palaeoecology*, **114**, 101–10.

Weissert, H. (1989). C-isotope stratigraphy, a monitor of palaeoenvironmental change: a case study from the Early Cretaceous. *Surveys in Geophysics*, **10**, 1–61.

Westrop, S.R. (1989). Trilobite mass extinction near the Cambrian–Ordovician boundary in North America. In *Mass extinctions: processes and evidence* (ed. S.K. Donovan), pp. 89–103. Belhaven Press, London.

Westrop, S.R. and Ludvigsen, R. (1987). Biogeographic control of trilobite mass extinction at an Upper Cambrian 'biomere' boundary. *Paleobiology*, **13**, 84–99.

Whatley, R.C. (1991). The platycopid signal: a means of detecting kenoxic events using Ostracoda. *Journal of Micropalaeontology*, **10**, 181–5.

Whatley, R.C. (1988). Patterns and rates of evolution among Mesozoic Ostracoda. In *Evolutionary biology of Ostracoda* (ed. T. Hanai, N. Ikeya, and K. Ishizaki), pp. 1021–40. Kodansha, Tokyo.

Whatley, R.C., Siveter, D.J., and Boomer, I.D. (1993). Arthropoda (Crustacea: Ostracoda). In *The fossil record 2* (ed. M.J. Benton), pp. 321–56. Chapman & Hall, London.

Wicander, E.R. (1975). Fluctuations in a late Devonian–early Mississippian phytoplankton flora of Ohio, USA. *Palaeogeography, Palaeoclimatology, Palaeoecology*, **17**, 89–108.

Wiedmann, J. (1973). Ammonoid (r)evolution at the Permian–Triassic boundary. In *The Permian and Triassic systems and their mutual boundary* (ed. A. Logan and L.V. Hills), pp. 513–21. Canadian Society of Petroleum Geologists, Memoir, **2**.

Wignall, P.B. (1990). Ostracod and foraminifera micropaleontology and its bearing on biostratigraphy: a case study from the Kimmeridgian (Late Jurassic) of north west Europe. *Palaios*, **5**, 219–26.

Wignall, P.B. (1994). *Black shales*. Oxford University Press, Oxford.

Wignall, P.B. and Hallam, A. (1992). Anoxia as a cause of the Permian/Triassic extinction: facies as evidence from northern Italy and the western United States. *Palaeogeography, Palaeoclimatology, Palaeoecology*, **93**, 21–46.

Wignall, P.B. and Hallam, A. (1993). Griesbachian (earliest Triassic) palaeoenvironmental changes in the Salt Range, Pakistan and southwest China and their bearing on the Permo-Triassic mass extinction. *Palaeogeography, Palaeoclimatology, Palaeoecology*, **102**, 215–37.

Wignall, P.B. and Hallam, A. (1996). Facies change and the end-Permian mass extinction in S.E. Sichuan, China. *Palaios*, **11**, 587–96.

Wignall, P.B. and Twitchett, R.J. (1996). Oceanic anoxia and the end Permian mass extinction. *Science*, **272**, 1155–8.

Wignall, P.B., Hallam, A., Lai, X., and Yang, F. (1995). Palaeoenvironmental changes across the Permian/Triassic boundary at Shangsi (N. Sichuan, China). *Historical Biology*, **10**, 175–89.

Wignall, P.B., Kozur, H., and Hallam, A. (1996). The timing of palaeoenvironmental changes at the Permo-Triassic (P/Tr) boundary using conodont biostratigraphy. *Historical Biology*, **12**, 39–62.

Wilde, P. and Berry, W.B.N. (1984). Destabilization of the oceanic density structure and its significance to marine 'extinction' events. *Palaeogeography, Palaeoclimatology, Palaeoecology*, **48**, 143–62.

Wilde, P. and Berry, W.B.N. (1986). The role of oceanographic factors in the generation of global bio-events. In *Global bio-events* (ed. O.H. Walliser), pp. 75–91. Springer-Verlag, Berlin.

Wilde, P., Berry, W.B.N., Quinby-Hunt, M.S., Orth, C.J., Quintana, L.R., and Gilmore, J.S. (1986). Iridium abundances across the Ordovician–Silurian stratotype. *Science*, **233**, 339–41.

Wilde, P., Quinby-Hunt, M.S., and Erdtmann, B.-D. (1996). The whole-rock cerium anomaly: a potential indicator of eustatic sea-level changes in shales of the anoxic facies. *Sedimentary Geology*, **101**, 3–53.

Wilkinson, I.P. and Riley, N.J. (1990). Namurian entomozoacean Ostracoda and eustatic events. In *Ostracoda and global events* (ed. R. Whatley and C. Maybury), pp. 161–72. British Micropalaeontological Society Publications Series. Chapman & Hall, London.

Wille, W. (1982). Evolution and ecology of upper Liassic dinoflagellates from SW Germany. *Neues Jahrbuch für Geologie und Paläontologie, Abhandlungen*, **164**, 74–132.

Williams, S.H. (1988). Dob's Linn – the Ordovician–Silurian boundary stratotype. *Bulletin of the British Museum of Natural History (Geology)*, **43**, 17–30.

Wilson, E.O. (1992). *The diversity of life*. Penguin Books, London.

Wolbach, W.S., Lewis, R.S., and Anders, E. (1985). Cretaceous extinctions: evidence for wildfires and search for meteoritic material. *Science*, **230**, 167–70.

Wolbach, W.S., Gilmour, I., and Anders, E. (1990). Major wildfires at the Cretaceous/ Tertiary boundary. *Geological Society of America Special Paper*, **247**, 391–400.

Wolbach, W.S., Roegge, D.R., and Gilmour, I. (1994). The Permian–Triassic of the Gartnerkofel-1 core (Carnic Alps, Austria): organic carbon isotope variation. In *New developments regarding the K/T event and other catastrophes in Earth history*, pp. 133–134. Abstract volume, Houston, Texas.

Wolfe, J.A. (1992). Climatic, floristic, and vegetational changes near the Eocene/Oligocene boundary in North America. In *Eocene–Oligocene climatic and biotic evolution* (ed. D.R. Prothero and W.A. Berggren), pp. 464–93. Princeton University Press, Princeton.

Wolfe, J.A. and Upchurch, G.R. (1987). North American non-marine climates and vegetation during the Late Cretaceous. *Palaeogeography, Palaeoclimatology, Palaeoecology*, **61**, 33–77.

Worsley, T.R., Moore, T.L., Fraticelli, C.M., and Scotese, C.R. (1994). Phanerozic CO_2 levels and global temperatures inferred from changing paleogeography. *Geological Society of America Special Paper*, **288**, 57–73.

Wright, J., Seymour, R.S., and Shaw, H.F. (1984). REE and Nd isotopes in conodont apatite: variations with geological age and depositional environment. In *Conodont biofacies and provincialism* (ed. D.L. Clark), pp. 325–40. Geological Society of America Special Paper, **196**, 325–40.

Wright, J., Schrader, H., and Holser, W.T. (1987). Paleoredox variations in ancient oceans recorded by rare earth elements in fossil apatite. *Geochimica et Cosmochimica Acta*, **51**, 631–44.

Wu, S., Liu, J., and Zhu, Q. (1993). The beginning climax and amplitude of transgression. In *Permo-Triassic events of South China* (ed. Z. Yang, S. Wu, H. Yin, G. Xu, K. Zhang and X. Bi), pp. 9–15. Geological Publishing House, Beijing.

Wyatt, A.R. (1987). Shallow water areas in space and time. *Journal of the Geological Society*, **144**, 115–20.

Wyatt, A.R. (1993). Phanerozoic shallow water diversity driven by changes in sea-level. *Geologische Rundschau*, **82**, 203–11.

Wyatt, A.R. (1995). Late Ordovician extinctions and sea-level change. *Journal of the Geological Society*, **152**, 899–902.

Xu, C. (1984). Influence of the Late Ordovician glaciation on basin configuration of the Yangtze Platform in China. *Lethaia*, **17**, 51–9.

Xu, D. and Yan, Z. (1993). Carbon isotope and iridium event markers near the Permian/ Triassic boundary in the Meishan section, Zhejiang Province, China. *Palaeogeography, Palaeoclimatology, Palaeoecology*, **104**, 171–5.

Xu, D., Ma, L., Chai, Z., Mao, X., Su, Y., Zhang, Q., and Yong, Z. (1985). Abundance of iridium and trace metals at the Permian/Triassic boundary at Shangsi in China. *Nature*, **314**, 154–6.

Xu, D., Yan, Z., Zhang, Q., Shen, Z., Sun, Y. and Ye, L. (1986). Significance of a $\delta^{13}C$ anomaly near the Devonian/Carboniferous boundary at the Nuhua Section, South China. *Nature*, **321**, 854–5.

Xu, G. and Grant, R.G. (1992). Permo-Triassic brachiopod succession and events in South China. In *Permo-Triassic boundary events in the eastern Tethys* (ed. W.C. Sweet, Z. Yang, J.M. Dickins and H. Yin), pp. 98–108. Cambridge University Press, Cambridge.

Xu, G. and Grant, R.E. (1994). Brachiopods near the Permian–Triassic boundary in South China. *Smithsonian Contributions to Paleobiology*, **76**.

Yan, Z., Hou, H., and Ye, L. (1993). Carbon and oxygen isotope event markers near the Frasnian–Famennian boundary, Luoxiu Section, South China. *Palaeogeography, Palaeoclimatology, Palaeoecology*, **104**, 97–104.

Yang, F. (1993). Biotic mass extinction and biotic alteration at the Permo-Triassic boundary. Ammonoids. In *Permo-Triassic events of South China* (ed. Z. Yang, S. Wu, H. Yin, G. Xu, K. Zhang and X. Bi), pp. 102–8. Geological Publishing House, Beijing.

Yang, Z. and Li, Z. (1992). Permo-Triassic boundary relations in South China. In *Permo-Triassic events in eastern Tethys* (ed. W.C. Sweet, Z. Yang, J.M. Dickins and H. Yin), pp. 9–20. Cambridge University Press, Cambridge.

Yang, Z., Yin, H., Wu, S., Yang, F., Ding, M., and Xu, G. (1987). Permian–Triassic boundary stratigraphy and fauna of South China. *Geological Memoirs*, Series 2, **6**. Geological Publishing House, Beijing.

Yin, H. (1982). Uppermost Permian (Changxingian) Pectinacea from South China. *Rivista Italiana Palaeontologia e Stratigrafia*, **88**, 337–86.

Yin, H. (1985a). On the transitional bed and the Permian–Triassic boundary in South China. *Newsletters in Stratigraphy*, **15**, 13–27.

Yin, H. (1985b). Bivalves near the Permian–Triassic boundary in South China. *Journal of Paleontology*, **59**, 572–600.

Yin, H. (1990). Paleogeographic distribution and stratigraphic range of the Lower Triassic *Claraia, Pseudoclaraia* and *Eumorphotis* (Bivalvia). *Journal of the China University of Geosciences*, **1**, 98–110.

Yin, H., Huang, S., Zhang, K., Hansen, H.J., Yang, F., Ding, M. and Bie, X. (1992). The effects of volcanism on the Permo-Triassic mass extinction in South China. In *Permo-Triassic events in eastern Tethys* (ed. W.C. Sweet, Z. Yang, J.M. Dickins and H. Yin), pp. 146–57. Cambrudge University Press, Cambridge.

Yin, H., Wu, S., Ding, M., Zhang, K., Tong, J., and Yang, F. (1994). The Meishan Section – candidate of the Global Stratotype Section and Point (GSSP) of the Permian–Triassic Boundary (PTB). *Albertiana*, **14**, 15–30.

Zachos, J.C., Arthur, M.A., and Dean, W.E. (1989). Geochemical evidence for suppression of pelagic marine productivity at the Cretaceous/Tertiary boundary. *Nature*, **337**, 61–4.

Zakharov, V.A. and Yanine, B.T. (1975). Les bivalves à la fin du Jurassique et au début du Crétacé. *Bulletin de Recherches Géologiques et Minières, Mémoire*, **86**, 221–8.

Zakharov, V.A., Lapukhov, A.S., and Shenfil, O.V. (1992). Are the iridium anomalies always evidence of impact character of Earth biosphere turnovers? *5th International Conference on Global Bioevents, Göttingen 1992. Abstract Volume*, 125.

Zhou L. and Kyte, F. (1988). The Permian–Triassic boundary event: a geochemical study of three Chinese sections. *Earth and Planetary Science Letters*, **90**, 411–21.

Zhuravlev, A.Y. (1996). Reef ecosystem recovery after the Early Cambrian extinction. In

Biotic recovery from mass extinction events (ed. M.B. Hart), pp. 79–96. Geological Society of London Special Publication, **102**.

Zhuravlev, A.Y. and Wood, R.A. (1996). Anoxia as the cause of the mid-Early Cambrian (Botomian) extinction event. *Geology*, **24**, 311–14.

Ziegler, A.M., Parrish, J.M., Yao, J., Gyllenhaal, E.D., Rowley, D.B., Parrish, J.T., Nie, S., Bekker, A., and Hulver, M.L. (1994). Early Mesozoic phytogeography and climate. In *Palaeoclimates and their modelling* (ed. J.R.L. Allen, B.J. Hoskins, B.W. Sellwood, R.A. Spicer and P.J. Valdes), pp. 89–98. Chapman & Hall, London.

Ziegler, W. and Lane, H.R. (1987). Cycles in conodont evolution from Devonian to mid-Carboniferous. In *Palaeobiology of conodonts* (ed. R.J. Aldridge), 147–63. Horwood Press Chichester.

Ziegler, W. and Sandberg, C.A. (1990). The Late Devonian standard conodont zonation. *Courier Forschungsinstitut Senckenberg*, **121**, 1–115.

Ziegler, W., Ji, Q., and Wang C. (1988). Devonian–Carboniferous boundary – final candidates for a stratotype section. *Courier Forschungsinstitut Senckenberg*, **100**, 15–19.

Zinsmeister, W.J., Feldmann, R.M., Woodburne, M.O. and Elliot, D.H. (1989). Latest Cretaceous/earliest Tertiary transition on Seymour Island. *Journal of Paleontology*, **63**, 731–8.

Index